SECOND EDITION

MATH

FOR INDUSTRIAL PROFESSIONALS

DOUGLAS STULTZ
Clover Park Technical College

Kendall Hunt
publishing company

Cover © Shutterstock.com

Kendall Hunt
publishing company

www.kendallhunt.com
Send all inquiries to:
4050 Westmark Drive
Dubuque, IA 52004-1840

ISBN: 978-1-5249-9186-9

Published in the United States of America

Table of Contents

Foreword

The text, Math for Industrial Professionals, is designed to help students solidify a mathematics foundation as they face their industrial professional program. It is important the student has confidence in the use of the basic mathematic and Algebraic principles and properties as well as an understanding on how he or she can apply them in creating and solving Algebraic expressions and equations and in using formulas, geometric shapes, and measurements. Each chapter provides instruction, demonstrations, practice work, and assignments for additional practice.

It is extremely important students have confidence in mathematical basics, such as fractions, decimals, order of operations, and power and root laws, as they are used in much of problem solving and calculations in industrial measuring. As such, Chapter One provides instruction on mathematical operations (adding, subtraction, multiplication, division, raising to a power, taking a root.) There is instruction and practice working with fractions, decimals, and proper order of operations. There is also demonstration and practice using the scientific calculator.

As the mathematics used in problem solving in the industrial profession often require using variable formulas, a basic understanding of algebraic concepts is needed. Chapter two provides instruction on creating and simplifying algebraic expressions, and when these algebraic expressions are used in equations, there is instruction on how to using mathematic properties of equality to solve for the variable in the equation. Also covered are ratio and proportions, rational variable equations, and, of course, evaluating and solving formulas.

It is critical that students in the industrial profession programs understand and be able to use standard units of measure. Chapter Three covers systems of measurement, focusing on the Customary System utilized primarily in the United States and the Metric System favored by the rest of the world. Measurements covered will include length, volume, mass/weight, area, temperature, force and power. Description of the various units of measure will be demonstrated and instruction will be given on conversions within measuring families and between different measuring families.

A basic understanding of two dimensional, or plane, geometry is essential. Chapter four covers the geometry of special closed plane figures and how they are measured for perimeter and area. Figures dealt with include polygonal shapes such as triangles, quadrilaterals, regular hexagons, regular octagons, as well as Circles, Circle arcs, Ellipses, and annulus. Also covered is the Pythagorean Theorem and its use in finding lengths of sides of right triangles.

The industrial professional student needs to be familiar with three dimensional, or solid, geometry as well. Chapter five covers special enclosed solid figures and how they are measured for surface area and volume. Figures dealt with include Prisms, Cylinders, Pyramids, Cones, Spheres, and Frustums. In addition, there is instruction on density of matter and it's use.

Finally, Chapter Six, covers some Trigonometry that is useful in the industrial profession. It provides instruction on the basic trigonometric function ratios for Sine, Cosine and Tangent and how they and the inverse trigonometric functions are used to find lengths and angle measures of right triangles. Also covered are laws of Sine and laws of Cosine and their use in finding sides and angles of any triangle.

It is hoped that the review and instruction provided by this text will assist the student in their course of study by giving them tools to use and the confidence they need to be successful.

CHAPTER 1

Order of Operations

1-1 FRACTIONS/DECIMALS

ALL FRACTIONS ARE CONSIDERED TO BE EXACT.

A fraction is a number that describes a part of a whole. It is also the relationship between two values by division. The fraction appears in this way: one of the values, called the numerator, is placed over the other value, called the denominator, with the fraction line between them acting like a division sign.

For example $\frac{2}{3}$, with 2 the numerator and 3 the denominator, is the relationship of 2 to 3 by division, or represents 2 divided by 3.

When the fraction "number" itself is left undivided, we say the number is EXACT.

When the fraction "number" is then divided, resulting in a decimal value, we consider the resulting decimal at best an APPROXIMATE value. For example, $\frac{2}{3}$ is exact, but when 2 is divided by 3 we get .6666… or $.\overline{6}$, a decimal that repeats the number 6 indefinitely. To come up with a number value, we can, at best, round the number to some decimal place value. We can say $\frac{2}{3} = .6666\ldots \approx .67$ rounded to the nearest hundredth …. So the best we can say is the decimal value is approximate. Even when fractions result in whole numbers or decimals that terminate which appear "exact", such as $\frac{4}{2} = 2.0$ or $\frac{1}{2} = .50$, we still say the decimal is approximate because there is still a call for some determination of rounding to a value. As in, 2.0 rounded to the nearest tenth, 2.00 rounded to the nearest hundredth, and so on.

There are other operation forms that have exactness versus approximation reasoning, such as

- Roots, example $\sqrt{2}$ is exact; it's number value is 1.414… which is approximate
- logarithms, example log 50 is exact; its number value is 1.698…. which is approximate
- trig functions, example sin 50° is exact; .7660…. is approximate

WHEN TO USE FRACTIONS (or other operation) AND WHEN TO USE DECIMAL IN SIMPLIFYING EXPRESSIONS AND/OR SOLVING EQUATIONS.

The Rule of thumb says "ANSWER IN THE SAME FORM YOU ARE GIVEN!"

Example, Joe mixes ½ quarts of skim milk with 1/4 quarts of whole milk. How much new milk does he have?

½ + ¼ = ¾, so he would have ¾ quarts of new milk.

However, it also true that ½ + ¼ = .5 + .25 = .75. So it would correct to say he has .75 quarts of new milk, but it would be BETTER and proper to say 3/4 quarts instead.

1-1 Exercises

1. Which of the following are EXACT numbers and which are APPROXIMATE numbers
   a) .56 ____________
   b) $.17\overline{8}$ ____________
   c) $\frac{5}{6}$ ____________
   d) $\frac{21345}{14992}$ ____________
   e) .5000 ____________
   f) $\frac{1}{2}$ ____________
   g) $\sqrt{7}$ ____________
   h) $2\frac{3}{4}+5+.0004$ ____________
2. True or false
   a) Some decimal numbers can be considered exact. ____________
   b) All rational fractional numbers are considered exact. ____________
   c) Ideally, if a problem uses fractions with math operations (adding, subtracting,..), it doesn't matter if your answer solution is in a fraction or a decimal form ____________
   d) If a number is given as a fraction plus a decimal, then the number is not exact. ____________
   e) Unless otherwise instructed, the sum $\frac{1}{2}+3\frac{1}{4}$ should be recorded as $3\frac{3}{4}$ and not 3.75 _______

Answers

1. a) APPROXIMATE
   b) APPROXIMATE
   c) EXACT

d) EXACT

e) APPROXIMATE

f) EXACT

g) EXACT

h) APPROXIMATE

2. a) FALSE

b) TRUE

c) FALSE

d) TRUE

e) TRUE

1-2 ADDITION AND SUBTRACTION OF FRACTIONS

ONLY LIKE QUANTITIES CAN BE ADDED OR SUBTRACTED.

We can add (or subtract) fractions only when they have the SAME denominator. We say these fractions with common denominators are LIKE fractions.

If you are adding (or subtracting) fractions with the same denominator it is easy: you simply add (or subtract) the numerators and put that sum (or difference) over the common denominator.

Such as $\frac{2}{7}+\frac{4}{7}=\frac{2+4}{7}=\frac{6}{7}$

Or $\frac{9}{15}-\frac{3}{15}=\frac{9-3}{15}=\frac{6}{15}$

When you want to add (or subtract) fractions with different denominators, we have to create equivalent fractions with the same denominator so they are like fractions. We can do so by multiplying the fraction(s) by fractional "identities". A fractional identity is a fraction that contains the same value in the numerator and denominator, making the fraction equal to 1. This works because any number divided by itself is equal to 1.

Examples of fractional identities are $\frac{3}{3}, \frac{500}{500}, \frac{y}{y}, \frac{a+b}{a+b}$ and so on.

We can change our fractions in our sum (or difference) to those with common denominators.

Example add $\frac{3}{4}+\frac{4}{5}$. We have uncommon denominators. 4 and 5. But if we multiply 4 by 5 we get 20 and multiply 5 by 4 we get 20, so we can get a common denominator of 20 by multiplying the numerator and denominator of $\frac{3}{4}$ by 5 and multiplying the numerator and denominator of $\frac{4}{5}$ by 4.

Here is what it will look like:

$$\frac{3}{4}+\frac{4}{5}=\frac{3}{4}\cdot\frac{5}{5}+\frac{4}{5}\cdot\frac{4}{4}=\frac{15}{20}+\frac{16}{20}$$

Do you see the fraction identities, $\frac{5}{5}$ *and* $\frac{4}{4}$?

Now we have a common denominator of 20, and adding

$\frac{15}{20}+\frac{16}{20}=\frac{15+16}{20}=\frac{31}{20}$

The same thing can be done with variable fractions.

Example: we want to subtract $\frac{5}{x}-\frac{2}{3y}$.

Since x times 3y = $3xy$ and 3y times $x = 3xy$, we can get a common denominator of $3xy$ by multiplying $\frac{5}{x}$ by $\frac{3y}{3y}$ and $\frac{2}{3y}$ by $\frac{x}{x}$,

so $\frac{5}{x}-\frac{3}{3y}=\frac{5}{x}\cdot\frac{3y}{3y}-\frac{2}{3y}\cdot\frac{x}{x}=\frac{15y}{3xy}-\frac{2x}{3xy}=\frac{15y-2x}{3xy}$

Practice

$\frac{2}{9}+\frac{5}{9}=$

$\frac{3}{5}-\frac{1}{7}=$

$\frac{3}{y}+\frac{4}{y}=$

$\frac{2}{3y}-\frac{4}{5x}=$

REDUCING FRACTION

The fractional identity also can be used to reduce fractions.

When the numerator and denominator share a common integer factor, one can "divide" that integer in both numerator and denominator, or dividing by a fraction identity.

Example, reduce $\frac{6}{15}$

3 is a factor of both 6 and 15, so we can write $\frac{6}{15}=\frac{6\div 3}{15\div 3}=\frac{2}{5}$

When there are no other common integer factors, we say the fraction is reduced.

HERE ARE SOME DIVISIBILTY RULES TO HELP YOU

Number	Rule
2	If the number is even
3	If you can add all of the digits of a number and that number is divisible by three. Example: 123. Add 1 + 2 + 3 = 6. Because 6 is divisible by 3, 123 must be divisible by 3
4	The last 2 digits are divisible by 4, or it is divisible by 2 twice. 124 is divisible by 4 because 24 is divisible by 4.
5	Ends in a 0 or 5

6	It is divisible by 2 and 3. An even number that follows the divisibility rule for 3.
7	No good rule. Ignore it and use the nearest number method discussed in the mental math section.
8	Same rule as 4 but for the last 3 digits, or it is divisible by 2 three times. 1,240 is divisible by 8 because 240 is divisible by 8
9	Same rule as 3, but the sum of the digits must be divisible by 9
10	If it ends in a 0.

Practice

Reduce these fractions

$\frac{10}{25}$ ________ $\frac{8}{12}$ ________ $\frac{16}{64}$ ________ $\frac{27}{45}$ ________ $\frac{132}{1731}$ ________

CONVERTING MIXED NUMBERS TO IMPROPER FRACTIONS

A **mixed number** is a combination of a whole number and a fraction and represents the sum of the whole number and the fraction. For example, $2\frac{1}{2}$ is the sum $2+\frac{1}{2}$.

A fraction in which the numerator is smaller than the denominator, like $^1/3$ or $^2/5$ is called a **proper fraction.** A fraction in which the numerator is larger than or equal to the denominator, like $^5/2$, $^{17}/3$, or $^6/6$ is called an **improper fraction.** To put it another way, a positive fraction with a value less than 1 is a proper fraction. A fraction with a value greater than or equal to 1 is an improper fraction.

To turn a mixed number into an improper fraction, multiply the whole number by the denominator (the bottom part), and add the result to the numerator (the top part). Then this sum goes over the original denominator.

Examples $4\frac{2}{3} = \frac{4 \cdot 3 + 2}{3} = \frac{14}{3}$; $10\frac{5}{7} = \frac{10 \cdot 7 + 5}{7} = \frac{75}{7}$

CONVERTING AN IMPROPER FRACTION TO A MIXED NUMBER

To write an improper fraction as a mixed number, divide the numerator (top part) by the denominator (bottom part). The quotient is the whole number, and the remainder is the numerator.

Examples $\frac{14}{3} \rightarrow 3\overline{)\,14}$ (quotient 4)

-12

2, so $\frac{14}{3} = 4\frac{2}{3}$

ADDING(SUBTRACTING) MIXED NUMBERS

When you add (or subtract) mixed numbers, you can add the whole numbers then add the fractions using fraction rules then convert the result into a new mixed number.

For Example: $2\frac{3}{4} + 5\frac{1}{7}$

We can write $2\frac{3}{4}+5\frac{1}{7}$ as $2+\frac{3}{4}+5+\frac{1}{7}$

We can rearrange the trems $2+\frac{3}{4}+5+\frac{1}{7}$ as $2+5+\frac{3}{4}+\frac{1}{7}$

Adding the whole numbers we get $2 + 5 = 7$

Adding the fractions $\frac{3}{4}+\frac{1}{7}=\frac{3\cdot 7}{4\cdot 7}+\frac{1\cdot 4}{7\cdot 4}=\frac{21}{28}+\frac{4}{28}=\frac{21+4}{28}=\frac{25}{28}$

So $2\frac{3}{4}+5\frac{1}{7}=7+\frac{25}{28}=7\frac{25}{28}$

What about subtracting mixed number $12\frac{4}{5}-9\frac{2}{3}$

We can write this as $12+\frac{4}{5}-(9+\frac{2}{3})=12+\frac{4}{5}-9-\frac{2}{3}$

We can rearrange the terms of $12+\frac{4}{5}-9-\frac{2}{3}$ as $12-9+\frac{4}{5}-\frac{2}{3}$

Now the whole number is $12-9=3$

and the fractions, $\frac{4}{5}-\frac{2}{3}=\frac{4\cdot 3}{5\cdot 3}-\frac{2\cdot 5}{3\cdot 5}=\frac{12}{5}-\frac{10}{15}=\frac{12-10}{15}=\frac{2}{15}$

So $12\frac{4}{5}-9\frac{2}{3}=3+\frac{2}{15}=3\frac{2}{15}$

Practice

$7\frac{3}{8}+4\frac{2}{7}$ ________ $15\frac{3}{4}-6\frac{1}{9}$ ________

ESTIMATING USING MIXED NUMBERS

When you add (or subtract) mixed numbers, you can use the whole number part to estimate your result.

Example $12\frac{3}{8}+15\frac{2}{7}$ can be estimated to $12 + 15 = 27$. Since the fraction parts can be anywhere from 0 to 2, So we say that this results would be about 27 or 28.

We also use estimating when we round numbers that are decimals. Becoming proficient in quickly estimating is important when we want to check our final answers, get a feel for what our answer should be around, etc.

1-2 Exercises

Reduce these fractions

1. $\frac{8}{16} =$
2. $\frac{14}{18} =$
3. $\frac{28}{35} =$
4. $\frac{145}{150} =$
5. $\frac{54}{108} =$
6. $\frac{51}{63} =$
7. $\frac{924}{1016} =$
8. $\frac{4545}{9999} =$

Add or subtract fractions, simplify final fractions

9. $\frac{1}{9} + \frac{5}{18} =$
10. $\frac{1}{9} + \frac{7}{20} =$
11. $\frac{3}{7} + \frac{1}{15} =$
12. $\frac{1}{12} + \frac{1}{9} =$
13. $\frac{7}{8} + \frac{6}{7} =$
14. $\frac{11}{20} + \frac{11}{40} =$
15. $\frac{2}{5} - \frac{1}{5} =$
16. $\frac{2}{5} - \frac{5}{16} =$
17. $\frac{3}{4} - \frac{2}{15} =$
18. $\frac{7}{8} - \frac{7}{16} =$
19. $\frac{1}{2} - \frac{3}{16} =$
20. $\frac{9}{20} - \frac{1}{7} =$
21. $\frac{1}{b} + \frac{5}{a} =$
22. $\frac{1}{xy} + \frac{7}{2w} =$
23. $\frac{3t}{7} + \frac{5t}{14} =$
24. $\frac{t}{p} - \frac{y}{x} =$
25. $\frac{ab}{cd} + \frac{ef}{gh} =$
26. $\frac{1}{f} + \frac{h}{4} =$
27. $2\frac{1}{7} + 9\frac{11}{15} =$
28. $4\frac{14}{15} + 1\frac{7}{12} =$
29. $7\frac{3}{10} + 7\frac{1}{16} =$
30. $7\frac{1}{7} + 1\frac{2}{9} =$
31. $5\frac{4}{9} + 8\frac{3}{7} =$
32. $9\frac{8}{9} + 6\frac{4}{7} =$
33. $8\frac{5}{7} - 2\frac{2}{15} =$
34. $9\frac{8}{9} - 6\frac{1}{2} =$
35. $10\frac{1}{12} - 7\frac{3}{4} =$
36. $5\frac{2}{5} - 1\frac{1}{2} =$
37. $8\frac{7}{20} - 4\frac{1}{4} =$
38. $7\frac{5}{9} - 6\frac{3}{4} =$

Answers

1. $\frac{8}{16} = \frac{1}{2}$
2. $\frac{14}{18} = \frac{7}{9}$
3. $\frac{28}{35} = \frac{4}{5}$
4. $\frac{145}{150} = \frac{29}{30}$
5. $\frac{54}{108} = \frac{1}{2}$
6. $\frac{51}{63} = \frac{17}{21}$
7. $\frac{924}{1016} = \frac{231}{254}$
8. $\frac{4545}{9999} = \frac{505}{1111}$
9. $\frac{1}{9} + \frac{5}{18} = \frac{7}{18}$

10. $\frac{1}{9}+\frac{7}{20}=\frac{83}{180}$

11. $\frac{3}{7}+\frac{1}{15}=\frac{52}{105}$

12. $\frac{1}{12}+\frac{1}{9}=\frac{7}{36}$

13. $\frac{7}{8}+\frac{6}{7}=1\frac{41}{56}$

14. $\frac{11}{20}+\frac{11}{40}=\frac{33}{40}$

15. $\frac{2}{5}-\frac{1}{5}=\frac{1}{5}$

16. $\frac{2}{5}-\frac{5}{16}=\frac{7}{80}$

17. $\frac{3}{4}-\frac{2}{15}=\frac{37}{60}$

18. $\frac{7}{8}-\frac{7}{16}=\frac{7}{16}$

19. $\frac{1}{2}-\frac{3}{16}=\frac{5}{16}$

20. $\frac{9}{20}-\frac{1}{7}=\frac{43}{140}$

21. $\frac{1}{b}+\frac{5}{a}=\frac{a+5b}{ab}$

22. $\frac{1}{xy}+\frac{7}{2w}=\frac{2w+7xy}{2xyw}$

23. $\frac{3t}{7}+\frac{5t}{14}=\frac{11t}{14}$

24. $\frac{t}{p}-\frac{y}{x}=\frac{tx-py}{px}$

25. $\frac{ab}{cd}+\frac{ef}{gh}=\frac{abgh+cdef}{cdgh}$

26. $\frac{1}{f}+\frac{h}{4}=\frac{4+fh}{4f}$

27. $2\frac{1}{7}+9\frac{11}{15}=11\frac{92}{105}$

28. $4\frac{14}{15}+1\frac{7}{12}=6\frac{31}{60}$

29. $7\frac{3}{10}+7\frac{1}{16}=14\frac{29}{80}$

30. $7\frac{1}{7}+1\frac{2}{9}=8\frac{23}{63}$

31. $5\frac{4}{9}+8\frac{3}{7}=13\frac{55}{63}$

32. $9\frac{8}{9}+6\frac{4}{7}=16\frac{29}{63}$

33. $8\frac{5}{7}-2\frac{2}{15}=6\frac{61}{105}$

34. $9\frac{8}{9}-6\frac{1}{2}=3\frac{7}{18}$

35. $10\frac{1}{12}-7\frac{3}{4}=2\frac{1}{3}$

36. $5\frac{2}{5}-1\frac{1}{2}=3\frac{9}{10}$

37. $8\frac{7}{20}-4\frac{1}{4}=4\frac{1}{10}$

38. $7\frac{5}{9}-6\frac{3}{4}=\frac{29}{36}$

1-3 MULTIPLYING FRACTIONS

When multiplying fractions, the rule is

MULTIPLY NUMERATORS ACROSS, MULTIPLY DENOMINATORS ACROSS

Example: $\frac{2}{7}\cdot\frac{6}{11}=\frac{2\cdot 6}{7\cdot 11}=\frac{12}{77}$

MULTIPLYING MIXED NUMBERS

When multiply mixed numbers, convert the mixed numbers to improper fractions then use fraction multiplication rule. Then reconvert the resulting improper fraction to a mixed number.

Example: $5\frac{1}{6}\cdot 3\frac{2}{9}=\frac{5\cdot 6+1}{6}\cdot\frac{3\cdot 9+2}{9}=\frac{31}{6}\cdot\frac{29}{9}=\frac{31\cdot 29}{6\cdot 9}=\frac{899}{54}$

Converting this improper fraction to a mixed number

$$\begin{array}{r} 16 \\ 54\overline{)899} \\ -54 \\ \hline 359 \\ -324 \\ \hline 35 \end{array} \quad \text{so } \frac{899}{54} = 16\frac{35}{54} \text{ so } 5\frac{1}{6} \cdot 3\frac{2}{9} = 16\frac{35}{54}$$

CANCELLING OUT FACTORS

When we multiply fractions, we can "reduce" the multiplication by cancelling our common factors in numerators and denominators across the fraction factors, making the calculation easier, as well as provide a simplified fraction result.

Example: $\frac{4}{11} \cdot \frac{33}{2}$ as the 4 in the first fraction numerator and the 2 in the second fraction denominator share a factor of 2 we can cancel out 2 from each. Also the 11 in the first fraction denominator and the 33 in the second fraction numerator share a factor of 11, we can cancel 11 from each. So we get

$$\frac{\overset{2}{\cancel{4}}}{\underset{1}{\cancel{11}}} \cdot \frac{\overset{3}{\cancel{33}}}{\underset{1}{\cancel{2}}} = \frac{2}{1} \cdot \frac{3}{1} = \frac{2 \cdot 3}{1 \cdot 1} = \frac{6}{1} = 6$$

Practice: Multiply

$\frac{3}{11} \cdot \frac{1}{2}$ ________, $\frac{7}{12} \cdot \frac{36}{21}$ ________, $2\frac{5}{6} \cdot 3\frac{4}{5}$ ________, $\frac{2}{3} \cdot 7\frac{6}{7}$ ________.

DIVISION BY QUANTITIES

A DENOMINATE NUMBER is made up of a numerical value followed by a unit of measure, such as 12 ft., or 345 lb., or 789 yr.

We can divide denominate numbers. What results when dividing denominate numbers is either a constant value or a fractional rate. When the unit of measure is the same unit, you get a constant value. For example, 6 ft. divided by 3 ft. = 3 ft. = $\frac{6 \text{ ft.}}{3 \text{ ft.}}$ = 2. The unit of measure cancels out which leaves $\frac{6}{3}$ = 2.

Fractional rates, or division by quantities, occur in real life all the time, especially in the industrial profession. For example, 10 mi. divided by 1 gal. results in the fractional rate 10 mi./1 gal. Other examples are 2,000 people/1 yr, slices/1 pizza, and 20 in./5 min.

As for other operations . . .

You can only add (or subtract) *LIKE* quantities, that is, only denominate numbers in the same unit of measure. But any quantities, like or not, can be multiplied together or divided. Remember, when you multiply quantities, you must multiply BOTH the numerical values and the unit of measures.

However, when multiplying unlike quantities, it is often more practical, when possible, to convert to like quantities, and create a new quantity raised to power(s).

Addition/Subtraction: For example, to add 12 ft. and 6 in., you must convert one of the quantities to either all feet or all inches, so 6 in. is 0.5 ft., so 12 ft. + 6 in. = 12 ft. + 0.5 in. = 12.5 ft.

Or 12 ft. is 144 in., so 12 ft. + 6 in. = 144 in. + 6 in. = 150 in.

Multiplication: To multiply 12 ft. by 6 in., you can get (12 ft.) (6 in.) = 72 "ft. in." This may fit a purpose in answering some question that uses "feet inches," but a more practical measure would be converting inches to feet then multiplying.

As 6 in. = 0.5 ft., (12 ft.) (6in.) = (12 ft.) (0.5 ft.) = 6 ft · ft. = $6\,\text{ft.}^2$ (or 6 sq. ft.), a more common measure for area.

And we could also figure as 12 ft. = 144 in., then (12 ft.) (6 in.) = (144 in.) (6 in.) = 864 in. · 864 in. = 864 in^2 (or 864 sq. in.).

FRACTION OF A QUANTITY

Taking a fraction "of" a quantity means we multiply the fraction times the quantity.

Example: 1/3 of $27 = \frac{1}{3}\cdot 27 = \frac{1}{3}\cdot\frac{27}{1} = \cdot\frac{1\cdot 27}{3\cdot 1} = \frac{27}{3} = 9$

Example: 3/4 of $10\frac{6}{7} = \frac{3}{4}\cdot\frac{10\cdot 7+6}{7} = \frac{3}{4}\cdot\frac{76}{7} = \frac{3\cdot 76}{4\cdot 7} = \frac{228}{28}$

$$\begin{array}{r} 8 \\ 28\overline{)\,228} \\ -224 \\ \hline 4 \end{array}$$

$\frac{228}{28} = 8\frac{4}{28} = 8\frac{1}{7}$ so 3/4 of $10\frac{6}{7} = 8\frac{1}{7}$

Practice

$\frac{4}{5}$ of 70 ________, $\frac{4}{9}$ of 22 ________, $\frac{2}{11}$ of $1\frac{6}{7}$ ________.

ESTIMATION FRACTION OF QUANTITY

If we are taking a fraction of a mixed number, multiply the fraction time the closest whole number part to get the estimate of what the result should be.

Example: $\frac{2}{5}$ of $20\frac{1}{3}$ estimates 2/5 of 20 to get 8. So the result should be between 8 and 9.

When using two mixed number, round each to the closest whole number part, then multiply.

Example: $4\frac{1}{3}$ of $9\frac{6}{7}\cdot 4\frac{1}{3}$ is closest to 4, $9\frac{6}{7}$ is closest to 10, so four times $10 = 40$. So $4\frac{1}{3}$ of $9\frac{6}{7}$ is estimated to be about 40 (it is actually $42\frac{5}{7}$). Estimating is a good way to check to see if your exact answer makes sense.

1-3 Exercises

Multiply these fractions, simplify final fractions

1. $\frac{1}{9}\cdot\frac{5}{7}=$

2. $\frac{4}{9}\cdot\frac{7}{11}=$

3. $\frac{3}{7}\cdot\frac{1}{15}=$

4. $\frac{1}{12}\cdot\frac{1}{9}=$

5. $\frac{7}{8}\cdot\frac{6}{7}=$

6. $\frac{11}{20}\cdot\frac{11}{40}=$

7. $\frac{2}{5}\cdot\frac{1}{5}=$

8. $\frac{2}{5}\cdot\frac{5}{16}=$

9. $\frac{3}{4}\cdot\frac{14}{15}=$

10. $\frac{7}{8}\cdot\frac{7}{16}=$

11. $\frac{1}{2}\cdot\frac{3}{16}=$

12. $\frac{9}{20}\cdot\frac{1}{18}=$

13. $\frac{1}{b}\cdot\frac{5}{a}=$

14. $\frac{1}{xy}\cdot\frac{7}{2w}=$

15. $\frac{3t}{7}\cdot\frac{5t}{14}=$

16. $\frac{t}{p}\cdot\frac{y}{x}=$

17. $\frac{ab}{cd}\cdot\frac{ef}{gh}=$

18. $\frac{1}{f}\cdot\frac{h}{4f}=$

19. $2\frac{1}{7}\cdot 9\frac{5}{6}=$

20. $4\cdot 1\frac{7}{12}=$

21. $7\frac{3}{10}\cdot 7\frac{1}{16}=$

22. $7\frac{1}{7}\cdot 1\frac{2}{9}=$

23. $5\frac{4}{9}\cdot 8\frac{3}{7}=$

24. $9\frac{1}{9}\cdot 6\frac{4}{41}=$

25. $8\frac{1}{2}\cdot 2\frac{2}{15}=$

26. $9\frac{8}{9}\cdot 6$

27. $10\frac{1}{12}\cdot 7\frac{3}{4}=$

28. $\frac{2}{3}$ *of* $500=$

29. $\frac{7}{8}$ of $2121=$

30. $\frac{1}{22}$ of $44000=$

31. $\frac{4}{7}$ *of* $5\frac{1}{3}=$

32. $2\frac{5}{6}$ *of* $1100=$

33. $1\frac{7}{8}$ *of* $4\frac{9}{13}=$

Answers

1. $\frac{1}{9}\cdot\frac{5}{7}=\frac{5}{63}$

2. $\frac{4}{9}\cdot\frac{7}{11}=\frac{28}{99}$

3. $\frac{3}{7}\cdot\frac{1}{15}=\frac{1}{35}$

4. $\frac{1}{12}\cdot\frac{1}{9}=\frac{1}{108}$

5. $\frac{7}{8}\cdot\frac{6}{7}=\frac{3}{4}$

6. $\frac{11}{20}\cdot\frac{11}{40}=\frac{121}{800}$

7. $\frac{2}{5}\cdot\frac{1}{5}=\frac{2}{25}$

8. $\frac{2}{5}\cdot\frac{5}{16}=\frac{1}{8}$

9. $\frac{3}{4}\cdot\frac{14}{15}=\frac{7}{10}$

10. $\frac{7}{8}\cdot\frac{7}{16}=\frac{49}{128}$

11. $\frac{1}{2}\cdot\frac{3}{16}=\frac{3}{32}$

12. $\frac{9}{20}\cdot\frac{1}{18}=\frac{1}{40}$

13. $\frac{1}{b}\cdot\frac{5}{a}=\frac{5}{ba}$

14. $\frac{1}{xy}\cdot\frac{7}{2w}=\frac{7}{2xyw}$

15. $\frac{3t}{7}\cdot\frac{5t}{14}=\frac{15t^2}{98}$

16. $\frac{t}{p} \cdot \frac{y}{x} = \frac{ty}{px}$

17. $\frac{ab}{cd} \cdot \frac{ef}{gh} = \frac{abef}{cdgh}$

18. $\frac{1}{f} \cdot \frac{h}{4f} = \frac{h}{4f^2}$

19. $2\frac{1}{7} \cdot 9\frac{5}{6} = 21\frac{1}{14}$

20. $4 \cdot 1\frac{7}{12} = 6\frac{1}{3}$

21. $7\frac{3}{10} \cdot 7\frac{1}{16} = 51\frac{89}{160}$

22. $7\frac{1}{7} \cdot 1\frac{2}{9} = 8\frac{46}{63}$

23. $5\frac{4}{9} \cdot 8\frac{3}{7} = 45\frac{8}{9}$

24. $9\frac{1}{9} \cdot 6\frac{4}{41} = 55\frac{5}{9}$

25. $8\frac{1}{2} \cdot 2\frac{2}{15} = 18\frac{2}{15}$

26. $9\frac{8}{9} \cdot 6 = 59\frac{1}{3}$

27. $10\frac{1}{12} \cdot 7\frac{3}{4} = 78\frac{7}{48}$

28. $\frac{2}{3}$ *of* $500 = 333\frac{1}{3}$

29. $\frac{7}{8}$ of $2121 = 1855\frac{7}{8}$

30. $\frac{1}{22}$ of $44000 = 2000$

31. $\frac{4}{7}$ *of* $5\frac{1}{3} = 3\frac{1}{21}$

32. $2\frac{5}{6}$ *of* $1100 = 3116\frac{2}{3}$

33. $1\frac{7}{8}$ *of* $4\frac{9}{13} = 8\frac{83}{104}$

1-4 RECIPROCALS/DIVIDING FRACTIONS

RECIPROCAL

A reciprocal of a number is a value when multiplied the number to equal 1. We can easily turn any fraction by "flipping" the numerator and denominator.

Example: The reciprocal of $\frac{3}{4}$ is $\frac{4}{3}$ and we know $\frac{3}{4} \cdot \frac{4}{3} = 1$

The reciprocal of $\frac{1}{9}$ is $\frac{9}{1}$ or 9 and we know that $\frac{1}{9} \cdot 9 = 1$.

The reciprocal of 11 is $\frac{1}{11}$ (note 11 can be written in fraction form as $\frac{11}{1}$).
Given a mixed number, simply convert it to improper form and then do your "flipping."

Example to find the reciprocal of $9\frac{4}{7}$ make $9\frac{4}{7} = \frac{9 \cdot 7 + 4}{7} = \frac{67}{7}$ so the reciprocal of $9\frac{4}{7}$ is $\frac{7}{67}$.

DIVISION BY FRACTION

When one fraction (or number) is divided by a second fraction, find the second fraction's reciprocal, then multiply it by the first fraction (or number).

Example: $\frac{2}{7} \div \frac{5}{6}$ $\frac{6}{5}$ is the reciprocal of $\frac{5}{6}$ so we multiply $\frac{2}{7} \cdot \frac{6}{5} = \frac{12}{35}$

Example: $5 \div \frac{3}{4} = 5 \cdot \frac{4}{3} = \frac{20}{3} = 6\frac{2}{3}$

Example: $\dfrac{\frac{2}{3}}{\frac{9}{11}} = \frac{2}{3} \div \frac{9}{11} = \frac{2}{3} \cdot \frac{11}{9} = \frac{22}{27}$

For division by a mixed number, convert to improper fraction and apply division rule.

Example: $5\frac{1}{2} \div 7\frac{1}{3}$ becomes $\frac{11}{2} \div \frac{22}{3} = \frac{11}{2} \cdot \frac{3}{22} = \frac{33}{44} = \frac{3}{4}$

Practice

$\frac{3}{7} \div \frac{5}{16}$ ________, $4 \div 2\frac{5}{6}$ ________, $4\frac{2}{3} \div 1\frac{9}{11}$ ________.

ESTIMATION OF DIVISION OF FRACTIONS

Estimating division of fractions can be tricky. One technique is to round fractions to the nearest whole value, then divide the whole numbers to find a good estimate.

Example: Estimate $5\frac{3}{4} \div 2\frac{1}{3}$ · $5\frac{3}{4}$ rounds to 6 and $2\frac{1}{3}$ rounds to 2, so $6 \div 2 = 3$. So $5\frac{3}{4} \div 2\frac{1}{3}$ is about 3 (it is actually $2\frac{13}{28}$)

Note: The problem is when a small value fraction, like ¼ or 1/11, which rounds to 0. So would require division.

1-4 Exercises

Divide these fractions, simplify final fractions

1. $\frac{1}{9} \div \frac{2}{3} =$
2. $\frac{4}{21} \div \frac{2}{7} =$
3. $\frac{3}{7} \div \frac{1}{15} =$
4. $\frac{1}{12} \div \frac{1}{8} =$
5. $\frac{7}{8} \div \frac{6}{7} =$
6. $\frac{8}{20} \div \frac{11}{30} =$
7. $\frac{2}{5} \div \frac{1}{5} =$
8. $\frac{2}{5} \div \frac{15}{16} =$
9. $\frac{3}{13} \div \frac{14}{15} =$
10. $\frac{7}{8} \div \frac{7}{16} =$
11. $\frac{1}{2} \div \frac{7}{16} =$
12. $\frac{9}{20} \div \frac{1}{12} =$
13. $2\frac{1}{7} \div 4\frac{1}{6} =$
14. $8 \div 1\frac{7}{12} =$
15. $14\frac{3}{10} \div 7\frac{1}{16} =$
16. $7\frac{4}{7} \div 5\frac{8}{9} =$
17. $8\frac{4}{9} \div 8\frac{3}{7} =$
18. $19\frac{7}{10} \div 6\frac{3}{5} =$
19. $8\frac{1}{2} \div 20\frac{5}{14} =$
20. $9\frac{2}{9} \div 6 =$
21. $2\frac{1}{16} \div 7\frac{3}{4} =$

Answers

1. $\frac{1}{9} \div \frac{2}{3} = \frac{1}{6}$

2. $\frac{4}{21} \div \frac{2}{7} = \frac{2}{3}$

3. $\frac{3}{7} \div \frac{1}{15} = \frac{45}{7} = 6\frac{3}{7}$

4. $\frac{1}{12} \div \frac{1}{8} = \frac{2}{3}$

5. $\frac{7}{8} \div \frac{6}{7} = \frac{49}{48} = 1\frac{1}{48}$

6. $\frac{8}{20} \div \frac{11}{30} = \frac{12}{11}$

7. $\frac{2}{5} \div \frac{1}{5} = 2$

8. $\frac{2}{5} \div \frac{15}{16} = \frac{32}{75}$

9. $\frac{3}{13} \div \frac{14}{15} = \frac{45}{182}$

10. $\frac{7}{8} \div \frac{7}{16} = 2$

11. $\frac{1}{2} \div \frac{7}{16} = \frac{8}{7} = 1\frac{1}{7}$

12. $\frac{9}{20} \div \frac{1}{12} = \frac{27}{5} = 5\frac{2}{5}$

13. $2\frac{1}{7} \div 4\frac{1}{6} = \frac{18}{35}$

14. $8 \div 1\frac{7}{12} = 5\frac{1}{19}$

15. $14\frac{3}{10} \div 7\frac{1}{16} = 2\frac{14}{565}$

16. $7\frac{4}{7} \div 5\frac{8}{9} = 1\frac{2}{7}$

17. $8\frac{4}{9} \div 8\frac{3}{7} = 1\frac{1}{531}$

18. $19\frac{7}{10} \div 6\frac{3}{5} = 2\frac{65}{66}$

19. $8\frac{1}{2} \div 20\frac{5}{14} = \frac{119}{285}$

20. $9\frac{2}{9} \div 6 = 1\frac{29}{54}$

21. $2\frac{1}{16} \div 7\frac{3}{4} = \frac{33}{124}$

1-5 POWERS AND ROOTS

POWERS

Exponents: Exponents are used to more easily write repeated multiplication of a number.

When we have $\boldsymbol{b^x}$, we have b, the base, and x, the exponent. This means we multiply b times itself x times. And we say "b raise to the power of x."

Example: 3^4; 3 is the base, 4 is the exponent, and this is 3 raised to the power of 4.

So, we multiply 3 by itself four times. Or $3^4 = 3 \cdot 3 \cdot 3 \cdot 3 = 81$

$$x^7 = x \cdot x \cdot x \cdot x \cdot x \cdot x \cdot x$$

When the exponent is 2, we have a special name for this operation. We say we "SQUARE" the base number, so, for example, 10^2 is 10 squared $= 10 \cdot 10 = 100$

When the exponent is 3, we say we "CUBE" the base number, so, for example, 4^3 is 4 cubed $= 4 \cdot 4 \cdot 4 = 64$.

USING THE CALCULATOR

On the scientific calculator, we use the ^ button (called the POWER key) when we want to apply an exponent. We enter the base number, followed by ^, then the exponent value, then ENTER.

So, to enter 5^3, we hit 5 ^ 3 ENTER to calculate (answer is 125).

For powers of 2, or squares, there is a special x^2 button you can use. You enter the base number followed by x^2 then ENTER and it will calculate the square of the number.

So, to find the square of 6, we enter 6 x^2 ENTER to calculate (answer is 36).

Practice

Use your calculator to find 7^4 ________ 13^3 __________ 2.11^5 _________

ESTIMATING POWERS

You can estimate exponential values by rounding to an easier number, then apply the repeated multiplication.

For example, 31^2 can be estimated as 30^2 or $30 \cdot 30 = 900$. So, 31^2 is just more than 900, (it is actually 961.)

ROOTS

A root is the inverse operation of a power, so, it "undoes" a power. Where $5^2 = 25$, we say 5 is the second root of 25, since five times 5 = 25. When we take a second root, we also call this a "square root." So, 5 is the "square root" of 25. We utilize a sign, $\sqrt{\ }$, called a radical sign, to represent a square root. So, $5 = \sqrt{25}$.

Some square roots are easy to calculate—if we are finding roots of "perfect Square." Perfect squares are whole numbers that are squared, and it is useful to memorize as many perfect squares as one can. So, taking the square root of perfect squares is the whole number itself. For example, as $1^2 = 1$, then $\sqrt{1} = 1$, and $2^2 = 4$, then $\sqrt{4} = 2$, and so on.

Here are perfect squares for the first 30 whole numbers.

Chart of Perfect Squares 1 to 30		
$1^2 = 1$	$11^2 = 121$	$21^2 = 441$
$2^2 = 4$	$12^2 = 144$	$22^2 = 484$
$3^2 = 9$	$13^2 = 169$	$23^2 = 529$
$4^2 = 16$	$14^2 = 196$	$24^2 = 576$
$5^2 = 25$	$15^2 = 225$	$25^2 = 625$
$6^2 = 36$	$16^2 = 256$	$26^2 = 676$
$7^2 = 49$	$17^2 = 289$	$27^2 = 729$
$8^2 = 64$	$18^2 = 324$	$28^2 = 784$
$9^2 = 81$	$19^2 = 361$	$29^2 = 841$
$10^2 = 100$	$20^2 = 400$	$30^2 = 900$

USING THE CALCULATOR

For any number, the scientific calculator can calculate a square root in two ways:

A) Hit 2nd then x^2 followed by the number followed by ENTER. If not a perfect square, you will get a decimal answer.

B) You can enter the number, followed by ^ key then 0.5 (or [1 ÷ 2]). THIS IS BECAUSE TAKING A SQUARE ROOT IS THE SAME AS RAISING A NUMBER TO THE ½ EXPONENT POWER. So, $\sqrt{22} = 22^{.5} = 22^{\frac{1}{2}}$ and $\sqrt{1254} = 1254^{.5} = 1254^{\frac{1}{2}}$, and so on.

So to find the square root of 121, we enter 2nd x^2 121 ENTER (answer is 11).

Or we enter 121 ^ 0.5 ENTER or 121 ^ (1 ÷ 2) ENTER.

Practice

Find square roots for 169 ___________, 5625 ___________, and 455 __________.

CUBE ROOT AND OTHER ROOTS

Taking the third root is the inverse operation of raising to the third power.

If $4^3 = 64$, then the third root of 64 is 4. We call a third root the "cube root" of the number. We use the radical $\sqrt[3]{\ }$, So the cube root of 64 is written as $\sqrt[3]{64}$ (Note: Only with square roots, we do not place a number at the top of the radical sign.)

Just like square roots knowing perfect cubes makes calculating cube roots easy. Here are a few perfect cubes

$1^3 = 1$	$5^3 = 125$	$9^3 = 729$
$2^3 = 8$	$6^3 = 216$	$10^3 = 1000$
$3^3 = 27$	$7^3 = 343$	$11^3 = 1331$
$4^3 = 64$	$8^3 = 512$	$12^3 = 1728$

USING THE CALCUATOR

For any number, then the scientific calculator can calculate a cube root in two ways:

A) Enter 3 then 2nd, then ^ key, then the number, followed by ENTER. If not a perfect cube, you will get a decimal answer.

B) You can enter the number, followed by ^ key then (1 ÷ 3). THIS IS BECAUSE TAKING A CUBE ROOT IS THE SAME AS RAISING A NUMBER TO THE 1/3 EXPONENT POWER.

So $\sqrt[3]{234} = 234^{\frac{1}{3}}$ and $\sqrt[3]{76} = 76^{\frac{1}{3}}$, and so on.

Practice

Find the cube roots for 345 ____________, 216 ________, and 78.9 ________.

OTHER ROOTS: Just as we can raise numbers to powers of 4, 5, 6, and so on, we can undo these by taking fourth, fifth, sixth, and on roots. There is no special names for these roots. We use the radical sign with the root number at the top, such as $\sqrt[4]{\ }$, $\sqrt[5]{\ }$, $\sqrt[6]{\ }$, and so on.

Use the calculator to find these roots, two ways

A) Type in the root number, then 2nd, then ^ key, then the number, followed by ENTER

B) You can enter the number, followed by ^ key then (1 ÷ root number) make sure the fraction is in parentheses when entering in calculator!

So $\sqrt[4]{23} = 23^{\frac{1}{4}}$ and $\sqrt[5]{12311} = 12311^{\frac{1}{5}}$, and so on.

ESTIMATING SQUARE ROOTS

Knowing the perfect squares can help you estimate a square root.

For example, take the square root of 10, or $\sqrt{10}$. Think of perfect squares that "sandwich" 10. That would be 9 and 16. We know the square root of 9 is 3, and the square root of 16 is 4. So the square root of 10 is between 3 and 4, and as 10 is just a little more than 9, we can estimate that $\sqrt{10}$ is a little more than 3 ($\sqrt{10}$ actually = 3.16 . . .).

Estimate $\sqrt{130}$. 130 is between perfect squares 121 and 144. It is also about halfway between 121 and 144. And since $\sqrt{121} = 11$ and $\sqrt{144} = 12$, then $\sqrt{130}$ should be about halfway between 11 and 12, say 11.5 ($\sqrt{130}$ is actually 11.402 . . .).

Practice

Use estimation to find the square root $\sqrt{67}$ ____________.

Chapter 1-5 Exercises

Find the following to the nearest hundredth.

1. 19^2 ______________
2. 3^7 __________
3. 12^3 ______________
4. 4^4 __________________
5. 2.5^6 ________________

6. 405^2 ________________

7. 2.344^4 ____________

8. $(-5)^2$ ________________

9. $(-7)^3$ ______________

10. $36^{\frac{1}{2}}$ ______________

11. $8^{\frac{1}{3}}$ __________________

12. $81^{\frac{1}{4}}$ ______________

13. $\sqrt{64}$ ________________

14. $\sqrt{225}$ ______________

15. $\sqrt{210}$ _____________

16. $\sqrt{10045}$ _____________

17. $\sqrt{.0041}$ _______________

18. $\sqrt[3]{216}$ ________________

19. $\sqrt[3]{2344}$ _____________

20. $\sqrt[4]{345}$ _______________

21. $\sqrt[5]{3000}$ _________________

22. $\sqrt[6]{3412}$ _____________

23. $\sqrt[11]{2189}$ ________________

24. $\sqrt[18]{65999}$ ______________

Answers

1. 361

2. 2187

3. 1728

4. 256

5. 244.14

6. 164025

7. 30.19

8. 25

9. – 343

10. 6

11. 2

12. 3

13. 8

14. 15

15. 14.49

16. 100.22

17. .06

18. 6

19. 13.28

20. 4.31

21. 4.96

22. 3.88

23. 2.01

24. 1.85

1-6 ORDER OF OPERATIONS

Order of Operations is a set of rules that dictate what operations to do first, then seconds, and so on, until we arrive at a single number, or simplified algebraic expression.

> When expressions have more than one operation, we follow
>
> 1. First do all operations in proper order (items 2–4) to expressions that lie inside
>   - Parentheses or brackets ( ), [ ], { }. Note: work innermost out.
>   - Under radical signs $\sqrt{\ }$, $\sqrt[3]{\ }$, $\sqrt[4]{\ }$, and so on.
>   - In numerators/denominator.
>   - Inside absolute value.
> 2. Next, do any work, left to right, with exponents or radicals.
> 3. Again, working from left to right, do all multiplication or division, whatever comes first.
> 4. Finally, working from left to right, do all addition or subtraction, whatever comes first.

PEM/DA/S is an acronym used to remember this order of operation.

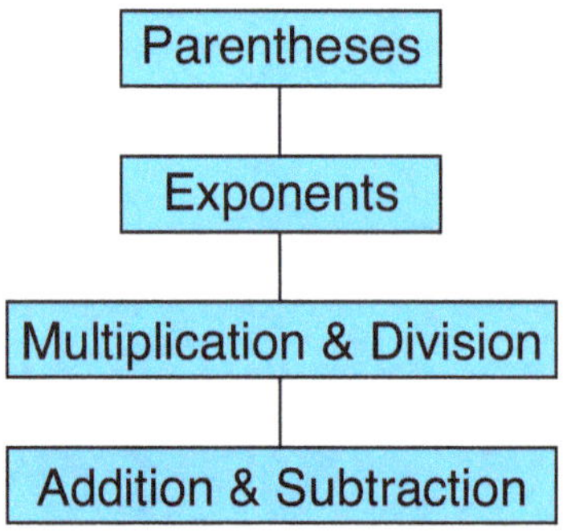

Practice

$$100 - 12 \div 3 + 32 \cdot 4 + 2$$

No parenthesis, no exponents (radicals, absolute value . . .) so look for multiplication/division, left to right.

$$100 - (12 \div 3) + 32 \cdot 4 + 2$$

Do division first, $12 \div 3 = 4$

$$100 - 4 + (32 \cdot 4) + 2$$

Do multiplication $32 \cdot 4 = 128$

$$(100 - 4) + 128 + 2$$

No more multiplication/division, look for addition or subtraction left to right, do $100 - 4 = 96$

$$(96 + 128) + 2$$

Do addition 96 + 128 = 224

$$224 + 2$$

Do addition 224 + 2 = 226

$$\text{so, } 100 - 12 \div 8 - 32 \cdot 4 + 2 = 226$$

Here's another one . . .

$$5^2 - 3[4 - 2(6 - 3)] \div 2$$

Do innermost parenthesis first

$$5^2 - 3[4 - 2(6 - 3)] \div 2$$

so do subtraction 6 – 3 = 3

$$5^2 - 3[4 - 2(3)] \div 2 \text{ or } 5^2 - 3[4 - 2 \cdot 3] \div 2$$

Now do [ ] brackets, multiply $2 \cdot 3 = 6$

$$5^2 - 3[4 - 6] \div 2$$

do subtraction 4 – 6 = –2

$$5^2 - 3[-2] \div 2 \text{ or } 5^2 - 3 \cdot -2 \div 2$$

no more parentheses, so do exponent $5^2 = 25$

$$25 - 3 \cdot -2 \div 2$$

No more exponents so do multiplication $3 \cdot -2 = -6$

$$25 - -6 \div 2$$

do division $-6 \div 2 = -3$

$$25 - -3$$

No more multiplication/division so do subtraction 25 – –3 = 25 + 3 = 28

$$\text{so } 5^2 - 3[4 - 2(6 - 3)] \div 2 = 28$$

Example: Let's try this more complicated example!!!!! simplify

$$12 - [4 \cdot (3^4 \div 9 + 6 \cdot \frac{6+4}{18+2})] + \sqrt{8+92}$$

$$12 - [4 \cdot (3^4 \div 9 + 6 \cdot \frac{6+4}{18+2})] + \sqrt{8+92}$$

Start with innermost parentheses because numerators and denominators of fractions are groupings, do numerator/denominator first, $\frac{6+4}{18+2} = \frac{10}{20} = \frac{1}{2}$

$$12 - [\,4 \cdot (3^4 \div 9 + 6 \cdot \frac{1}{2})] + \sqrt{8+92} \text{ next do exponent } 3^4 = 81$$

$$12 - [\,4 \cdot (81 \div 9 + 6 \cdot \frac{1}{2})] + \sqrt{8+92} \text{ do division } 81 \div 9 = 9$$

$$12 - [4 \cdot (9 + 6 \cdot \frac{1}{2})] + \sqrt{8+92} \text{ do multiplication } 6 \cdot \frac{1}{2} = \frac{6}{2} = 3$$

$$12 - [4 \cdot (9 + 3)] + \sqrt{8+92} \text{ do addition } 9 + 3 = 12\text{, drop inner parentheses}$$

$$12 - [4 \cdot 12] + \sqrt{8+92}$$

Do [ ] bracket, do multiplication $4 \cdot 12 = 48$ and remove brackets

$$12 - 48 + \sqrt{8+92} \text{ as under a radical is a grouping, do addition } 8 + 92 = 100$$

$$12 - 48 + \sqrt{100} \text{ do square root } \sqrt{100} = 10$$

$$12 - 48 + 10 \text{ do subtraction } 12 - 48 = -36$$

$$-36 + 10 \text{ do addition } -36 + 10 = -26$$

$$\text{so } 12 - [4 \cdot (3^4 \div 9 + 6 \cdot \frac{6+4}{18+2})] + \sqrt{8+92} = -26$$

Practice

1. $8 - 2 \cdot 4 - 12 \div 2 + 4 =$
2. $19 - [6 + 2\,(16 \div 8 \cdot 5 + 4)] \div 2 =$
3. $82 - 10\left(8 - \frac{6+2 \cdot 4}{18-11}\right)^2 \div \sqrt{25} =$
4. $12 \div 2 + 2\,[8 - (2 + 7) - \sqrt{4+12}\,]^2 =$

Answers: 1) –2, 2) 2, 3) 10, 4) 56.

Chapter 1-6 assignments

Evaluate these expressions using proper order of operation

1. $4\,(6) + 3 - 7\,(4) \div 2 =$
2. $5 \div 5 + 3 - 6 \cdot 2 =$
3. $3\,(7) - 11 + 15 \div 3 =$
4. $3 + 34 \div 17 + 4 \cdot 12 - 8 =$
5. $25 - (7 - 12 \div 6)\,(4) =$

6. $5[4 + (3)(2)] - (9 - 28 \div 7) \div 5 =$

7. $2^4 - 5(10 - 4^2 \div 2) + (30 - 3^3) =$

8. $(32 - 3^3 \div 9 \cdot 10)^5 - 4^2 \div 8 + 3^2$

9. $31 - \sqrt{36 \div 6 + 19} \cdot 2^2 - 11 =$

10. $3^3 + \dfrac{\sqrt{3 + 2 \cdot 3 + 27}}{4 - 1} - 4(5) =$

11. $4^2 + (14 - 48 \div \dfrac{8 + 6 \cdot 4}{1 + 7}) - 7 \times 2 =$

12. $\dfrac{6 + 3 \cdot 5 - 6^2 \div [2 + (21 \div 3)]}{2^3 - 3 \cdot 2} =$

Answers

1. 13
2. −8
3. 15
4. 45
5. 5
6. 49
7. 9
8. 39
9. 0
10. 9
11. 4
12. $\dfrac{17}{2}$

1-7 SCIENTIFIC CALCULATOR AND ORDER OF OPERATIONS.

NOTE: This course recommends the use of the scientific calculator, TI-30 xIIS. Demonstrations/ Instructions will be done using this calculator.

In the Industrial profession, it is critical that we have an understanding of how to enter calculations into our calculator. And it is important when we do any complicated calculation in the calculator we use proper order of operations, which means keeping the proper grouping in mind by utilizing the parentheses keys. Here are some helpful hints when using the calculator.

- If we want to square a negative number, say –4 to the 2nd power, we must remember to enter it as $(-4)^2$ and NOT as -4^2. Remember, when we multiply –4 by –4 we end up with a positive, as negative times negative is positive. So $(-4)^2$ will calculate the square of the negative and get the correct value, +16, whereas -4^2 will only square the 4, and will give an answer of –16.
- A mixed number needs to be treated like a grouped sum in the calculator. $3\frac{1}{2}$ would be written as $(3 + 1 \div 2)$
- Sums or differences in numerator or denominator. Let's say we have a fraction $\frac{2}{21+345}$ and we want to calculate it to a value on the calculator. Remember that the 21 + 345 in the denominator is a grouping and needs to be put in parenthesis. So it must be entered as $2 \div (21 + 345)$ which equals $2/366 \approx .00546$.

  NOTE: If it is entered in the calculator as $2 \div 21 + 345$, without the parentheses, the calculator will follow order of operation, and the resulting calculation is to divide 2 by 21 then add that to 345, and get ≈ 345.0952

  The same thing goes for a sums or difference s in numerators For $\frac{23+567}{7}$ it must be entered as $(23 + 567) \div 7 \approx 84.29$
- You need to practice with the calculator to become competent when to enter operations exact or when things need to be grouped. Examples:

  What about something like $3000\,(1 + .004 \cdot \frac{1}{4})^{20}$

  You could calculate the .004 times ¼ first to get .001, then add it to 1,
- Entering $3000\,(1 + .001)^{20} \approx 3060.5734\ldots$ to get the answer. However, you can enter your calculation steps as they come in one continuous flow, as long as you recognize proper grouping

  You can do so By entering 3000 (1 + .004 × 1 ÷ 4) ^ 20 which also yields about ≈ 3060. 5734 . . .

  In addition, here is a tip to make thing easier. As multiplying by 1 / 4 is really dividing by 4, this calculation can be done by dropping the X 1 and calculated as

  3000 (1 + .004 ÷ 4) ^ 20

  When we come to multiplying factors which are all fractions, such as $\frac{3}{4} \cdot \frac{5}{7} \cdot \frac{1}{8}$ you can do so in two ways

  1) we can put each fraction in parenthesis, then multiply, entering

     (3 ÷ 4) (5 ÷ 7) (1 ÷ 8) ≈ .0670

     Or

2) Remember when we multiply fractions, we multiply the numerators over multiplying the denominators, so we can use the numerator product over the denominator product, making sure we group them in parentheses, like,

(3 x 5 x 1) ÷ (4 x 7 x 8) ≈ .0670

➡ What about roots?

Lets say we want to calculate $\sqrt{8+22^4}$.

You could calculate $8+22^4$ first. As 22 ^ 4 = 234256, so $8+22^4$ = 8 + 234256 = 234264

Then for $\sqrt{234256}$, you would do the square root either as 2nd x^2 234264 or as 234264 ^ .5 or as 234264 ^ (1 ÷2) each which would give you ≈ 484.01

Or you can do it in one continuous flow using the operations as they come using parentheses.

This becomes Either as 2nd x^2 (8 + 22 ^ 4) ≈ 484.01

or as (8 + 22 ^ 4) ^ .5 ≈ 484.01

or as (8 + 22 ^ 4) ^ (1 / 2) ≈ 484.01

It is important what and when to put values in parentheses. And it comes from understanding order of operation and practice, practice, practice

PRACTICE, use scientific calculator and order of operation to calculate (so give answer in decimal to nearest hundredth):

a) $10\frac{3}{7} + 4\frac{2}{3}\left(6\frac{1}{6} - 5\frac{2}{3}\right)$

b) $\frac{4^2+5\cdot(11-2^3)}{7\cdot(-3)^2} + 6\frac{3}{11}$

c) $\sqrt{20^2 + 6\left(3\frac{1}{5}\right)}$

d) $\frac{3}{7}\cdot\sqrt{5+11^3} - \frac{(-26)^3 - 5^2}{27^3+8}$

Answers:

a) 12.76

b) 6.76

c) 20.47

d) 16.56

CHAPTER 2

Algebra: Expressions, Equations, and Formulas

2-1 VARIABLE EXPRESSIONS

In Mathematics, an **EXPRESSION** is a set of numbers and/or variables and operation signs that can set on one side of an equal sign.

Here are examples of expression that contain variable(s).

$$3x+2 \qquad -12x^2 5x-2 \qquad \sqrt{5y} \qquad \frac{2y+8}{y} \qquad \frac{23yxp+3x}{43} \qquad \log 23x+4z$$

An expression on its own with a **variable (or variables)** means, in most cases, that the unknown variable **COULD BE ANY REAL NUMBER VALUE**. For example, for $3x + 2$, x can be substituted with any real number value.

There are Exceptions. The value for a variable in some expressions can be limited when we are not able to do operations with certain numbers. For example, a single variable in a denominator can never be 0 since we cannot divide by 0. Another case is a variable under a radical sign for an even root can never be a negative.

EXAMPLE: given the expression $\frac{2y+8}{y}$, y can be any real number value except 0. We can see that $\frac{2y+8}{y} = \frac{2(0)+8}{0} = \frac{8}{0}$ which is undefined.

EXAMPLE: $\sqrt{5y}$ y can be any real number greater than or equal to 0. You can see if $y < 0$, then $5y$ would be negative, and you cannot have a negative under a square root radical.

Practice

Identify what values for the variable can be used in these expressions

$15x+29$ __________

$2y^2+35y-2$ _________

$\sqrt{w-5}$ _________

$\frac{2t+18}{t}$ ____________

$16h^{\cdot 5}$ __________

$5(3x+2)-7(x+7)$ __________

$\frac{5}{3-y}$ __________

2-1 Assignments

Identify what values for the variable can be used in these expressions. If any real number can satisfy the expression, write ALL REAL NUMBERS. If the set of values is limited, describe what that set is.

1. $32x$ __________
2. y^2+y-12 __________
3. $5x+45$ __________
4. $4t^3-6t^2-3t+4$ __________
5. $\sqrt{w}$ __________
6. $\sqrt{p+8}$ __________
7. $\frac{22}{x-7}$ __________
8. $\frac{5x+7}{3x+8}$ __________
9. $(x+8)-6(x+66)$ __________
10. $\frac{x^2+3x-4}{x-11}$ __________

Answers

1. All real numbers
2. All real numbers
3. All real numbers
4. All real numbers
5. All real numbers greater than or equal to 0, or $w \geq 0$
6. All real number greater than or equal to –8, or $p \geq -8$
7. All real number except 7, or $x \neq 7$

8. All real number except $-\frac{8}{3}$, or $x \neq -\frac{8}{3}$

9. All real number

10. All real number except 11, or $x \neq 11$

2-2 SIMPLIFYING VARIABLE EXPRESSION

An EQUATION is a statement consisting of making two expressions equal, or placing an = sign between the expressions. When we take two expressions with at least one of them with a variable (or variables) and make them equal, we can then try to solve for the variable values using properties of mathematic equality.

Before we can solve any variable equations, it is essential that both expression sides of the equation are SIMPLIFIED.

When we simplify expressions, we need to follow these rules through completion

1. remove parentheses by multiplying factors.
2. Combine like terms so that each variable to a specific power appear just once in the expression
3. combine the constants.
4. All constants and/or coefficients are simplified, including fractions.

SIMPLIFYING EXPRESSIONS TOOLS

Combining LIKE TERMS

LIKE TERMS are terms that have the same variable(s) to the same exponent power.

Examples of like terms are x, $4x$, $11x$, $.5x$, $-34x$. Note, each share a variable factor of x

Another set of like terms t^2, $4t^2, -t^2$, $10000t^2$. Note each share a variable factor of t^2

Another set of like terms are x^2y, $4x^2y$, $-.01x^2y$. Note each share a variable factor of x^2y

Also, numbers by themselves, called CONSTANTS, are consider like terms as well.

However, $4w$ and $10w^2$ are **not like terms**. They share a variable w, but each term has a different exponent value.

Other examples of terms that are not like are $2y^3$ *and* $9y^9$; 5 and x ; tr^3 *and* t^3r

ADDITION/SUBTRACTION RULE SAYS:

- We can add (or subtract) like terms and simplify.
- We cannot add (or subtract) terms that are not like to simplify. We need to leave the sum as it is.

HOW TO ADD (OR SUBTRACT) LIKE TERMS.

When the variable factor has a number in front of it, such as $3x$ or $7y^2$, we are multiplying the variable by that number. So $3x$ is 3 times x and $7y^2$ is 7 times y^2. The number in front is called the variable's COEFFICIENT. So for $3x$, 3 is the coefficient of x, For $7y^2$, 7 is the coefficient of y^2. When our term is just a variable, we are assuming multiplication by 1, so the coefficient is 1. So for the variable expression x, 1 is the coefficient.

Take the expression

$$4x^2 + 5x^2 + x^2$$

This expression is not simplified, as x^2 shows up 3 times in the expression.

We need to add, or combine, the like terms by adding the coefficients of each term and make that sum a single coefficient for our variable.

As 4 is the coefficient of $4x^2$, 5 is the coefficient of $5x^2$, and 1 is the coefficient of x^2, we can write this expression as

$$4x^2 + 5x^2 + x^2 = (4+5+1)x^2 = 10x^2$$

Here is an example of the difference of like terms:

$$10y^3 - 4y^3 = (10-4)y^3 = 6y^3$$

So the rule is

ADD/SUBTRACT LIKE TERMS BY ADDING/SUBTRACTING THE COEFFICIENTS AND KEEPING THE VARIABLE BASE AND ITS EXPONENT THE SAME.

Practice Simplifying these Expressions

$9a + 13a - 7a$ ________________

$13a + 9 + 9b + 12a - 7b$ ________________

$12z + 5 + 3z$ ________________

$10m + 12n - 5m + 3n + 5$ ________________

$4c^2 + 3c + 11 + 8c^2 - 9$ ________________

$13ab + 4b + 3ab - 2b^3 + 9a - 7b$ ________________

DISTRIBUTION PROPERTY

We need to use the DISTRIBUTIVE PROPERTY sometimes to simplify an algebraic expression. This gets rid of parentheses in the expression.

The **DISTRIBUTIVE PROPERTY** says

$$x(y + z) = xy + xz \text{ and } x(y - z) = xy - xz$$

Here are Examples of the distributive property

$3(x + 2) = 3(x) + 3(2) = 3x + 6$

$4(y - 2) = 4(y) + 4(-2) = 4y - 8$

$$x(x+7) = x(x) + x(7) = x^2 + 7x$$

$$x(w + x - 10) = x(w) + x(x) + x(-10) = xw + x^2 - 10x$$

$$\frac{2}{3}\left(\frac{4}{7}x + \frac{1}{2}\right) = \frac{2}{3}\left(\frac{4}{7}x\right) + \frac{2}{3}\left(\frac{1}{2}\right) = \frac{2 \cdot 4}{3 \cdot 7}x + \frac{2 \cdot 1}{3 \cdot 2} = \frac{8}{21}x + \frac{2}{6} = \frac{8}{21}x + \frac{1}{3}$$

What about $-(x - 8)$?

This also employs the distributive property, as we are multiplying –1 times $(x - 8)$

$$-(x - 8) \text{ is equal to } -1(x - 8) = -1(x) + -1(-8) = -x + 8.$$

Note: A quick way to simplify with multiplying by –1 in the distributive property is change the signs of the terms in the sum.

Example: $-1(d + 5) = -d - 5$ $-1(-w - 9) = w + 9$

Practice

$3(w + 11) =$ ____________

$5(k - 4) =$ ______________

$-12q(q - 9) =$ _________

$14(x + w + 3r) =$ ______________

$-123(-3x + 2.3y) =$ __________

$3t(t^2 - t + 2) =$ ____________

$\frac{1}{4}\left(\frac{5}{6}p + \frac{12}{5}\right) =$ ___________

$\frac{3}{4}x\left(b - \frac{4}{3}\right) =$ _________

2-2 Assignments

Simplify these expressions

1. $19x + 12x - 7x$ ______________
2. $5y - 6y + 11y - 5y$ _______________
3. $34g + 12g - 5g + 200g$ ___________
4. $10f + 9 + 29h + 12f - 7 + 11h$ ____________
5. $12.5k + 5 + 3.23k - 5$ ________________

6. $10x^2 + 12x - 35x + 3x^2 + 5$ ________________

7. $24d^2 + 3c + 11d + 22c^2 - 9d + c^2$ ________________

8. $3ab + 6b + 13ab - 2a^3 + 19a - 17b$ ____________

9. $100x^2 + 2x - 2 + 5x - 23x^2 + 5 - 77x^2 - 3$ ____________

10. $7(k + 10) =$ ______________

11. $-(3k - 8)$ _______________

12. $25(3h - 4) =$ ________________

13. $-r(r + 22) =$ __________

14. $18(2x + 3w + 4y) =$ ________________

15. $2(-7x + 1.3y) =$ ____________

16. $6y(y^2 + 2y - 12) =$ ______________

17. $\frac{1}{3}\left(\frac{5}{7}h + \frac{2}{5}\right) =$ _____________

18. $\frac{3}{8}x\left(16x - \frac{8}{3}\right) =$ ___________

19. $6(3x + 2) - 9(4x - 8) + 13x$ ___________________

20. $4t(k + 9h + p) - 6tk + 4th + 9tp$ ________________

21. $8(3m + 2) - 9(4n + 8) + 7(m + n)$ _______________

22. $7(4k + 6y - 10x) - 2(14k + 21y - 35x)$ ___________

Answers

1. $24x$
2. $5y$
3. $241g$
4. $22f + 40h + 2$
5. $15.73k$
6. $13x^2 - 23x + 5$
7. $24d^2 + 23c^2 + 2d + 3c$

8. $-2a^3 + 16ab + 19a - 11b$

9. $7x$

10. $7k + 70$

11. $-3k + 8$

12. $75h - 100$

13. $-r^2 - 22r$

14. $36x + 54w + 72y$

15. $-14x + 2.6y$

16. $6y^3 + 12y^2 - 72y$

17. $\frac{5}{21}h + \frac{2}{15}$

18. $6x^2 - x$

19. $-5x + 84$

20. $-2tk + 40th + 13tp$

21. $31m - 29n - 56$

22. 0

2-3 SOLVING SIMPLE VARIABLE EQUATIONS

Remember, we have to SIMPLIFY OUR ALGEBRAIC EXPRESSIONS in order to SOLVE ALGEBRAIC EQUATIONS. Here are the rules for simplifying expressions again.

Remove parentheses by multiplying factors.

Combine like terms so that each variable to a specific power appear just once in the expression

Combine the constants.

All constants and/or coefficients are simplified.

Therefore: An expression is simplified if

- there are no parentheses,
- all variables raised to a particular power appears only once,
- all constants are reduced to one constant

Only when we have simplified expressions on both sides of the equation can we then solve our variable in an equation.

SOLVING FOR A VARIABLE IN AN EQUATION

We need some tools to arrive at a solution. We can use the principle OF EQUALITY:

> WHATEVER YOU DO TO ONE SIDE OF THE EQUATION, YOU DO TO THE OTHER SIDE TO RETAIN EQUALITY.

So You can do the following to both sides of an equation

- Add
- Subtract
- Multiply
- Divide
- Raise to the same exponent power
- Take the square root, cube root, 4th root, ...
- Take the logarithm

And your equation remains EQUAL.

The goal is to do these operations in order to **ISOLATE YOUR VARIABLE** so it equals eventually a number value. And the strategy is to use inverse operations on both sides of the equation to undo operations already there.

Solve $11x + 6x = 204$

The left side is not simplified as the x appears twice. We need to combine terms on the left sides.

$11x + 6x = (11 + 6)x = 17x$

So $11x + 6x = 204$ becomes

$17x = 204$. Now both sides of the equation are simplified.

Now to isolate the x, we need to "undo" the multiplication of $17x$ by dividing both sides of the equation by 17

$$17x = 204$$

$\div 17 \quad \div 17$ we get

$x = 12$ which is our solution

Solve $4(x + 5) - 2x + 9 + 2(x + 4) = 233$

We need to use the distributive property twice on the left side

$$4(x + 5) - 2x + 9 + 2(x + 4) = 4(x) + 4(5) - 2x + 9 + 2(x) + 2(4) = 4x + 20 - 2x + 9 + 2x + 8$$

We need to combine like terms so there is only one x and one constant on the left side.

We can rearrange our terms so that like terms are together

$$4x + 20 - 2x + 9 + 2x + 8 = 4x - 2x + 2x + 20 + 9 + 8,$$

now we can combine like terms

$$4x - 2x + 2x + 20 + 9 + 8 = (4 - 2 + 2)\,x + (20 + 9 + 8) = 4x + 37$$

So $4(x + 5) - 2x + 9 + 2(x + 4) = 233$ becomes $4x + 37 = 233$

We need to isolate the x on the left side. We can use subtraction then division to undo the operations

$$\begin{array}{rcl} 4x + 37 & = & 233 \\ -37 & & -37 \\ \hline 4x & = & 196 \\ \div 4 & & \div 4 \\ \hline x & = & 196 \end{array}$$

subtract 37 both sides

divide both sides by 4

which is our solution

What if there are variables on both sides of an equation? We still need to simplify each side first, then we use properties of equality to get numbers one side and the variable on the other side of the equation.

Solve $3(x + 2) = 7(x - 6)$, We need to simplify both sides.

We use the distributive property both sides, $3(x) + 3(2) = 7(x) + 7(-6)$ which becomes $3x + 6 = 7x - 42$

The equation is now simplified on both sides

$$\begin{array}{rcl} 3x + 6 & = & 7x - 42 \\ -6 & & -6 \\ \hline 3x & = & 7x - 48 \\ -7x & & -7x \\ \hline -4x & = & -48 \\ \div -4 & & \div -4 \\ \hline x & = & 12 \end{array}$$

we can subtract 6 from both sides

we can subtract $7x$ from both sides

we divide both sides by -4

our solution

Solve this one: $.25C + 3 = C$

Remember, when there is no number in front of our variable C, we assume that the coefficient is 1 and we are multiplying by C by 1.

$$\begin{array}{rcl} .25C + 3 & = & 1.C \\ -.25C & & -.25C \\ \hline 3 & = & (1 - .25)C \\ 3 & = & .75\,C \\ \div .75 & & \div .75 \\ \hline 4 & = & C \end{array}$$

both sides are simplified. So we can subtract $.25C$ from both sides

As $1 - .25 = .75$

Divide both sides by .75

our solution

Practice

$3x + 12 = 33$

$\frac{1}{2}(m+4) = 49$

$3y + 8(y - 4) = -43$

$4p + 9p + 4(p - 6) = 19(p + 2)$

$.5x + 2(x - 4) = 4(x + .75)$

$9(g + 2) - 4g + 12 = 2(2g + 15)$

2-3 Assignments

Solve for the variable.

1. $4x = 48$
2. $5y - 1 = 19$
3. $32t + 100 = -60$
4. $5t + 43 = -68$
5. $23x + 14 = 111$
6. $2.5t + 3.5 = 12$
7. $.5(m - 6) = 40$
8. $11y + 10(y - 8) = 100$
9. $1.5x + 2.5(3x - 8) = 24$
10. $w + 7w - 4(w - 22) = 11(w + 3)$
11. $.5r + 6(r - 70) = 2(r + 1.75)$
12. $8(d + 8) + 7(d - 2) + 6(4d - 9) = 7(d - 1)$
13. $7(2g + 4) - 8g + 6 = 2(9g + 19)$
14. $7(11r - 4r) + 4 = 34r$
15. $8(p + 2) - 6(p - 5) + 8(p - 5) = 6(p - 2) + 8(p - 3) + 2(p + 6)$

Answers

1. $x = 12$
2. $y = 4$

3. $t = -5$
4. $t = -\frac{111}{5}$
5. $x = \frac{97}{23}$
6. $t = 3.4$
7. $m = 86$
8. $y = \frac{60}{7}$
9. $x = 4.\overline{8}$
10. $w = \frac{55}{7}$
11. $r = 94.\overline{1}$
12. $d = -\frac{3}{32}$
13. $g = -\frac{1}{3}$
14. $r = -\frac{4}{15}$
15. $p = 5$

2-4 RATIOS AND PROPORTIONS

Another name for a fraction is a RATIO. When one fraction equaling another fraction it is called a PROPORTION. For example, $\frac{1}{2} = \frac{5}{10}$, $\frac{13}{39} = \frac{1}{3}$, or $\frac{x}{3} = \frac{7}{35}$ are all proportions. We say the numerator to denominator changes in the same way for both fraction/ratios in proportions.

When we have ratios (or fractions) with a variable in our equation, we call these RATIONAL EQUATIONS. When the rational equation involves a proportion, we can solve it by using what is called **CROSS MULTIPLICATION**. It is also called the **MEANS/EXTREMES** property.

In any proportion, the means/extremes property says for

$$\text{if } \frac{a}{b} = \frac{c}{d}, \text{ then } ad = bc$$

The numerator a and denominator d are considered the "extremes" of the proportion, and the denominator b and numerator c are considered the "means" of the proportion. So given the proportion, the product of the extremes equals the product of the means.

You should see the "cross multiplying" in this property

$$\frac{a}{b} = \frac{c}{d} \text{ becomes } ad = bc$$

Solve $\frac{3}{5} = \frac{x}{10}$ for x we cross multiply to get $(3)(10) = (5)(x)$ or

$30 = 5x$ divide both sides by 5

$\div 5 \quad \div 5$

$6 = x$ our solution

So $\frac{3}{5} = \frac{6}{10}$

Example: Solve for y $\frac{3}{y} = \frac{11}{430}$ cross multiply $430(3) = y(11)$

$1290 = 11y$ divide both sides by 11

$\div 11 \quad \div 11$

$\frac{1290}{11} = y$

Practice solve for variable $\frac{2}{3} = \frac{w}{50}$ $\quad \frac{7}{11} = \frac{2}{h}$

$\frac{2q}{5} = \frac{4}{9}$ $\quad \frac{5}{9.6} = \frac{2}{4.5x}$

WHAT IF YOU HAVE A SUM OR DIFFERENCE IN A NUMERATOR OR DENOMINATOR IN A TWO RATIONAL EXPRESSION EQUATION

Example: $\frac{x+3}{x} = \frac{3}{4}$ Because we have single term fractions (and the $x + 3$ forms a single value in the numerator), we, therefore, have a proportion, and we can use cross multiplication to solve for x.

$\frac{x+3}{x} = \frac{3}{4}$ becomes $4(x + 3) = 3(x)$ using distributive property on the left, we get

$4x + 12 = 3x$ subtract $4x$ from both sides

$-4x \quad -4x$

$12 = -x$ divide both sides by -1

$\div -1 \quad \div -1$

$-12 = x$

$\frac{4}{h+3} = \frac{11}{2h-3}$ This is a proportion, so cross multiply

$\frac{4}{h+3} = \frac{11}{2h-3}$ becomes $4(2h - 3) = 11(h + 3)$ distribute both sides

$8h - 12 = 11h + 33$ add 12 both sides

$+12 \quad +12$

$8h = 11h + 45$ subtract $11h$ both sides

$$\begin{aligned} \underline{-11h \quad -11h} \\ -3h = 45 \\ \underline{\div -3 \quad \div -3} \\ h = -15 \end{aligned}$$

divide both sides by -3

Practice

$\frac{x}{x+1} = \frac{13}{15}$ $\qquad$ $\frac{p-5}{2} = \frac{p}{200}$

$\frac{r+2}{5} = \frac{r-10}{4}$ $\qquad$ $\frac{m-5}{7-m} = \frac{9}{13}$

REMEMBER YOU CANNOT CANCEL ACROSS A SUM OR DIFFERENCE IN A FRACTION

we know that $\frac{24+8}{8} = \frac{32}{8} = 4$

This is not the same as $\frac{24+\not{8}}{\not{8}} = \frac{24+1}{1} = 25.$

You cannot cancel out the 8s across the sum in the numerator.

For $\frac{20x+11}{5}$, you cannot cancel the 5 out of just the $20x$ to get $\frac{4x+11}{1}$. The only way you can cancel out the 5 is if 5 is a common factor of both terms in the numerator. Because 5 is not a factor of 11, the best we can do is to keep it as $\frac{20x+11}{5}$. It is simplified!

If you had $\frac{20x+15}{5}$, the 5 is a factor of both $20x$ and 15, so we can cancel out a 5 in each term to get $\frac{20x+15}{5} = 4x+3$. So it is now simplified.

YOU CAN ONLY CROSS MULTIPLY WHEN YOU HAVE TWO TERMS IN A RATIONAL EQUATION

So . . . You cannot simply cross multiply when you have more than two terms in a rational equation. For example, $\frac{y}{2} + 8 = \frac{4}{11}$ has three terms, so you cannot cross multiply as is. However, you can change this into a two term proportion and then use cross multiplication. This will be discussed in Section 2-5.

We can use cross multiplication to solve proportion word problems involving conversions.

Example: Richie Rich is vacationing in Europe. He needs to convert \$450 to Euros. Richie checked the exchange rate and found it to be 1 Euro = \$1.23

We want to find how many Euros Richie will get. So first let x = number of Euros.

The relationship between Euros and Dollars is the ratio $\frac{1 \text{ Euro}}{\$1.23}$

We can now set up a proportion to find the equivalent Euro value for $450.

We get $\frac{1 \text{ Euro}}{\$1.23} = \frac{x \text{ Euros}}{\$450.00}$ We can write this without units of measure as $\frac{1}{1.23} = \frac{x}{450}$ Now we cross multiply $(1)(450) = (1.23)(x)$ or $450 = 1.23x$

$$\begin{array}{cc} \div 1.23 & \div 1.23 \\ \hline 365.85 = x & \end{array}$$

Rounding to the nearest hundredth of a Euro, $450 = 365.85 Euros

Example: Mr. Magoo wants to go to Canada. He has $2900.03 and needs to convert it to Canadian Dollars.

Currently the exchange ratio is 1.00 Canadian $ = $0.81 US

Let C = Canadian dollar amount

Set up exchange ratio $\frac{\$1.00 \text{ Canadian}}{\$0.81}$

Set it equal to the ratio $\frac{C \text{ \$ Canadian}}{\$2900.03}$ or $\frac{1.00}{0.81} = \frac{C}{2900.03}$

Solve $\frac{1.00}{0.81} = \frac{C}{2900.03}$ by cross multiplying

$$\begin{array}{l} (1.00)(2900.03) = (0.81)(C) \\ 2900.03 = 0.81C \\ \div 0.81 \quad \div 0.81 \\ \hline 3580.28 = \quad C \end{array}$$

So $2900.03 is equal to $3580.28 Canadian

Example: Wellington weights 30 kg. What is his weight in pounds?

1 kg is about 2.2 lb.

Let P = number of pounds.

Conversion ratio is $\frac{1 \text{ kg}}{2.2 \text{ lb.}}$, set it equal to $\frac{30 \text{ kg}}{P \text{ lb.}}$ or $\frac{1}{2.2} = \frac{30}{P}$

Solve $\frac{1}{2.2} = \frac{30}{P}$ by cross multiplying $(1)(P) = (2.2)(30)$ or $P = 66$

So 30 kg is equal to about 66 lb. Therefore, Wellington weight is about 66 lb.

Example: Fast Freddy drives his motor car at 88 mph. How many kilometers does he travel in 300 min?

We need to do two conversions, turning 88 mi. into km and turning 300 min into hr.

The conversion factor for miles to kilometers is 1 km = 0.62 mi.

Let k = number of kilometers

So $\frac{1 \text{ km}}{0.62 \text{ mi.}} = \frac{k \text{ km}}{88 \text{ mi.}}$ or $\frac{1}{0.62} = \frac{k}{88}$

Solve $\frac{1}{0.62} = \frac{k}{88}$ by cross multiplying $(1)(88) = (0.62)(k)$

$$\begin{array}{c} 88 = 0.62k \\ \div 0.62 \quad \div 0.62 \\ \hline 141.9354\ldots = k \end{array}$$

Rounding to the nearest hundredth of a kilometer, 88 mi. = 141.94 km. So if Fast Freddy drives 88 mi. in 1 hr, then he drives 141.94 km in 1 hr.

Now we need to convert 300 min to hr

Let h = number of hours

Since there is 60 min in 1 hr, then $\frac{1 \text{ hr}}{60 \text{ min}} = \frac{h \text{ hr}}{300 \text{ min}}$ or $\frac{1}{60} = \frac{h}{300}$

Solve $\frac{1}{60} = \frac{h}{300}$ by cross multiplying $(1)(300) = (60)(h)$

$$\begin{array}{c} 300 = 60\text{h} \\ \div 60 \quad \div 60 \\ \hline .5 \;=\; h \end{array}$$

So, 300 min = 5 hr.

We now need the number of kilometers driven in 5 hr, let x = kilometers driven in 5 hr, and we can say $\frac{141.94 \text{ km}}{1 \text{ hr}} = \frac{x \text{ km}}{5 \text{ hr}}$ or $\frac{141.94}{1} = \frac{x}{5}$

Cross multiply $(141.94)(5) = 1(x)$

$$709.70 = x,$$

so Fast Freddy drive to get $709.70 = x$ km in 300 min

Practice

Quagmire has 410 g of mercury. He needs to convert this to ounces. 1 g = 0.04 oz.

Pep Cee has 123 fl. oz. of a tasty beverage. Pep needs to know how many liters of the beverage he has.

I L = 33.81 fl. oz.

Watson's Elementary Dairy sells milk for $1.19 per 16 fl. oz.

What would 3.2 gal. of milk cost if this rate applies.

There are 128 fl. oz. in 1 gal.

2-4 Assignments

Solve for the variable

1. $\frac{5}{6}=\frac{w}{60}$ __________

2. $\frac{q}{12}=\frac{5}{9}$ __________

3. $\frac{y}{17}=\frac{23}{51}$ __________

4. $\frac{3}{7}=\frac{72}{f}$ __________

5. $\frac{7}{11.5}=\frac{6}{3.5x}$ __________

6. $\frac{9}{13}=\frac{3}{1.7h}$ __________

7. $\frac{3m}{7}=\frac{m-2}{31}$ __________

8. $\frac{x}{x+2}=\frac{19}{35}$ __________

9. $\frac{r-9}{3}=\frac{r}{100}$ __________

10. $\frac{5+2d}{9}=\frac{d-11}{4}$ __________

11. $\frac{k-5}{9-k}=\frac{8}{13}$ __________

12. $\frac{h+15}{h-20}=\frac{12}{13}$ __________

Solve the following using the conversion equation in proportions

13. Henry has 56 Liters of Lemonade. He needs to know how many quarts of Lemonade he has. 1.00 quart = .95 Liters.

14. Delores has 456 grams of toothpaste. She needs to know how many ounces of toothpaste she has. 1.00 ounce = 28.35 grams

15. Paula walked 4.5 miles. How many kilometers did Paula walk? 1 mile = 1.61 kilometers

16. The package has a volume of 460 cubic feet. How many cubic meters is the package? 1.0 cubic meter = 35.32 cubic feet

17. The new Howdy Doody postage stamp has an area of 2.08 in^2. What is the area of this stamp in cm^2? 1.0 $in^2 = 6.45\ cm^2$

18. Mom and Pop's Shopette ordered 910.5 pints of half and half cream. How many gallons of half and half did they order? 1.0 gallon = 8.0 pints

19. Speedie ran the 200 meter race in 25.5 seconds. How much of an hour did it take Speedie to run the race? 1.0 hour = 3600.0 seconds

20. A great whale weighed 9560.4 kilograms. What is the weight in tons? 1.0 ton = 907.03 kilograms

21. Wendell' Plumbing Supply sold metal pipping at $36.50 per 15 yard length. What would 75.4 meter of piping cost. 1.0 meter = 1.09 yards

22. It cost $2 per gallon of pool cleaner to keep the pool antiseptic. It takes 1 gallon of pool cleaner to properly clean 9 gallons of pool water. The swimming pool currently has 39600 cubic inches of water. a) How many gallons of water are in the pool? b) And how much does it cost for pool cleaner? 1 gallon = 231 cubic inches.

Answers

1. $w = 50$
2. $q = \frac{20}{3}$
3. $y = \frac{23}{3}$
4. $f = 168$
5. $x = 2.82$ or $\frac{138}{49}$
6. $h = 2.55$ or $\frac{130}{51}$
7. $m = -\frac{7}{43}$
8. $x = \frac{19}{8}$
9. $r = \frac{900}{97}$
10. $d = 119$
11. $k = \frac{137}{21}$
12. $h = -435$
13. $\frac{x\ quart}{56\ Liters} = \frac{1\ quart}{.95\ Liters}$, 58.95 quarts
14. $\frac{x\ ounce}{456\ grams} = \frac{1\ ounce}{28.35\ grams}$, 16.08 ounces
15. $\frac{x\ kilometers}{4.5\ miles} = \frac{1.61\ kilometers}{1\ mile}$, 7.25 kilometers

16. $\dfrac{x\ cubic\ meters}{460\ cubic\ feet} = \dfrac{1\ cubic\ meters}{35.32\ cubic\ feet}$, 13.02 cubic meters

17. $\dfrac{x\ cm^2}{2.08\ in^2} = \dfrac{6.45\ cm^2}{1\ in^2}$, 13.42 cm^2

18. $\dfrac{x\ gallons}{910.5\ pints} = \dfrac{1\ gallon}{8\ pints}$, 113.81 gallons

19. $\dfrac{x\ hours}{25.5\ seconds} = \dfrac{1\ hours}{3600\ seconds}$, .007 hour

20. $\dfrac{x\ tons}{9560.4\ kilograms} = \dfrac{1\ ton}{907.03\ kilograms}$ 10.54 tons

21. $\dfrac{x\ yd}{75.4\ meters} = \dfrac{1.09\ yd}{1\ meter}$, 82.19 yards. 82.19 yards (\$36.50/15 yds.) = \$200.00

22. a) $\dfrac{x\ gallons}{39600\ cubic\ inches} = \dfrac{1\ gallon}{231\ cubic\ inches}$, so $x = 171.43$ gallons of pool water,

b) $\dfrac{y\ gallons\text{ of pool cleaner}}{171.43\ gallons\text{ of pool water}} = \dfrac{1\ gallon\text{ of pool cleaner}}{9\ gallons\text{ of pool water}}$, so $y = 19.05$ gallons of pool cleaner, and $\left(\dfrac{\$2.00}{1\ gallon\text{ of pool cleaner}}\right)(19.05\ gallons\text{ of pool cleaner}) = \$\,38.10$

2-5 SOLVING RATIONAL EQUATION

In Section 2-4, we dealt with rational equations when you have a two term variable proportion, using means-extremes property, or cross multiplication to solve them. And it was noted that **YOU CAN ONLY CROSS MULTIPLY WHEN YOU HAVE TWO TERMS IN A RATIONAL EQUATION.**

What if you have more than two terms in a rational equation? We will show two methods to solve these: (A) CONVERTING INTO A PROPORTION and (B) MULTIPLYING BY A COMMON DENOMINATOR MULTIPLE.

A) CONVERTING INTO A PROPORTION

For Example: $\dfrac{y}{2} + 8 = \dfrac{4}{11}$

The goal is to get the $\dfrac{y}{2}$ on one side and a fraction on the other side. We can accomplish this by subtracting 8 both sides.

$$\frac{y}{2} + 8 = \frac{4}{11} \quad \text{subtracting 8 both sides}$$

$$\underline{-8 \qquad -8}$$

$$\frac{y}{2} = \frac{4}{11} - 8 \quad \text{For the left side using the fraction subtraction rules,}$$

$$\frac{4}{11} - 8 = \frac{4}{11} - \frac{8}{1} = \frac{4}{11} - \frac{8 \cdot 11}{1 \cdot 11} = \frac{4}{11} - \frac{88}{11} = \frac{4-88}{11} = \frac{-84}{11}$$

So $\frac{y}{2} = \frac{-84}{11}$ Now we have a proportion, and we can <u>cross multiply</u>

$\frac{y}{2} = \frac{-84}{11}$ Becomes $11y = 2(-84)$ or $11y = -168$ divide both sides by 11

$$\underline{\div 11 \quad \div 11}$$

$$y = \frac{-168}{11}$$

Example: $\frac{5}{y} + \frac{1}{2} = \frac{13}{4}$ subtract $\frac{1}{2}$ both sides

$$\underline{-\frac{1}{2} \quad -\frac{1}{2}}$$

$$\frac{5}{y} = \frac{13}{4} - \frac{1}{2} = \frac{13}{4} - \frac{2}{4} = \frac{13-2}{4} = \frac{11}{4}$$

So $\frac{5}{y} = \frac{11}{4}$ Now we have a proportion, so we cross multiply

$5(4) = y(11)$

$20 = 11y$ divide both sides by 11

$\underline{\div 11 \div 11}$

$\frac{20}{11} = y$

Practice using conversion to proportion

$$\frac{2x}{3} + \frac{5}{6} = \frac{10}{6}$$

$$\frac{5}{2y} + 2 = \frac{1}{2}$$

$$\frac{5x}{11} - \frac{7}{9} = \frac{1}{9}$$

$$1 + \frac{11}{3t} = 2\frac{12}{13}$$

B) MULTIPLYING BY A COMMON DENOMINATOR MULTIPLE

Another handy technique to solve a rational is one designed to eliminate all the denominators, so NO MORE FRACTION!!!! YEAH! This is accomplished by find a common multiple of each of the denominators and then multiplying both sides of your equation.

Here are the steps

- Simplify each fraction in our rational equations, if possible.
- Then multiply both sides of the equation by a COMMON MULTIPLE OF ALL THE DENOMINATORS. Ideally, the least common multiple (LCM) of the denominators is the best one to use.
- Then use algebra steps, factoring, and so on, to solve.

NOTE: If you don't know the LCM of the denominators, you can always find the product of all the denominators, and use that as your multiplier. It will cancel out the denominators, but may require simplifying your answer at the end.

Example:

Solve $\frac{x}{2} + \frac{x}{3} = 10$

The denominators of our three terms are 2, 3, and 1.

The LCM of the denominators 2, 3, (and 1) is 6. So we want to multiply both sides of the equation by 6 to eliminate the denominators.

$6\left(\frac{x}{2} + \frac{x}{3}\right) = 6(10)$ use distributive property right side

$6\left(\frac{x}{2}\right) + 6\left(\frac{x}{3}\right) = 6(10)$ multiply through both sides

$3x + 2x = 60$ Note: Fractions eliminated!! Now combine like terms right side

$5x = 60$ divide both sides by 5

$x = 12$

Another example: $\frac{x+2}{3} = \frac{x-4}{7}$

The LCM of 3 and 7 is 21, so multiply both sides by 21,

$$21 \cdot \frac{x+2}{3} = 21 \cdot \frac{x-4}{7}$$

$7(x + 2) = 3(x - 4)$ use distributive property both sides

$7x + 14 = 3x - 12$ subtract 14 both sides

$7x = 3x - 26$ subtract $3x$ both sides

$4x = -26$ divide 4 both sides

$x = -\frac{26}{4} = -\frac{13}{2}$

Example: solve for x $\frac{2}{3} + \frac{3}{4x} = \frac{7}{12}$ the LCM of 3, $4x$, and 12 is $12x$

So multiply both sides by $12x$

$$12x\left(\frac{2}{3} + \frac{3}{4x}\right) = 12x\left(\frac{7}{12}\right) \text{ use distribute property left side}$$

$12x\left(\frac{2}{3}\right) + 12x\left(\frac{3}{4x}\right) = 12x\left(\frac{7}{12}\right)$ now carry out multiplication canceling out denominators!!

$4x(2) + 3(3) = x(7)$ carry out the multiplication both sides

$8x + 9 = 7x$ Subtract $8x$ both sides

$\underline{-8x \quad -8x}$

$9 = -x$ or $x = -9$

Practice using multiplying by Common Denominator

Solve for variable

$$\frac{p}{8} + \frac{p}{4} = 100$$

$$\frac{w}{5} + \frac{2}{3} = \frac{w}{3}$$

$$\frac{x}{7} + \frac{4x}{5} = \frac{13}{35}$$

$$\frac{2}{5} + \frac{9}{y} = \frac{7}{6}$$

$$\frac{3}{d} + \frac{5}{6} = \frac{1}{6d}$$

$$\frac{p+9}{3} = \frac{p-4}{8} + \frac{1}{2}$$

2-5 Assignments, Use any Method You Prefer

1. $\frac{r}{7} + \frac{r}{21} = 336$

2. $\frac{w}{9} + \frac{w}{81} = 20$

3. $\frac{c}{24} + \frac{5}{12} = \frac{c}{2}$

4. $\frac{k}{15} + \frac{9}{30} = \frac{7k}{90}$

5. $\frac{p}{17} + \frac{6p}{51} = \frac{1}{3}$

6. $\frac{4}{5} + \frac{22}{t} = \frac{5}{6}$

7. $\frac{30}{n} + \frac{25}{11} = \frac{9}{22n}$

8. $\frac{h+5}{4} = \frac{h-4}{12} + \frac{1}{6}$

9. $\frac{5}{p} + \frac{7}{45} = \frac{7}{15}$

10. $\frac{13g}{14} + \frac{g}{7} = \frac{11}{35}$

11. $\frac{a+8}{11} + \frac{1}{4} = \frac{a+49}{44}$

12. $\frac{b+9}{7} = \frac{b-4}{8} + \frac{1}{2}$

13. $\frac{15}{x} + \frac{31}{3x} = \frac{7}{15}$

Answers

1. $r = 1764$
2. $w = 162$
3. $c = \frac{10}{11}$
4. $k = 27$
5. $p = \frac{17}{9}$
6. $t = 660$
7. $n = -\frac{651}{50}$
8. $h = -\frac{17}{2}$
9. $p = \frac{225}{14}$
10. $g = \frac{22}{75}$
11. $a = 2$
12. $b = -72$
13. $x = \frac{380}{7}$

2-6 FORMULAS

A FORMULA is an equation that uses multiple unknowns to describe a relationship of fact. For our purposes, a formula will be any equation with more than one unknown (or variable).

The mathematics of Industry uses many formulas to find answers. So we need to know how to use formulas to our best advantage!

Which of these five equations are formulas given our definition??

1. $2x + 3y = 12$
2. $V = \dfrac{QR}{3}$
3. $2K - (5 + K) = 121$
4. $3x^2 + 2x + 1 = 100$
5. $I = Prt$

Equations 1), 2), and 5) are formulas.

Equation 1) has two unknowns, x and y.

Equation 2) has three unknowns, V, Q, and R.

Equation 5) has four unknowns, I, P, r, and t.

Equations 3) and 4) each has one unknown, note: Equation 3) has K as an unknown, even though it shows up twice. It can be reduced, when distributing and combining like terms, to $K - 5 = 121$. So there is only one value to find for the unknown K. Equation 4) has one unknown, x, even though it appears twice, and one time as a square. There is still only one value to find for the x. Therefore, equations 3) and 4) are not formulas.

The First use of formulas is to substitute known values in all but one of the unknown values, then use the algebra steps to solve for the remaining variable.

For example, the formula $V = IR$ is used by electricians, where

- V = Volts (for voltage),
- I = Amps of current through a circuit, and
- R = Ohms of resistance in the circuit.

What is the Volts given .23 Amps and 400 Ohms,

Lets make $I = .23$ and $R = 400$, and substitute into $V = IR$

So $V = IR = (.23)\,(400) = 92$. So the Volts is 92 Volts

What If there are 100 Volts, and .67 Amps, and we want to find the Ohms of resistance?

we make $V = 100$, $I = .67$ then substitute into $V = IR$ as

so $V = IR$ becomes

$$\begin{array}{l} 100 \;=\; .67\,R. \\ \div .67 \div .67 \\ \hline 149.25 = \;R, \end{array}$$

We need to then divide both sides by .67

so there is 149.25 Ohms of resistance.

So the strategy is solve in formulas is

a. substitute for all but one of the unknowns (variables),

b. use reduction and equality properties in correct order of operation so that the unknown (variable) is isolated on one side of the equation.

Example: A simple one: $y = 3x + 7$ is a linear equation, where (x, y) represents points on the line when graphed,

Lets Solve for x, given $y = 13$ Substitute 13 for y in $y = 3x + 7$ we get

$13 = 3x + 7$ now subtract 7 both side

$\underline{-7 \quad\quad -7}$

$6 = 3x,$ divide both sides by 3

$\underline{\div 3 \quad \div 3}$

$2 = x$ So $x = 2$ when $y = 13$ in $y = 3x + 7$

Example:

In business, the formula $I = Prt$ is an important one. It is called the SIMPLE INTEREST FORMULA, where I = Interest accumulated, P = money borrowed (or principal) r = interest rate, t = time of investment.

Find I if $P = 2500$, $r = .06$, and $t = \frac{1}{12}$. We substitute these values into the formula as

$$I = Prt = 2500\,(.06)\left(\frac{1}{12}\right) = \frac{2500\,(.06)\,(1)}{12} = 12.50$$

Find r if $I = 45$, $P = 4600$ and $t = .25$. We substitute these values into the formula $I = Prt$ as

$45 = 4600\,(r)\,.25$ or $45 = 1150\,r$, divide both sides by 1150

$\underline{\div 1150 \div 1150}$

$.012 = r$

A formula for the relationship between Celsius and Fahrenheit temperature degrees is $C = \frac{5}{9}(F - 32)$ with C the degree Celcius and F the degree Fahrenheit

If $F = 95$ degrees, what is Celsius? Our formula becomes

$C = \frac{5}{9}(95 - 32) = \frac{5}{9}(F - 32) = \frac{5}{9}(63) = 35$ so 95 degree Fahrenheit = 35 degree Celsius

If $C = -20$, what is Fahrenheit? Our formula becomes

$-20 = \frac{5}{9}(F - 32)$ use distributive property right side

$$-20 = \frac{5}{9}F - \frac{5}{9}(32) \qquad \text{or}$$

$$-20 = \frac{5}{9}F - 17\frac{7}{9} \qquad \text{add } 17\frac{7}{9} \text{ both sides}$$

$$+17\frac{7}{9} \quad + \quad 17\frac{7}{9}$$

$$-2\frac{2}{9} = \frac{5}{9}F \qquad \text{multiply both sides by } \frac{9}{5}$$

$$\frac{9}{5}\left(-2\frac{2}{9}\right) = \frac{9}{5}\left(\frac{5}{9}F\right) \text{ or } \frac{9}{5}\left(-\frac{20}{9}\right) = \frac{9}{5}\left(\frac{5}{9}F\right) \text{ carry out the multiplication}$$

$-4 = F$ so -20 degrees Celsius is -4 degree Fahrenheit

Practice

For the formula $C = 2\,(3.14r)$ find C if $r = 2.3$

Find r if $C = 200$

For $I = Prt$ find I if $P = 3000$, $r = .03$, $t = 20$

Find t if $P = 11000$, $r = .09$, $I = 4620$

For $B = \frac{A}{(1+rt)^2}$ Find B if $A = 1000$, $r = .02$ and $t = 12$

Find A if $B = 1234$, $r = .065$ and $t = 365$

For $P = \frac{y}{2q} + 1$ Find P if $y = 88$, $q = 22$

Find q if $P = 10$, $y = 18$

2-6 Assignments

Solve these formulas

1. for $3xy + p = 12$
   a) find y when $x = 8$ and $p = 2$ ___________
   b) find p when $x = 5$ and $y = -8$ ___________

2. for $2.3r - y + 6 = 21$
   a) find r when $y = 7$ ___________
   b) find y when $r = 4.23$ ___________

3. for $\frac{5+2x}{3y} = 4$
   a) find x when $y = 2$ ___________
   b) find y when $x = -17$ ___________

4. for $2q - 8p + t^2 = 2p$
   a) find q when $p = 3$ and $t = 4$ ___________
   b) find p when $q = .5$ and $t = 7$ ___________

5. for $2x + 2y = A$
   a) find y when $x = 5$ and $A = 7.8$ ___________
   b) find A when $x = -3$ and $y = 3$ ___________

6. for $I = prt$
   a) find r when $I = 600$, $p = 2340$ and $t = 3$ ___________
   b) find p when $I = 212$, $r = .04$, and $t = 12$ ___________

7. for $A = 6.28\,r^2 + 6.28\,rh$
   a) find h when $A = 600$ and $r = 4$ ___________
   b) find A when $r = 3$ and $h = 6.5$ ___________

8. for $\frac{g-h}{k-m} = 45$
   a) find h when $g = 4$, $k = 12$, and $m = 7$ ___________
   b) find k when $g = 110$, $h = 20$, and $m = 31$ ___________

9. for $40 - 3y + x - 8w = 34 + 6y - 5x + w$
   a) find w when $y = 11$ and $x = 2$ ___________
   b) find y when $x = 20$ and $w = 10$ ___________
   c) find x if $y = 2$ and $w = 1$ ___________

10. for $V = \frac{1 + rp}{\left(1 - \frac{r}{t}\right)^2}$
   a) find V for $r = .06$, $p = 100$ and $t = 12$ ___________
   b) find p for $V = 2300$, $r = .8$, and $t = 2$ ___________

Answers

1. a) $y = \frac{5}{12}$ b) $p = 132$

2. a) $r \approx 9.57$ b) $y = -5.27$

3. a) $x = \frac{19}{2}$ **b)** $y = -\frac{29}{12}$

4. a) $q = 7$ **b)** $p = 5$

5. a) $y = -1.1$ **b)** $A = 0$

6. a) $r \approx .085$ **b)** $p \approx 441.67$

7. a) $h \approx 19.89$ **b)** $A \approx 178.98$

8. a) $h = -221$ **b)** $k = 33$

9. a) $w = -9$ **b)** $y = 4$ **c)** $x = 3.5 \text{ or } = \frac{7}{2}$

10. a) $V \approx 7.07$ **b)** $p = 1033.75$

2-7 LITERAL EQUATIONS

Sometimes you have a formula, and you need to solve for some variable other than the "standard" one. For instance, the formula for the perimeter P of a square with sides of length s is $P = 4s$. You might need to solve this equation for s, so you can plug in a perimeter and figure out the side length.

This process of solving a formula for a given variable is called "solving literal equations." One of the dictionary definitions of "literal" is "related to or being comprised of letters," and variables are sometimes referred to as literals. So "solving literal equations" seems to be another way of saying "taking an equation with lots of letters, and solving for **one letter in particular**."

At first glance, these exercises appear to be much worse than your usual solving exercises, but they really aren't that bad. You pretty much do what you've done all along for solving linear equations and other sorts of equation; the only substantial difference is that, due to all the variables, you won't be able to simplify your answers as much as you're used to. Here's an example of how "solving literal equations" works.

DISTANCE, SPEED, AND TIME

A classis science formula is $d = r \cdot t$ where d = distance traveled, r = speed or rate of travel, t = time

Suppose the rate traveled is $\frac{65 \text{ mi.}}{\text{hr}}$ and the time traveled is 10 hr. We let $r = \frac{65 \text{ mi.}}{\text{hr}}$ and $t = 10$ hr.

So $d = r\,t$ becomes $d = \frac{65 \text{ mi.}}{\text{hr}}(10 \text{ hr}) = 65(10) \text{ mi.} = 650 \text{ mi.}$

What if we need a formula for the time if we know the distance traveled and the speed.

So we want a formula for t "in terms of" d and r.

So we start with $d = r\,t$, to isolate t on one side, we divide both sides by r

$$\div r \quad \div r$$

$$\frac{d}{r} = t$$

So now we take the distance traveled and divide by the speed to get the time of travel

Example, the distance, d, traveled is 50 mi., and the speed, r is 5 miph.

$$\frac{d}{r} = t \text{ becomes } \frac{50 \text{ mi.}}{\left(\frac{5 \text{ mi.}}{1 \text{ hr}}\right)} = 10 \text{ hr} = t.$$

Now, figure a formula for r in terms of d and t . . . .

Another example:

Solve $4x + 5y = 100$ for y

We want to isolate y on one side, so

$4x + 5y = 100$ subtract $4x$ both sides

$\underline{-4x \qquad\qquad -4x}$

$5y = 100 - 4x$ divide both sides by 5

$\underline{\div 5 \qquad\qquad \div 5}$

$y = \frac{100}{5} - \frac{4x}{5}$ simplifies to $y = 20 - \frac{4x}{5}$

Geometry example:

The formula for area of a rectangle is $A = bh$, with A = area, b = base measure, and h = height measure.

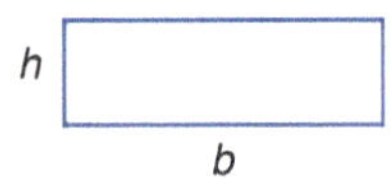

What is the formula for the base, b, in terms of the area, A, and the height, h?

In other words, Let's solve $A = bh$ for b in terms of A and h

$A = bh$ divide both sides by h

$\frac{A}{h} = \frac{bh}{h}$ Cancel out the h on the right side

$\frac{A}{h} = b,$

What is the formula for the height in terms of the area and base?

PERIMETER OF A RECTANGLE

➡ Perimeter of a rectangle is the formula $P = 2l + 2w$ with P = perimeter, l = length, and w = width.

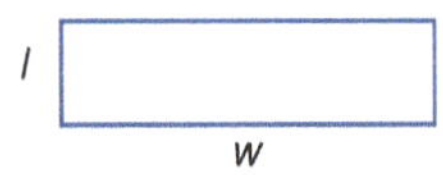

Let's solve $P = 2l + 2w$ for w in terms of P and l.

$P = 2l + 2w$ subtract $2l$ both sides

$P - 2l = 2l + 2w - 2l$

$P - 2l = 2w$ divide by 2 both sides

$$\frac{P-2l}{2} = \frac{2w}{2}$$

$$\frac{P-2l}{2} = w$$

What is l in terms of P and w?

Solve $Q = \frac{c+d}{2}$ for d in terms of Q and d

$$Q = \frac{c+d}{2} \quad \text{multiply both sides by 2}$$

$$(2)\ Q = (2)\ \frac{c+d}{2}$$

$$2Q = c + d \quad \text{subtract } c \text{ both sides}$$

$$2Q - c = d$$

What is c in terms of Q and d?

Solve $12d + 13h - 4(d + 3) - 100 = 32d - 45h + 2r - 90$

For h in terms or d and r

Simplify left side, distributive property

$$12d + 13h - 4d - 12 - 100 = 32d - 45h + 2r - 90$$

Combine like terms left side

$$8d + 13h - 112 = 32d - 45h + 2r - 90$$

Use properties of equality to isolate h one side.

$$8d + 13h - 112 = 32\text{d} - 45h + 2r - 90$$

$$\underline{-8d \qquad\qquad\qquad\qquad -8d}$$

$$13h - 112 = 24d - 45h + 2r - 90$$

$$\underline{+112 \qquad\qquad\qquad\qquad +112}$$

$$13h = 24d - 45h + 2r + 22$$

$$\underline{+45h \qquad\qquad +45h}$$

$$58h = 24d + 2r + 22$$

$$\underline{\div 58 \quad \div 58}$$

$$h = (24d + 2r + 22) \div 58 = \frac{24d + 2r + 22}{58}$$

$$h = \frac{12d + r + 11}{29}$$

Practice

Solve $T - W = 11$ for W in terms of T

For T in terms of W

Solve $12x = 4y + 30$ for y in terms of x

Solve $2R + 8B + 3C = 54 + 3R$ for B in terms of R and C

Solve $F = \frac{ma}{q}$ for a in terms of F, m, and q

For q in terms of Fm and a

2-7 Homework Assignments

Solve these formulas for the variable indicated in terms of the other variables.

1. Solve $y = 3x + 2$ for x ________________
2. Solve $r - x + 2y = 1$ for y ________________
3. Solve $A = .5bh$ for b ________________
4. Solve $6t + ry = \frac{n}{2}$ for y ________________
5. Solve $48t + 12k = 6(p - 2)$ for p ________________
6. Solve $2(x - h) - t = 3t$ for x ________________
7. Solve $\frac{1}{8}x + 4w = 3y$ for w ________________
8. Solve $V = 3.14r^2h$ for h ________________
9. Solve $.5\,h(k + 2t) = x$ for t ________________
10. Solve $y - 2a - 2b = -x + a + b$ for x ________________
11. Solve $\frac{3d + 7b - 9k}{2g + m} = 7$ for m ________________
12. Solve $\frac{a}{4c} + \frac{b}{2c} = 12$ for c ________________

Answers

1. $x = \frac{y - 2}{3}$
2. $y = \frac{1 - r + x}{2}$
3. $b = \frac{A}{.5h} = \frac{2A}{h}$

4. $y = \dfrac{n - 12t}{2r}$

5. $p = 8t + 2K + 2$

6. $x = 2t + h$

7. $w = \dfrac{(24y - x)}{32}$

8. $h = \dfrac{V}{3.14r^2}$

9. $t = \dfrac{2x - kh}{2h}$ or $= \dfrac{x}{h} - \dfrac{k}{2}$

10. $x = 3a + 3b - y$

11. $m = \dfrac{3d + 7b - 9k - 14g}{7}$

12. $c = \dfrac{a + 2b}{48}$

2-8 QUADRATIC EQUATIONS

Many formulas have variables raised to **whole number exponent powers.** We will use some of them in the course.

Example of this type of equation is the **QUADRATIC EQUATION.** This is an equation where we raise one of the unknown (variable) to the second power (or squaring the variable) and it is the highest exponent in the expression

$$\text{Such as } y = x^2 + 3x + 1$$

This is a formula. So, we can solve the formula by giving one of the variables a value. If we make $x = 2$, we simply substitute the 2 for x in the formula and get

$$y = x^2 + 3x + 1 = 2^2 + 3(2) + 1 = 4 + 6 + 1 = 11$$

However, what if we make $y = 2$? We get $2 = x^2 + 3x + 1$

So how do we solve this for x????

We will need to make use of the **SQUARE ROOT** in the work of solution. We also need to recognize that there will technically be **<u>two</u> solutions** to this quadratic, although at times each of the two solutions can be the same value.

Let's first solve a simple one variable quadratic.

Example, solve for x in $x^2 = 16$ to "undo" the squaring of the x to isolate the x, we take the square root of x^2. Remember, we need to do the same operation to both sides of the equation, so we take the square root of both sides of our equation $x^2 = 16$ we get $\sqrt{x^2} = \sqrt{16}$ or $x = 4$

Checking this, when we take 4^2, it does equals 16. However, it is partially correct!!

Because when we square a negative number, we get a positive value, $(-4)^2$ is also equal to 16. So, the real solution to $x^2 = 16$ is + 4 or −4. See, we actually have two solutions for the quadratic.

The rule for taking a square root says

> if $x^2 = k$, where k is some real number, then $\sqrt{x^2} = x = \pm\sqrt{k}$.

Solving for x in $x^2 = 81$, take the square root both sides we get $\sqrt{x^2} = \pm\sqrt{81}$ or $x = \pm 9$.

Solving for t in $t^2 = 144$, take the square root both sides we get $\sqrt{t^2} = \pm\sqrt{144}$ or $t = \pm 12$.

We call the positive root value the PRINCIPAL ROOT. Calculators only calculate principal square roots, so when you calculate it in the calculator using the square root tool, or raise to the ½ (or .5) power, you will only see the positive root value. Just keep in mind, that the negative will also be a solution.

In addition, unless you have a perfect square which will result in a ± integer square root value, you will receive a decimal answer on your calculator. Often, the decimal will have numbers that don't repeat endlessly. The best you can do is to write these decimal to some rounded place value. Sometimes if we want an EXACT value, we leave the root value in the square root radical.

We can now solve simple quadratic equations, using the root operation as well as other inverse operations, doing them in reverse order of operations, as shown before. The goal is to isolate our variable squared, then take the square root of both sides.

Example: Solve for M

$3.4M^2 = 126.7.$ We need to undo the multiplication of 3.4 by dividing both sides by 3.4

$\div 3.4 \qquad \div 3.4$

$M^2 = 37.2647....$ now we undo the square by taking the square root both sides. Remember, we need to take a + and −

$\sqrt{M^2} = \pm\sqrt{37.2647....}$ or $M = \pm 6.10$ to the nearest hundredth value.

An exact answer for M would be $\pm\sqrt{\frac{126.7}{3.4}}$ and a rounded answer would be ± 6.10

Example: Solve for x for $3x^2 + 22 = 100$ first we need to undo the + 22 by subtracting 22 both sides

$-22 \qquad -22$

$3x^2 = 78$ now undo three times by dividing by 3 both sides

$\div 3 \quad \div 3$

$x^2 = 26$ now we undo the square by taking the square root both sides.
Remember, we need to take + and − of the number side

$\sqrt{x^2} = \pm\sqrt{26}$ or $x = \pm\sqrt{26}$.

In a decimal rounded to the nearest hundredth, $\pm\sqrt{26} = \pm 5.10$ so an exact answer for x is $\pm\sqrt{26}$ and a rounded answer is ± 5.10.

We can do this with literal equations as well.

Solve for c in

$E = mc^2$ We need to isolate the c^2 so we divide both sides by m

$\div m \quad \div m$

$\frac{E}{m} = c^2$ now we undo the square by taking the square root both sides.

Remember, we need to take + and − of the "number" (the $\frac{E}{m}$) side

$$\pm\sqrt{\frac{E}{m}} = \sqrt{c^2} \text{ or } \pm\sqrt{\frac{E}{m}} = c$$

SOLVING QUADRATICS IN THE FORM $ax^2 + bx + c = 0$ using THE QUADRATIC FORMULA

When a quadratic equation has both x^2 and x terms, we cannot simply take the square roots of both sides. We need another way—we will introduce a method called the QUADRATIC FORMULA.

An equation in the form $\boldsymbol{ax^2 + bx + c} = 0$ where $\boldsymbol{a} \neq 0$, b and c are real numbers is a STANDARD FORM quadratic equation.

When our quadratic is written in standard form, solutions to the quadratic follows this QUADRATIC FORMULA

$$x = \frac{-b \pm \sqrt{b^2 - 4ac}}{2a}$$

We take our standard form equation set equal to 0, determine the values for a, b, and c, then substitute for a, b, and c into the formula to solve for x.

Example: solve for x in $x^2 - 7x + 12 = 0$

$a = 1$, $b = -7$, $c = 12$ so plugging these values in we get

$$x = \frac{-b \pm \sqrt{b^2 - 4ac}}{2a} = \frac{--7 \pm \sqrt{(-7)^2 - 4(1)(12)}}{2(1)} = \frac{7 \pm \sqrt{49 - 48}}{2} = \frac{7 \pm \sqrt{1}}{2} = \frac{7 \pm 1}{2}$$

$$\text{or } x = \frac{7+1}{2} = \frac{8}{2} = 4 \quad \text{or} \quad x = \frac{7-1}{2} = \frac{6}{2} = 3.$$

So solutions to $x^2 - 7x + 12 = 0$ are $x = 4$ or 3.

We can do a quick check to show this by plugging 4 and 3 into our equation

For $x = 4$, $x^2 - 7x + 12 = (4)^2 - 7(4) + 12 = 16 - 28 + 12 = 0$

For $x = 3$, $x^2 - 7x + 12 = (3)^2 - 7(3) + 12 = 9 - 21 + 12 = 0$ so they check

Example: Use the quadratic formula to solve for x in $x^2 + 6x + 4 = 0$

$a = 1, b = 6, c = 4$, so plugging these values in we get

$$x = \frac{-b \pm \sqrt{b^2 - 4ac}}{2a} = \frac{-6 \pm \sqrt{6^2 - 4(1)(4)}}{2(1)} = \frac{-6 \pm \sqrt{36 - 16}}{2} = \frac{-6 \pm \sqrt{20}}{2} = \text{approximately } \frac{-6 \pm 4.47}{2}$$

$$\text{or } x = \frac{-6 + 4.47}{2} = -.765 \text{ or } x = \frac{-6 - 4.47}{2} = -5.235$$

Example: Use the quadratic formula to solve for x in $-2x^2 + 15x = 10$

This is not in the standard form, so we must subtract 10 both sides to get

$$-2x^2 + 15x - 10 = 10 - 10 \textit{ or } -2x^2 + 15x - 10 = 0 \quad \textit{now}$$

$a = -2, b = 15, c = -10$, so plugging these values in we get

$$x = \frac{-b \pm \sqrt{b^2 - 4ac}}{2a} = \frac{-15 \pm \sqrt{15^2 - 4(-2)(-10)}}{2(-2)} = \frac{-15 \pm \sqrt{225 - 80}}{-4} = \frac{-15 \pm \sqrt{145}}{-4} =$$

$$\text{approximately } \frac{-15 \pm 12.04}{-4} \text{ so}$$

$$x = \frac{-15 + 12.04}{-4} = .74 \text{ or } x = \frac{-15 - 12.04}{-4} = 6.76$$

Practice: Solve for the variable (give answers to nearest hundredth)

$x^2 = 121$	$r^2 = 22801$
$y^2 = 23$	$p^2 - 100 = 0$
$x^2 + 8x + 9 = 0$	$3w^2 + 6x - 2 = 0$
$x^2 + 3x + 1 = 2$	$11x^2 + 4x - 2 = 0$

2-8 Assignments

Solve for all solutions to these quadratic equations. Give answer to nearest hundredth.

1. $x^2 = 256$
2. $r^2 = 3451$
3. $y^2 - 400 = 0$
4. $p^2 - 2100 = 0$

5. $x^2 + 13x + 42 = 0$

6. $w^2 - 6w + 8 = 0$

7. $2x^2 + 7x + 4 = 2$

8. $9x^2 + 12x - 2 = 0$

9. $20w^2 - 8w - 2 = 0$

10. $p^2 - 9p - 112 = 0$

11. $5x^2 - 4x = 17$

12. $t^2 + 12t + 36 = 0$

Answers

1. $x = \pm\sqrt{256} = \pm 16$

2. $r = \pm\sqrt{3451} \approx \pm 58.75$

3. $y = \pm\sqrt{400} = \pm 20$

4. $p = \pm\sqrt{2100} \approx \pm 45.83$

5. $x = -7$ or -6

6. $w = 4$ or 2

7. $x = \dfrac{-7 \pm\sqrt{33}}{4} \approx -.31$ or -3.19

8. $x = \dfrac{-12 \pm\sqrt{216}}{18} \approx .15$ or -1.48

9. $w = \dfrac{8 \pm\sqrt{224}}{40} \approx .57$ or $-.17$

10. $p = 16$ or -7

11. $x = \dfrac{4 \pm\sqrt{356}}{10} \approx 2.28$ or -1.49

12. $t = -6$

CHAPTER 3

Systems of Measurement and Unit Conversion

3-1 MEASUREMENT SYSTEMS, CUSTOMARY AND METRIC

MEASUREMENT SYSTEMS: There have been thousands of measuring systems used throughout history in different civilizations. Today, there has evolved two main standard forms of measuring: The Customary (also known as the old English) System and The Metric System.

When writing quantities of measuring units we use a DENOMINATE NUMBER. A denominate number consists of a number value and a unit of measure. Examples of denominate numbers are 5 inches, 10 meters, 3456 pounds, .08 milligrams, 450 days, and 12 degrees.

Let's look at these two main measuring systems.

ENGLISH/CUSTOMARY – ***USED PRIMARILY IN THE U.S. (and not many other places!!!!)***

Origins of Measurements of length

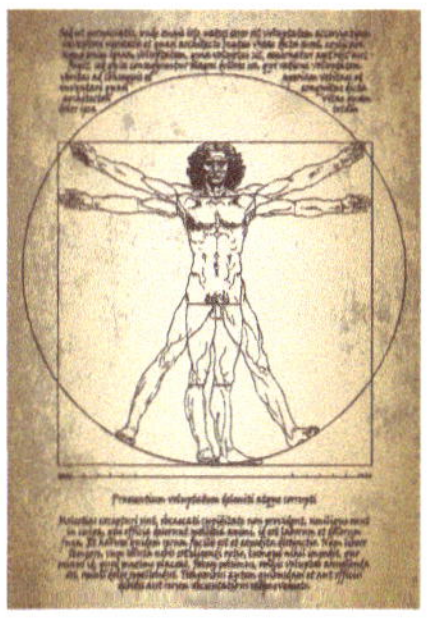

©Kwirry/Shutterstock.com

In ancient times, the **BODY** ruled when it came to measuring.

The length of a **foot**, the **width** of a finger, and the **distance of a step** were all accepted measurements.

©GCapture/Shutterstock.com

INCH: At first an inch was the width of a man's thumb.

©Swapan Photography/ Shutterstock.com

In the 14th century, King Edward II of England ruled that 1 inch equaled 3 grains of barley placed end to end lengthwise.

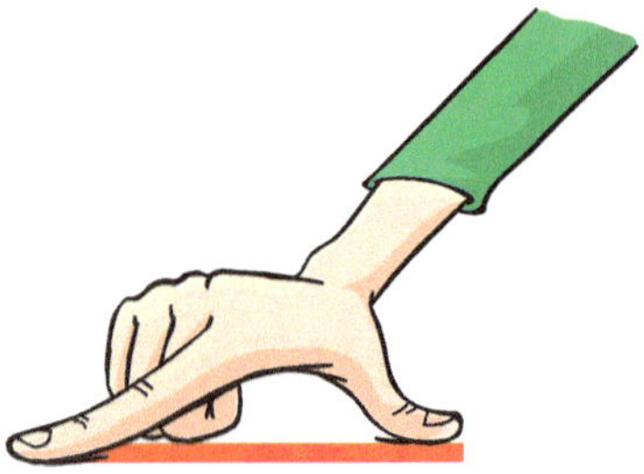

©Dusan Pavlic/Shutterstock.com

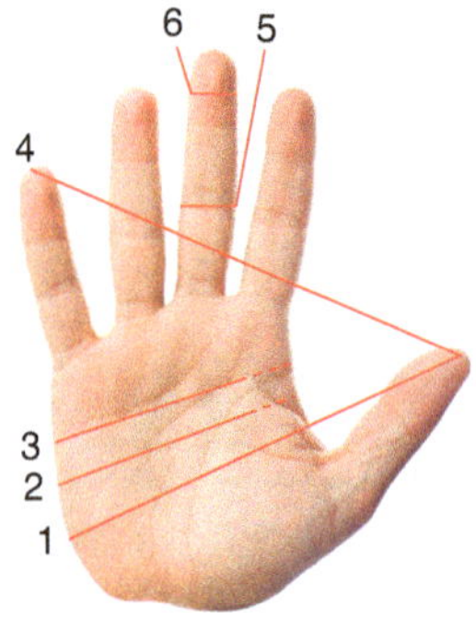

Adapted from ©Kues/Shutterstock.com; © Kendall Hunt Publishing

A hand was approximately 5 inches or 5 digits (fingers) across. Today, a hand is 4 inches and is used to measure horses (from the ground to the horse's withers, or shoulder).

SPAN: A span was the length of the hand stretched out, about 9 inches.

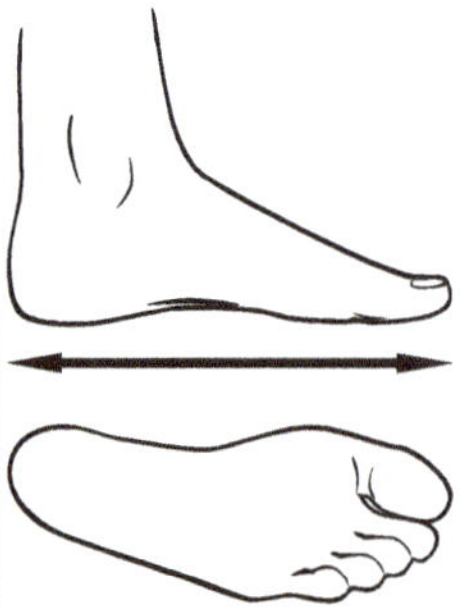

©BIELOUS NATALIIA/ Shutterstock.com

FOOT: In ancient times, the foot was $11^1/_4$ inches. Today it is 12 inches, the length of the average man's foot.

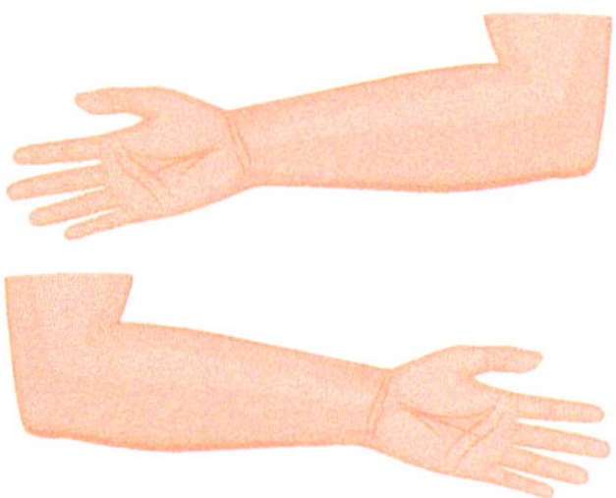

©Aksanaku/Shutterstock.com

YARD: A yard was originally the length of a man's belt or girdle, as it was called.

In the 12th century, King Henry I of England, (because her could, he was the king!!!), fixed the yard as the distance from his nose to the thumb of his out-stretched arm. Today it is 36 inches.

CUBIT: [use top image only and remove labels]

In ancient Egypt, a cubit was the distance from the elbow to the fingertips. Today a cubit is about 18 inches.

[Only use the image on the left and put a line below his feet to measure between his feet]

©Shendart/Shutterstock.com

PACE: The ancient Roman soldiers marched in paces, which were the length of a double step, . . about 5 feet; 1,000 paces was a mile. Today, a pace is the length of one step, $2^1/_2$ to 3 feet.

These measurements would be ok, except - everyone's body is different. So, depending on who was doing the measuring, measurements could vary. There was a need to create a **STANDARD SYSTEM OF MEASUREMENTS**:

And the English or Customary system we use today predominantly in the US has these basic standard units of measures,

- Inch. in. = 0.083 or 1/12 feet
- Foot, ft = 12 inches
- Yard, yd = 3 feet or 36 inches
- Mile, mi. = 5,280 feet or 1,760 yards

METRIC SYSTEM – used worldwide

The metric system was developed to utilize the base ten numbering system that was popular about the world.

METRIC LENGTH

The basic unit of measure for length in the metric system is the METER. A meter is about 3 feet (or a yard) in U.S. Measurement. We use m. to abbreviate meters.

The Meter was deemed an official measurement length by the French back in the 1700's. The officials in charge of measurements set it equal to be 10^{-7} of the distance from the equator to the pole on the meridian of the Earth though Paris.

That is $\frac{1}{10,000,000}$ of the distance you would walk in a straight line from the north pole to the equator. How they measured that distance in the 1700's - who knows, but they did it, then divided it by 10,000,000 and, voila, we have the meter length!!!!!!. Whew!

The metric system has prefix modifiers that apply to the basic meter unit, and create multiples of ten for new units. Here are the most common prefixes used. And **<u>We find these prefixes apply not only to measure of length, but to other measures, such as volume and weight.</u>**

Here are the most common prefix/multiples of ten used.

Prefix	Multiplier	Power of 10	Symbol
mega	1,000,000	10^6	M
kilo	1,000	10^3	k
meter	1		
centi	1/100	10^{-2}	c
milli	1/1,000	10^{-3}	m
micro	1/1,000,000	10^{-6}	mc or μ

Here is the breakdown for length measures:

- A megameter, Mm, is 1,000,000 (or 10^6) meters
- A kilometer, km, is 1000 (or 10^3) meters
- A meter, m, is the basic unit of length
- A centimeter, cm is 1/100 (or 10^{-2}) meter
- A millimeter mm, is 1/1000 (or 10^{-3}) meter
- A micrometer, mcm, is 1/1,000,000 (or 10^{-6}) meter

In reality, the kilometer (to measure really long distances), the meter, the centimeter (for small distances) and the millimeter (for even smaller distances) are the most common metric length measures.

For your information, Here is a more complete prefix, multiplier, power of ten and symbol chart.

Prefix	Symbol	Multiplication factor		
exa	E	10^{18}	=	1 000 000 000 000 000 000
peta	P	10^{15}	=	1 000 000 000 000 000
tera	T	10^{12}	=	1 000 000 000 000
giga	G	10^{9}	=	1 000 000 000
mega	M	10^{6}	=	1 000 000
kilo	K	10^{3}	=	1 000
hecto	h	10^{2}	=	100
deca	da	10^{1}	=	10
deci	d	10^{-1}	=	0.1
centi	c	10^{-2}	=	0.01
milli	m	10^{-3}	=	0.001
micro	μ	10^{-6}	=	0.000 001
nano	n	10^{-9}	=	0.000 000 001
Pico	p	10^{-12}	=	0.000 000 000 001
Femto	f	10^{-15}	=	0.000 000 000 000 001
Atto	a	10^{-18}	=	0.000 000 000 000 000 001

To give you an idea of how metric lengths compare to US measures, here is a sampling

For example:

1 kilometer (km) = 0.62 mile (mi)
1 kilometer (km) = 3280.8 feet (ft)
1 meter (m) = 3.28 feet (ft)
1 centimeter (cm) = 0.39 inch (in)
1 millimeter (mm) = 0.039 inch (in)
1 inch (in) = 2.54 centimeters (cm)
1 inch (in) = 25.4 millimeters (mm)
1 foot (ft) = 0.30 meter (m)
1 yard (yd) = 0.91 meter (m)
1 yard (yd) = 0.00091 kilometer (km)
1 mile (mi) = 1.61 kilometers (km)

The fraction would be:

$$\frac{1\text{ kilometer (km)}}{0.62\text{ miles (mi)}} \quad \frac{\text{(numerator)}}{\text{(denominator)}}$$

VOLUME: To measure the space taken up inside solid figure, we use volume. Standard measures have developed in these units.

CUSTOMARY (US) VOLUME MEASURES

The common measures of volume in the U.S. system of measurements are:

- Teaspoons, tsp,
- Tablespoons, Tsp. = 3 teaspoons
- fluid ounces, fl. oz. = 2 tablespoons, 6 teaspoons
- cups, c. = 8 fluid ounces, 16 tablespoons
- pints, pt. = 2 cups, 16 fluid ounces
- quarts, qt. = 2 pints, 4 cups
- gallons, gal = 4 quarts, 8 pints, 16 cups

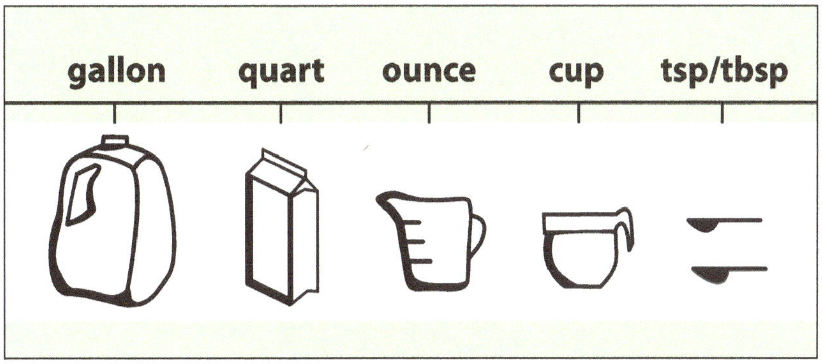

©Brookm/Shutterstock.com

METRIC VOLUMES MEASURES

The basic volume measure is the LITER. One liter is slightly more than one Quart. We use the L. to abbreviate liters.

Applying the metric prefixes we get

Here is the breakdown for volume measures:

- A megaliter, ML, is 1,000,000 (or 10^6) liters
- A kiloliter, kL, is 1000 (or 10^3) liters
- A liter, L., is the basic unit of volume
- A centiliter, cL. is 1/100 (or 10^{-2}) liter
- A milliliter, mL, is 1/1000 (or 10^{-3}) liter
- A microliter, mcL, is 1/1,000,000 (or 10^{-6}) liter

Usually, the most common measures used for volume are the liter, L, and the milliliter, mL..

To give you a comparison of normal customary to metric volume measures:

1 Quart = .946 Liters;	1 liter = 1.057 Quarts
1 teaspoon = 4.93 milliliters;	1 milliliter = .203 teaspoons
1 Tablespoon = 14.787 milliliters;	1 milliliter = .068 tablespoons
1 cup = .237 liters;	1 liter = 4.227 cups
1 gallon = 3.786 liters;	1 liter = .264 gallons

MASS/WEIGHT - Mass is a measurement of the amount of matter something contains, while Weight is the measurement of the pull of gravity on an object. We often use the same units of measure for each, but weight can be different given the gravitational pull on the object (like the difference between being on the earth and being on the moon.)

©Loomitz/Shutterstock.com

The CUSTOMARY UNITS of WEIGHT MEASURE are

- Ounces, oz. a slice of bread is about an ounce in mass/weight.
- Pounds, lbs. = 16 ounces
- Tons, = 2000 pounds.

(Note, **fluid** ounces and ounces are different. Fluid ounces measures volume while ounces measure mass and weight.)

METRIC MASS/WEIGHT MEASURES

The basic mass/weight measure is the GRAM. A gram uses g. for abbreviation. One gram is about .036 of an ounce. Also there are about 28.35 grams in one ounce.

Here is the breakdown for volume measures:

- A megagram, Mg, is 1,000,000 (or 10^6) grams
- A kilogram, kg, is 1000 (or 10^3) grams
- A gram, g., is the basic unit of mass/weight
- A centigram, cg. is 1/100 (or 10^{-2}) grams
- A milligram, mg, is 1/1000 (or 10^{-3}) grams
- A microgram, mcg, is 1/1,000,000 (or 10^{-6}) grams

Usually, the most common measures used for mass are kilograms, grams, and milligrams,

To give you a comparison of normal customary to metric volume measures:

1 ounce = 28.35 grams;	1 gram = .036 ounce
1 Ounce = 28349.5231 Milligrams;	1 milligram = 0.0000352 ounce
1 kilograms = 2.205 pounds	

3-1 Exercises

For 1 – 8, True or false

1. Upper and lower case letters are Important to prefixes and abbreviations in metric measurement. __________
2. The Customary System of measurement is base 12. __________
3. The Metric System of Measurement is base 10. ____________
4. The abbreviation μL means microliters. __________
5. The abbreviation μ g means milligrams. ______________
6. There are about four liters in one gallon. _____________
7. There are about 450 grams in one pound. ______________
8. The Customary Measurement System is the one used universally throughout the world. ____

Write the best customary unit of measurement and the best metric unit of measure for measuring the following items

9. The length of your thumb
10. The length of your kitchen table

11. The fluid volume of a glass of water

12. The weight of a new born.

13. The fluid volume in a carton of milk.

14. The fluid volume of a gas tank of a car.

15. The weight of hard cover textbook.

16. The thickness of a quarter.

17. The weight of a marshmallow.

18. The length of a human foot.

19. The weight of a cell phone.

20. The weight of an automobile.

For the following measure, indicate the metric prefix that should be used for the basic unit of measure.

21. .01 grams = 1 _________ grams

22. 1000 Liters = 1 ________ Liters

23. .001 meters = 1 ___________ meters

24. 10 grams = 1 _____ grams

25. 10^6 meters = 1 _______ meters

26. .000001 Liters = 1 ________ Liters

27. 100 grams = 1 ______ grams

28. 10^{-1} Liters = 1 ______ Liters

Answers

1. True

2. False

3. True

4. True

5. False
6. True
7. True
8. False
9. Inch, centimeter
10. Foot, meter
11. Fluid ounce, millileter
12. Pound, kilogram
13. Quarts, liters
14. Gallon, liter
15. Pound, kilogram
16. Inch, millimeter
17. Ounce, milligram
18. Foot, centimeter
19. Ounces, grams
20. Tons, megagram
21. Centi
22. Kilo
23. Milli
24. Deca
25. Mega
26. Micro or μ
27. Hecto
28. Deci

3-2 DIMENSIONS, DENOMINATE NUMBERS, AND OPERATIONS

A DIMENSION is a fundamental physical quantity in any system of measurement. There are seven fundamental dimensions in modern system of measurement. They are

1. LENGTH
2. MASS
3. TIME
4. TEMPERATURE
5. ELECTRICAL CURRENT
6. LUMINOSITY
7. AMOUNT OF SUBSTANCE

Units are measures used to quantify a dimension, and units associated with numbers quantify the units of measure. We do this when we create a DENOMINANT NUMBER. The number is coupled with the unit and the unit is coupled with the number.

So 12 ft., 5 lb., 98°, 45 hr, and so on, all describe some dimension, or identifies what IS BEING MEASURED OR COUNTER.

A denominate number is really an algebraic quantity.

We can think of

- The unit itself as sort of a variable.
- The number with the unit as the variables coefficient.

So "3 in." is like 3 in. or three times the inch measure.

43.5 mL is like 43.5 mL or 43.5 times the milliliter measure, and so on.

Thinking this way, we can apply the simplifying rules when adding (subtracting), multiplying (dividing), and using properties of equality with units of measures in equations.

Here are some units of measure rules that follow the variable rule:

- You can only add and subtract like units of measure.
- You can multiply and divide both like and unlike units of measure.

ADDING AND SUBTRACTING

Just like simplifying $3x + 4x$ to $(3 + 4)x$ to $7x$, we can make 3 in. + 4 in. equal to (3 + 4) in. equal to 7 in.

However, $3x + 4y$ is simplified as we cannot add these two unlike term.

Same goes for 3 in. + 4 ft. Inches and feet are unlike units, so we cannot add them as is into one unit of measure.

How about 43 qt. – 23 qt. As we have like units, it becomes (43 – 23) qt. = 20 qt.

But we cannot subtract 43 qt. – 23 pt. as is. into one unit of measure

The only way to add or subtract unlike units of measure is to change (or convert) one of them in terms of the other, such as changing inches into feet.

Example: Find 5 in. + 4 ft. in feet.

As there are 12 in. in 1 ft., 5 in. is $\frac{5}{12}$ of a foot. So we can write

$$5 \text{ in.} + 4 \text{ ft.} = \frac{5}{12} \text{ ft.} + 4 \text{ ft.} = 4\frac{5}{12} \text{ ft.}$$

Now, find 5 in. + 4 ft. in inches.

As 4 ft. is 4(12) in. = 48 in., 5 in. + 4 ft. = 5 in. + 48 in. = 53 in.

Example: 14 gal. – 13 qt. in gallons.

As there are 4 qt. in 1 gal., 12 qt. is 3 gal., and 1 qt. is $\frac{1}{4}$ gal, so 13 qt. is $3 + \frac{1}{4}$ gal. $= 3\frac{1}{4}$ gal.

$$14 \text{ gal.} - 13 \text{ qt.} = 14 \text{ gal.} - 3\frac{1}{4} \text{ gal.} = 14 - (3 + \frac{1}{4}) \text{ gal.}$$

$$= (14 - 3 - \frac{1}{4}) \text{ gal.} = (11 - \frac{1}{4}) \text{ gal.} = 10\frac{3}{4} \text{ gal.}$$

Now, find 14 gal. – 13 qt. in quarts.

As 14 gal. is 14(4) qt. = 56 qt., then 14 gal. – 13 qt.

$$= 56 \text{ qt.} - 13 \text{ qt.} = 43 \text{ qt.}$$

DOUBLE DENOMINATE NUMBERS AND ADDITION/SUBTRACTION

What if we have double denominate numbers in addition or subtraction? One way is to change them to one unit of measure then add or subtract.

For example, 4 ft. 3 in. + 2 ft. 11 in.

We can change both into feet then add.

$$4 \text{ ft. } 3 \text{ in.} + 2 \text{ ft. } 11 \text{ in.} = 4\frac{3}{12} \text{ ft.} + 2\frac{11}{12} \text{ ft.} = (4 + \frac{3}{12} + 2 + \frac{11}{12}) \text{ ft.}$$

$$= (6 + \frac{14}{12}) \text{ ft.} = (6 + 1 + \frac{2}{12}) = 7\frac{2}{12} \text{ ft.} = 7\frac{1}{6} \text{ ft.}$$

Or change both into inches then add

$$4 \text{ ft. } 3 \text{ in.} + 2 \text{ ft. } 11 \text{ in.} = 48 \text{ in.} + 3 \text{ in.} + 24 \text{ in.} + 11 \text{ in.}$$
$$= (48 + 3 + 24 + 11) \text{ in.} = 86 \text{ in.}$$

Another way we can add double denominate number is to add the respective units of measure and create a final double denominate number.

$$4 \text{ ft. } 3 \text{ in.} + 2 \text{ ft. } 11 \text{ in.} = 4 \text{ ft.} + 3 \text{ in.} + 2 \text{ ft.} + 11 \text{ in.}$$

$$= 4 \text{ ft.} + 2 \text{ ft.} + 3 \text{ in.} + 11 \text{ in.} = 6 \text{ ft.} + 14 \text{ in., and } 14 \text{ in.} = 1 \text{ ft. } 2 \text{ in.,}$$

Then 6 ft. + 14 in. = 6 ft. + 1 ft. + 2 in. = 7 ft. 2 in.

Practice

Add/subtract, give answers in both feet and inches

12 ft. + 7 in. =

14 ft. – 5 in. =

Add to quarts then to pints

6 qt. + 7 pt. =

8 qt. – 3 pt. =

Add/subtract and give as double denominate number

6 ft. 6 in. – 3 ft. 8 in. =

20 lb. 5 oz. + 5 lb. 12 oz. =

DIVIDING

$15y \div 3y$ can be simplified to $\dfrac{15y}{3y} = \dfrac{5}{1} = 5$

So is 15 yd. ÷ 3 yd. = $\dfrac{15 \text{ yd.}}{3 \text{ yd.}} = \dfrac{5}{1} = 5$. The units of measure cancel each other out, just like the variable y does in the previous expression.

Now $15x \div 3y = \dfrac{15x}{3y} = \dfrac{5x}{y}$

We cannot further simplify the $\dfrac{x}{y}$ as they are unlike

Same goes for 15 mi. ÷ 3 hr = $\dfrac{15 \text{ mi.}}{3 \text{ hr}} = \dfrac{5 \text{ mi.}}{\text{hr}}$. This forms what is called a rate. A rate is a comparison of two different measure denominate numbers by division. We use rates all the time in industrial profession, such as miles/gallon, inches/hour, people/year, milligram per minute, and so on.

PRACTICE: Find the rate for 117 gal. ÷ 9 mi.

For 678 in. ÷3 min

For 456 lb. ÷ 8 in.^{2}

MULTIPLYING

When multiplying $6t\ 7p$, we multiply the numbers then multiply the variables to get $6t\ 7p = 6 \cdot 7 \cdot t \cdot p = 42\,t\,p$.

When multiplying units of measure, it is ok to multiply unlike units. Like variable factors, we multiply the numbers and we multiply the units of measure. For example,

$$6 \text{ ft.} \cdot 7 \text{ in.} = 6 \cdot 7 \cdot \text{ft. in.} = 42 \text{ “ft. in.”}$$

However, it may not always produce a *sensible* new unit of measure. We normally do not use “feet inches” as a unit of measure.

In this case, it makes sense to convert one of the denominate number measure to the other unit of measure.

For 6 ft.· 7 in., we can change the 6 ft. into in. As there are 12 in. in 1 ft., 6 ft. is $(6 \cdot 12)$ in. = 72 in.

So 6 ft. · 7 in. = 72 in. · 7 in. = 72 · 7 · in. · in. = 504 in.^{2} or 504 sq. in. A square inch is a recognized unit of measure for area.

Example:

$$4 \text{ ft. } 3 \text{ in.} \cdot 7 \text{ ft. } 7 \text{ in.}$$

Change to feet then multiply

$$4 \text{ ft. } 3 \text{ in.} = 4\frac{3}{12} \text{ ft.}, 7 \text{ ft. } 7 \text{ in.} = 7\frac{7}{12} \text{ ft.}$$

$$\text{So } 4 \text{ ft. } 3 \text{ in.} \cdot 7 \text{ ft. } 7 \text{ in.} = 4\frac{3}{12} \text{ ft.} \cdot 7\frac{7}{12} \text{ ft.} = 4\frac{3}{12} \cdot 7\frac{7}{12} \text{ ft} \cdot \text{ft.}$$

$$= \frac{51}{12} \cdot \frac{91}{12} \text{ ft.}^2 = \frac{4641}{144} \text{ ft.}^2 = \frac{1547}{48} \text{ ft.}^2 = 32\frac{11}{48} \text{ ft.}^2$$

Or change to inches then multiply

$$4 \text{ ft. } 3 \text{ in.} = 48 \text{ in.} + 3 \text{ in.} = 51 \text{ in.}, 7 \text{ ft. } 7 \text{ in.} = 84 \text{ in.} + 7 \text{ in.} = 91 \text{ in.}$$

$$\text{So, } 4 \text{ ft. } 3 \text{ in.} \cdot 7 \text{ ft. } 7 \text{ in.} = 51 \text{ in.} \cdot 91 \text{ in.} = 51 \cdot 91 \cdot \text{in.} \cdot \text{in.} = 4641 \text{ in.}^2$$

NOTE: Conversion is not always needed, however, there are units of measures that use multiplication of individual units, such as light year (speed of light times years) or foot pounds (feet times pounds).

Practice

Multiply, give answers in both square feet and in square inches

7 ft. · 18 in. =

8 ft. 9 in. · 4 ft. 1 in. =

2 ft. · 7 in. =

5 ft. 10 in. 40 ft. 3 in. =

3-2 Exercises

Convert these double denominate numbers to one denominate number

1. 12 feet 3 inches = ____________ feet
2. 12 feet 3 inches = ____________ inches
3. 2 miles 1320 feet = __________ feet
4. 2 miles 1320 feet = ___________ miles
5. 3 gallons 12 fl ounces = _________ gallons
6. 3 gal 12 fl oz = ____________ fl oz
7. 3 hr 38 min = _________ min
8. 3 T 750 lb = _______T

Perform the indicated addition or subtraction, write answer in simplified double denominate number

9. 9 ft 5 in + 12 ft 8 in ____________
10. 4 mi 3880 ft + 4 mi 2690 ft ______________
11. 13 gal 1 qt + 4 gal 3 qt ______________
12. 15 gal 2 qt – 5 gal 3 qt _____________
13. 3 hr 34 min + 6 hr 55 min ___________
14. 55 min 45 sec – 30 min 55 sec ____________

Perform the indicated division, Give as fractional unit rate

15. 1566 miles ÷ 6 hours ___________
16. 980 pounds ÷ 14 in^2 ____________
17. 672 miles ÷ 14 gallons _________
18. 5676 yard ÷ 6 hours ___________

Perform the multiplication, answer in unit of measure indicated

19. 8 feet times 11 inches = _________ sq feet
20. 4 feet times 8 yard = ___________ sq yards

21. 6 ft 3in times 3 ft 3 in = ___________ sq inches

22. 2 ft 9in times 4 ft 3 in = ___________ sq ft

23. 5 yds 1 ft times 6yd 2 ft = _________ sq ft

24. 9 yards 2 feet times 2 yards 1 foot = ___________ sq yards

Answers

1. 12.25
2. 147
3. 11880
4. 2.25
5. 3.094
6. 396
7. 218
8. 3.375
9. 22 ft 1 in
10. 9 mi 1290 feet
11. 18 gal
12. 9 gal 3 qt
13. 10 hr 29 min
14. 24 min 50 sec
15. $\frac{\text{261 mile}}{\text{1 hour}}$
16. $\frac{\text{70 lbs}}{1\ \text{in}^2}$
17. $\frac{\text{48 mi}}{\text{1 gal}}$
18. $\frac{\text{946 yds}}{\text{1 hour}}$

19. $7.3\overline{3}$

20. $10.6\overline{6}$

21. 2925

22. 11.6875

23. 320

24. 22.56

3-3 CUSTOMARY SYSTEM UNIT CONVERSIONS

US CUSTOMARY SYSTEM OF MEASUREMENT

Here are the basic units of measure for the seven dimensions

Dimension	Basic unit of measure
LENGTH	INCH
MASS	POUND
TIME	SECOND
TEMPERATURE	DEGREE OF FAHRENHEIT
ELECTRICAL CURRENT	AMPERE
LUMINOSITY	CANDELA
AMOUNT OF SUBSTANCE	MOLE

We could have just stick to these basic units and figure everything in terms of inches, pounds, seconds, and so on. However, larger number of these units or small fraction of these units are just too cumbersome to write (like 20,000,000 in. or 1/64 of a pound) so to help ease this, other common units of measure evolve from multiples or fractional parts of these basic units.

For example, when we have 12 in., it is recognized that is equivalent to 1 ft.; when we have 2,000 lb., we say we have 1 tn., and so on.

As we have learned, we need to be able to convert between units of measure so we can apply the operations addition and subtraction, and in some cases multiplication, to like units of measure.

In order to be able to convert, we must first learn some basis conversion equations among measure in the US Customary System. Here are some .... These should be committed to memory!

US CUSTOMARY SYSYEM OF MEASUREMENT

LENGTH 1 ft. = 12 in.

1 yd. = 3 ft. = 36 in.

1 mi. = 5,280 ft. = 1,760 yd.

WEIGHT	1 lb. = 16 oz.
	1 tn. = 2,000 lb.
TIME	1 min = 60 sec
	1 hr = 60 min
	1 day = 24 hr
AREA	1 sq. ft. (ft.^{2}) = 144 sq. in. (in.^{2})
	1 sq. yd. (yd.^{2}) = 9 sq. ft.
VOLUME	1 cup (c) = 8 fl. oz.
	1 pt. = 2 cups = 16 fl. oz.
	1 qt. = 2 pt. = 4 cups = 32 fl. oz.
	1 gal. = 4 qt. = 8 pt. = 16 cups = 128 fl. oz.
	1 cu. ft. (ft.^{3}) = 1728 cu. in. (in.^{3})
	1 cu. yd. (yd.^{3}) = 27 cu. ft.
	1 gal. = 231 cu. in.

NOTE: For area, you can use sq. ft. for ft.^{2} and for volume, you can use cu. ft. for ft.^{3}

To convert from one unit of measure to another we can

- Write the quantity you desire to convert as a fraction.
- Multiply by a fraction formed by the conversion equation (called a conversion identity) so the units not in the final answer cancel out.
- Simplify your final number value if needed.

HERE IS AN EXAMPLE: Convert 31 in. to feet.

First, write 31 in. as $\frac{31 \text{ in.}}{1}$

Next, the conversion table shows that there are 12 in. in 1 ft. so 12 in. = 1 ft. As they are equivalent, we can write an identity fraction $\frac{1 \text{ ft.}}{12 \text{ in.}}$ and this is equal to 1. This makes this an identity conversion fraction.

We take our $\frac{31 \text{ in.}}{1}$ and multiply by $\frac{1 \text{ ft.}}{12 \text{ in.}}$ or

$$\frac{31 \text{ in.}}{1} \cdot \frac{1 \text{ ft.}}{12 \text{ in.}}$$ Now doing the multiplication

We get

$$\frac{31 \text{ in.}}{1} \cdot \frac{1 \text{ ft.}}{12 \text{ in.}} = \frac{31 \text{ in.} \cdot 1 \text{ ft.}}{1 \cdot 12 \text{ in.}} = \frac{31 \text{ in. ft.}}{12 \text{ in.}} = \frac{31}{12} \text{ ft.} = 2\frac{7}{12} \text{ ft.}$$

Note, the "inches" unit of measure cancel out in our final answer. This is why we made our identity conversion fraction $\frac{1\text{ ft.}}{12\text{ in.}}$ instead of $\frac{12\text{ in.}}{1\text{ ft.}}$.

EXAMPLE: Convert 3,260 min to hr

Write as fraction $\frac{3{,}260\text{ min}}{1}$. As 60 min = 1 hr, and we want to cancel out minutes, our identity conversion fraction is $\frac{1\text{ hr}}{60\text{ min}}$

Multiply

$$\frac{3{,}260\text{ min}}{1}\ \frac{1\text{ hr}}{60\text{ min}}=\frac{3{,}260\text{ hr}}{60}=54\frac{20}{60}\text{ hr}=54\frac{1}{3}\text{ hr}$$

EXAMPLE: Convert 15 gal. to cubic feet

Write as fraction $\frac{15\text{ gal.}}{1}$

We don't have a direct conversion from gallons to cubic feet. We do, however, have a conversion from gallons to cubic inches and then a conversion from cubic inches to cubic feet, so we will do a two-step conversion.

1 gal. = 231 cu. in. We want to cancel out gallons, so use the identity conversion fraction $\frac{231\text{ in.}^3}{1\text{ gal.}}$, and we multiply

$$\frac{15\text{ gal.}}{1}\cdot\frac{231\text{ in.}^3}{1\text{ gal.}}=3{,}465\text{ cu. in.}$$

Now, write as fraction $\frac{3{,}465\text{ in.}^3}{1}$

Since 1 cu. ft. = 1,728 cu. in. and we want to cancel out cubic inches we use the identity conversion fraction $\frac{1\text{ ft.}^3}{1{,}728\text{ in.}^3}$

$$\text{Multiply }\frac{3{,}465\text{ in.}^3}{1}\cdot\frac{1\text{ft.}^3}{1{,}728\text{ in.}^3}=\frac{3{,}465}{1{,}728}\text{ cu. ft.}=2\ \frac{9}{1{,}728}\text{ cu. ft.}=2\frac{1}{192}\text{ cu. ft.}$$

We can also do this in one continuous equation where we line up our identities and multiply through

$$\frac{15\text{ gal.}}{1}\cdot\frac{231\text{ in.}^3}{1\text{ gal.}}\cdot\frac{1\text{ft.}^3}{1{,}728\text{ in.}^3}=\frac{15\cdot 231\cdot 1\text{ft.}^3}{1\cdot 1\cdot 1728}=\frac{3{,}465}{1{,}728}\text{ cu. ft.}=2\ \frac{1}{192}\text{ cu. ft.}$$

PRACTICE: Convert

321 in. to yards

14 mi. to feet

600 oz. to pounds

2,134 sec to minutes

2,467 sq. in. to square yards

463 cu. in. to quarts

3-3 Exercises

Convert each quantity. Round the nearest hundredth of a unit when appropriate.

1. 192 inches = _________ feet
2. 11 pound = ______ ounces
3. 1 Ton = _______ ounces
4. 39 feet = ______ yards
5. 19.5 pints = _______ quarts
6. 4 miles = _________ feet
7. 2.3 miles = _________ yards
8. 656 ounces = _____ pounds
9. 9000 pounds = _____ Tons
10. 32 gallons = _____ cups
11. 428 gal = _____ qt
12. 9 ft^2 = _____ in^2
13. 1008 in^2 = _____ ft^2
14. 324 ft^3 = ______ in^3
15. 5 ft^3 = ______ qt.
16. 345611 oz. = ______ T.
17. 8 mi. = __________ in.
18. 324 yd^2 = __________ in^2
19. 1 yd^3 = ____________ gal.
20. 1 ft^3 = ______ qt.

21. 1 fl oz = _____ in^3

22. 1 in^3 = ______ fl oz.

23. 3450 min. = _______ hrs

24. 9201 hrs = ______ sec.

25. 45500 sec = ______ min

26. 10 qt/1 hr = ______ $ft^3/1min$

27. 5.5 $yd^3/1min$ = ____ $ft^3/1hr$

28. 6.1 yd^3/hr _________ gal./min

29. 5670 in^3/sec = _______ pt./min

30. 3.45 T./hr = ______________ oz./sec

Answers

1. 16
2. 176
3. 32000
4. 13
5. 9.75
6. 21120
7. 4048
8. 41
9. 4.5
10. 512
11. 1712
12. 1296
13. 7
14. 559872
15. 149.61

16. 10.80

17. 506880

18. 419904

19. 201.97

20. 29.92

21. 1.80

22. .56

23. 57.5

24. 33123600

25. 758.33

26. .0056

27. 8910

28. 20.53

29. 11781.82

30. 30.67

3-4 METRIC UNIT CONVERSIONS

Here are the basic unit of measure for METRIC in the seven dimensions

Dimension	Basic unit of measure
LENGTH	meter
MASS	gram
TIME	SECOND
TEMPERATURE	DEGREE OF Celsius
ELECTRICAL CURRENT	AMPERE
LUMINOSITY	CANDELA
AMOUNT OF SUBSTANCE	MOLE

In addition, for volume, the basic unit is the LITER.

Just like US customary, we could have just stick to these basic units and figure everything in terms of meters, grams, etc. However larger number of these units or small fraction of these units are just too cumbersome to write (like 20000000 meters or 1/1000 of a gram) Fortunately, Metric has a built in way of dealing with this, the POWERS OF TEN multipliers/prefix to represent larger or smaller units of the basic unit. So any conversion from metric to metric is either a multiplication or a division of a power of ten!!

Here are the most common prefix/powers of ten used in metric conversion.

prefix	multiplier	power of 10	symbol
mega	**1,000,000**	10^6	**M**
kilo	**1,000**	10^3	**k**
BASE UNIT	**1**		
centi	**1/100**	10^{-2}	**c**
milli	**1/1,000**	10^{-3}	**m**
micro	**1/1,000,000**	10^{-6}	**mc or μ**

Example:

Convert 50 grams to milligrams.

1 Milligram is 1/1000 of a gram. This also means there are 1000 milligrams in 1 gram.

We get $\frac{50\ grams}{1} \cdot \frac{1000\ milligrams}{1\ gram} = \frac{50000\ milligrams}{1} = 50000$ milligrams

Convert 8.4 centimeters to meters.

1 centimeter = 1/100 of a meter. This also means there are 100 centimeters in 1 meter.

We get $\frac{8.4\ centimeters}{1} \cdot \frac{1\ meter}{100\ centimeters} = \frac{8.4\ meters}{100} = .084$ meter

WE CAN USE THE DIFFERENCE IN POWERS TO FIGURE CORRECT IDENTITY FRACTION FOR CONVERSIONS

Convert 5 kiloliters to milliliters.

As kiloliters is 1000 liters or 10^3 liters and milliliters is 1/1000 liters or 10^{-3} *liters*, there is a difference 10^6 between kiloliters and milliliters. (3 to –3 is a difference of 6) So this means that there are 10^6 milliters in 1 kiloliters, or 1000000 milliliters = 1 kiloliter.

We now get $\frac{50\ kiloliters}{1} \cdot \frac{1000000\ milliliters}{1\ kiloliter} = \frac{5000000\ milliliters}{1} = 5000000$ milliliters.

Example, Convert 5768999 micrometers to centimeters.

As micrometers is 10^{-6} meters and centimeters is 10^{-2} meters, the difference of 10^4 between centimeters and micrometers ($-2 - -6 = 4$). So this means there are 10^4 micrometers in 1 centimeter, or 10000 micrometers in 1 centimeters.

Therefore $\frac{5768999\ mcm}{1} \cdot \frac{1\ cm}{10000\ mcm} = \frac{5768999}{1000}$ cm = 576.8999 cm

Practice

Convert 689 Liters to kiloliters

Convert 754 milligrams to Kilograms

Convert 23 Megameters to Centimeters

Convert 56 microliters to milliliters

Convert 100 centigrams to Megagrams

3-4 Exercises

Convert these metric measurements

1. .0005 kilometers = ________ meters
2. 2.3 centimeters = ______ meters
3. 624 millimeters = ________meters
4. .045 Liters = _______ milliLiters
5. 4350 grams = ______ milligrams
6. 307 kilogram = _______ grams
7. 6.5 cm = _______ m
8. .8 kL = _______ L

9. 434 mm = _______ cm

10. 50 mcg = ______ mg

11. 5960 mL = _______ kL

12. .005 kg = __________ mg

13. 3.31 L = ______ mL

14. 9000 μg = _______ g

15. 45000000 cm = ________ Mm

16. 23456 L = _________ kL

17. 12300000 mcm = ________ km

18. .92 m = ________ mm

19. 123 Mg = ________ mg

20. 345000000 mL = ________ kL

Answers

1. .5
2. .023
3. .624
4. 45
5. 4350000
6. 307000
7. .065
8. 800
9. 43.4
10. .05
11. .00596
12. 5000

13. 3310

14. .009

15. .45

16. 23.456

17. .0123

18. 920

19. 123000000000

20. 345

3-5 CONVERSIONS BETWEEN CUSTOMARY AND METRIC SYSTEMS

We are often called upon to convert between the Customary and Metric families. These should be committed to memory!

US CUSTOMARY TO METRIC SYSTEM OF MEASUREMENT

LENGTH	I in. = 2.54 cm
	3.281 ft. = 1 m
	1 mi. = 1.609 km
WEIGHT	2.205 lb. = 1 kg
AREA	1 sq. in. = 6.452 cm^2
	10.765 sq. ft. = 1m^2
VOLUME	1 cu. in. = 16.387 cm^3
	35.32 cu. ft. = 1 m^3
	1 fl. oz. = 29.57 mL
	1.057 qt. = 1 L
	1 gal. = 3.785 L

To convert from one unit of measure to another we can

- Write the quantity you desire to convert as a fraction.
- Multiply by a fraction formed by the conversion equation (called a conversion identity) so the units not in the final answer cancel out.
- Simplify your final number value if needed.

Convert 165 lb. to kilograms

As there are 2.205 lb. in 1 kg,

we get $\frac{165 \text{ lb.}}{1} \cdot \frac{1 \text{ kg}}{2.205 \text{ lb.}} = \frac{165 \text{ kg}}{2.205} = 74.83 \text{ kg}$

Convert 37 cm to inches

As there are 2.54 cm in 1 in.

We get $\frac{37 \text{ cm}}{1} \cdot \frac{1 \text{ in.}}{2.54 \text{ cm}} = \frac{37 \text{ in.}}{2.54} = 14.57 \text{ in.}$

Convert half a liter to fluid ounces

There are 1,000 mL in 1 L and there are 29.57 mL in 1 fl. oz., so

we get $\frac{1}{2} \cdot \frac{1 \text{ L}}{1} \cdot \frac{1{,}000 \text{ mL}}{1 \text{ L}} \cdot \frac{1 \text{ fl. oz.}}{29.57 \text{ mL}} = \frac{1{,}000 \text{ fl. oz.}}{2 \cdot 29.57} = 16.91 \text{ fl. oz.}$

Convert 60 mph to kilometers per second

There are 1.609 km in 1 mi., 60 min in 1 hr and 60 sec in 1 min.

So we get $\frac{60 \text{ mi.}}{1 \text{ hr}} \cdot \frac{1.609 \text{ km}}{1 \text{ mi.}} \cdot \frac{1 \text{ hr}}{60 \text{ min}} \cdot \frac{1 \text{ min}}{60 \text{ sec}} = \frac{60 \cdot 1.609 \text{ km}}{60 \cdot 60 \text{ sec}} = \frac{.0268 \text{ km}}{1 \text{ sec}}$ or about .0268 km per sec.

Practice

14 mi. to kilometers

2,134 kg to pounds

463 cu. in. to qt.

2,345 mL to gallons

$\frac{23 \text{ L}}{\text{day}}$ to $\frac{\text{fluid ounces}}{\text{min}}$

600 fl. oz. to milliliters

2,467 sq. in. to square centimeters

2,340 mi. to meters

13 mg to pounds

3-5 Exercises

Convert the following to the nearest hundredth when appropriate

1. $1 \text{ cm}^2 =$ ______ m^2

2. $1 \text{ cm}^3 =$ ______ m^3

3. $1 \text{ m}^2 =$ ______ cm^2

4. $1 \text{ m}^3 =$ ______ cm^3

5. $15 \text{ m}^2 =$ ______ cm^2

6. $34200 \text{ cm}^2 =$ ______ m^2

7. $56700000\ cm^3 =$ ______ m^3

8. $11\ m^3 =$ ______ cm^3

9. 480 mL = ______ fl oz

10. 242 lb = ______ kg

11. 6 ft = ______ m

12. 12 gal = ______ L

13. 20 mi = ______ km

14. 567 m = ______ ft

15. 5678 L = ______ qt

16. 367 kg = ______ lb

17. 12200 mL = ______ gal

18. 345 km = ______ ft

19. 20 gal = ______ mL

20. $18.4\ ft^2 =$ ______ m^2

21. $45\ m^2 =$ ______ ft^2

22. $43.2\ cm^3 =$ ______ in^3

23. 9.2 L = ______ pt

24. 40 mi/1 min = ______ km/1 sec

25. 40 mi/1 gal = ______ m/1 L

26. 120 m/1 hr = ______ ft/1 sec

27. $15\ lb/1\ in^2 =$ ______ $kg/1\ m^2$

28. $782\ kg/1\ m^3 =$ ______ $lb/1\ ft^3$

29. $1\ g/1\ cm^3 =$ ______ $oz/1\ in^3$

30. 2.5 L/1 day = ______ fl oz/1 hr

31. $1600\ cm^3/1\ hr =$ ______ $in^3/1\ min$

32. 7.5 km/1 hr = ______ mi/1 min

Answer

1. .0001
2. .000001
3. 10000
4. 1000000
5. 150000
6. 3.42
7. 56.7
8. 11000000
9. 16.23
10. 109.75
11. 1.83
12. 45.42
13. 32.18
14. 1860.33
15. 6001.65
16. 809.24
17. 3.22
18. 1132131.76
19. 75700
20. 1.71
21. 484.43
22. 2.64
23. 19.45
24. 1.07
25. 17003.96

26. .11

27. 10543.59

28. 48.82

29. .59

30. 3.52

31. 1.63

32. .08

3-6 UNITS OF FORCE, ENERGY, AND POWER

SPEED

Speed is how fast (or slow) something is moving and it is calculated by length divided by time, such as $\frac{55\text{ mi.}}{1\text{ hr}}$. If someone is driving a car at $\frac{55\text{ mi.}}{1\text{ hr}}$, it means that after 1 hr of driving, that person (vehicle) has traveled a total of 55 mi.

ACCELERATION

We know that not everything moves at a constant speed. Sometimes things go faster and sometimes things go slower. This changing of the speed is measured as ACCELERATION. Acceleration is speed divided by time.

©Iconic Bestiary/Shutterstock.com

Let's say Bosco gets into his car and from a stop sitting position, at 0 mph, speeds up the vehicle to 30 mph and it takes 20 sec to do so. What is the acceleration?

The change in MILES per hour is (30–0)/hr = 30 mph. We must now figure this change in miles per hour over the 20 sec it takes to change the speed. We also need a consistent time measure, so let's change 20 sec to hours. Using our conversion, 1 hr = 3,600 sec,

$$20 \text{ sec} \cdot \frac{1 \text{ hr}}{3{,}600 \text{ sec}} = \frac{20 \text{ hr}}{3{,}600} \approx .0055\bar{5} \text{ hr}$$

Acceleration is therefore $\dfrac{30 \text{ mi.}/\text{hr}}{.0055\bar{5} \text{ hr}} = \dfrac{30 \text{ mi.}}{.0055\bar{5} \text{ hr} \cdot \text{hr}} = \dfrac{5{,}400 \text{ mi.}}{\text{hr}^2}$

ACCELERATION is measured in length divided by *time squared*.

We can think of this acceleration as increasing 5,400 mph for each hour. Wow! That means after 1 hr, Bosco's speed is 5,400 mph, and after 2 hr, it is (5,400 + 5,400) mph or 10,800 mph, after 3 hr, a speed of (10,800 + 5,400) mph or 16,200 mph … say what????? He would need a space suit! However, keep in mind this same acceleration experienced through the 20 sec would just be going and going and going through seconds, minutes, hours … not realistic on a highway (at some time, he would level off his speed and or slow down …). It may make more sense to think of this acceleration in feet per seconds squared. Let's convert … We already know 5,280 ft. = 1 mi., and 1 hr = 3,600 sec, then 1 $\text{hr}^2 = (3{,}600 \text{ sec})^2 = 12{,}960{,}000 \text{ sec}^2$. Now we can convert

$$\frac{5{,}400 \text{ mi.}}{1 \text{ hr}^2} \cdot \frac{5{,}280 \text{ ft.}}{1 \text{ mi.}} \cdot \frac{1 \text{ hr}^2}{12{,}960{,}000 \text{ sec}^2} = \frac{5{,}400 \cdot 5{,}280 \text{ ft.}}{12{,}960{,}000 \text{ sec}^2} = \frac{2.2 \text{ ft.}}{\text{sec}^2}$$

So Bosco is increasing his speed 2.2 ft./sec for each second.

Let's look at it this way: just before Bosco hits the gas pedal, the speed is 0 ft per second. Now the car starts moving and after 1 sec, Bosco's speed is 2.2 feet per second. After the 2nd second, we add 2.2 to the feet per second so his speed is now (2.2 + 2.2) feet per second or 4.4 feet per second. After the third second, the speed is now (3 · 2.2) feet per second or 6.6 feet per second, and so on.

And in the 20th second, his acceleration has lead him to a speed of

$$\frac{20 \cdot 2.2 \text{ ft.}}{\text{sec}} = \frac{44 \text{ ft.}}{\text{sec}}$$

Here is the breakdown of speed by each second from 1 to 20

Seconds	0	1	2	3	4	5	6	7	8	9	10
Speed ft./sec	$\frac{0 \text{ ft.}}{\text{sec}}$	$\frac{2.2 \text{ ft.}}{\text{sec}}$	$\frac{4.4 \text{ ft.}}{\text{sec}}$	$\frac{6.6 \text{ ft.}}{\text{sec}}$	$\frac{8.8 \text{ ft.}}{\text{sec}}$	$\frac{11 \text{ ft.}}{\text{sec}}$	$\frac{13.2 \text{ ft.}}{\text{sec}}$	$\frac{15.4 \text{ ft.}}{\text{sec}}$	$\frac{17.6 \text{ ft.}}{\text{sec}}$	$\frac{19.8 \text{ ft.}}{\text{sec}}$	$\frac{22 \text{ ft.}}{\text{sec}}$
Seconds	11	12	13	14	15	16	17	18	19	20	
Speed ft./sec	$\frac{24.2 \text{ ft.}}{\text{sec}}$	$\frac{26.4 \text{ ft.}}{\text{sec}}$	$\frac{28.6 \text{ ft.}}{\text{sec}}$	$\frac{30.8 \text{ ft.}}{\text{sec}}$	$\frac{33 \text{ ft.}}{\text{sec}}$	$\frac{35.2 \text{ ft.}}{\text{sec}}$	$\frac{37.4 \text{ ft.}}{\text{sec}}$	$\frac{39.6 \text{ ft.}}{\text{sec}}$	$\frac{41.8 \text{ ft.}}{\text{sec}}$	$\frac{44 \text{ ft.}}{\text{sec}}$	

Does the $\dfrac{44 \text{ ft.}}{\text{sec}}$ equals the 30 mph speed Bosco attained in the 20th second? Let's check.

Converting to miles per hour

$$\frac{44 \text{ ft.}}{1 \text{ sec}} \cdot \frac{1 \text{ mi.}}{5{,}280 \text{ ft.}} \cdot \frac{3{,}600 \text{ sec}}{1 \text{ hr}} = \frac{44 \cdot 3{,}600 \text{ mi.}}{5280 \text{ hr}} = \frac{30 \text{ mi.}}{1 \text{ hr}}$$

We will next look at FORCE, but we first want to convert this acceleration in metric acceleration.

Changing $\frac{2.2 \text{ ft.}}{\sec^2}$ into m/sec^2

We know there are 3.281 ft. in 1 m, so

$$\frac{2.2 \text{ ft.}}{\sec^2} \cdot \frac{1 \text{ m}}{3.281 \text{ ft.}} = \frac{2.2 \text{ m}}{3.281 \sec^2} = \frac{.671 \text{ m}}{\sec^2}$$

So Bosco's acceleration was .671 m/sec^2.

FORCE

Force is the push or pull on something, which results in speed and acceleration. We can measure force using acceleration. It is formally figured as

$$\text{Force} = \text{mass times acceleration or } F = ma$$

where F = force, m = mass, and a = acceleration.

Demonstrating force in metric, let's use our Bosco car problem.

If Bosco's car weighs 2,205 lb.? What is the force provided to reach 30 mph in the 20th second?

We need acceleration and we need mass to figure force. And we want this as a metric calculation.

We calculated Bosco's car acceleration as = $\frac{.671 \text{ m}}{\sec^2}$

We now need the mass, m. As we are given the weight in pounds, we need to convert 2,205 lb. to kilograms.

There are 2.205 lb. in 1 kg

2,205 lb. $\cdot \frac{1 \text{ kg}}{2.205 \text{ lb.}}$ = 1,000 kg. Bosco's car weighs 1,000 kg.

Now Force = F = ma = 1000 kg $\cdot \frac{.671 \text{ m}}{\sec^2} = \frac{671 \text{ kg} \cdot \text{m}}{\sec^2}$

Do you see the derived unit of measure $\frac{\text{kg} \cdot \text{m}}{\sec^2}$. This is a metric unit of force and it is called a *NEWTON*, abbreviated N.

So 1 Newton = 1 N = $\frac{1 \text{ kg} \cdot \text{m}}{\sec^2}$

And the force need for Bosco's 2,205 lb. car to reach 30 mph in 20 sec is 671 N.

WEIGHT VERSUS MASS

In metric, mass is measured in grams, kilograms, milligrams, and so on.

This is different than weight. Weight is based on the gravitational force of the surface, so is really a FORCE based on a pull of push. In US Customary measures, we use ounces, pounds, and tons, and these weight measures are based on the Earth's gravitational force. So the concept of 1 lb. is really a force on the earth.

It has been calculated that the force of 1 lb. on the Earth's surface is

$$1 \text{ lb.} = 4.4482 \text{ N}$$

And this means 1 lb. is being pulled down at a force of $\dfrac{4.4482 \ \text{kg}\cdot\text{m}}{\text{sec}^2}$.

So if the force for Bosco's 2,205 lb. car to reach 30 mph in 20 sec is 671 N, that means there are

$$671 \text{ N}\cdot \frac{1 \text{ lb.}}{4.4482 \text{ N}} \approx 150.848 \text{ pounds of force.}$$

ENERGY

If force is the push or pull, then energy is what causes the force.

One calculation definition that is used for energy is

$$\text{Energy} = \text{Force times distance or Energy} = Fd$$

Example: Bosco's 2,205 lb. car travels 75 m during the 20 sec of acceleration to 30 mph. What is the energy used in accelerating?

We have calculated the force to be 671 N, or $\dfrac{671 \ \text{kg}\cdot\text{m}}{\text{sec}^2} = F$

Now the distance traveled, *d* equals 75 m, so Energy is

$$\text{Energy} = \text{F}d = \frac{671 \ \text{kg}\cdot\text{m}}{1 \text{ sec}^2}\cdot 75 \text{ m} = \frac{671\cdot 75\cdot\text{kg}\cdot\text{m}\cdot\text{m}}{1 \text{ sec}^2} = \frac{50{,}325 \text{ kg}\cdot\text{m}^2}{\text{sec}^2}$$

Note: The derived unit $\dfrac{\text{kg}\cdot\text{m}^2}{\text{sec}^2}$ is used for energy and it is called a *JOULE*, abbreviated J.

$$\text{So 1 Joule} = 1 \text{ J} = \frac{1 \ \text{kg}\cdot\text{m}^2}{\text{sec}^2}$$

Therefore, the energy needed for moving Bosco's car to reach 30 mph in 20 sec is

$$50{,}325 \ \frac{\text{kg}\cdot\text{m}^2}{\text{sec}^2} = 50{,}325 \text{ J}$$

NOTE: 1 Newton times 1 meter = 1 Joule, or 1 N times 1 m = 1 J

Energy is encountered in many forms. There is energy that comes from

- Moving objects
- Electrical current
- Wood-burning stove
- Sun light
- Food we eat … to name a few

What have evolved are several different units of measure.

However, the basics of all forms energy can be broken down into the metric form mass times length squared divided by time squared, as in the Joule = $\dfrac{\text{kg}\cdot\text{m}^2}{\text{sec}^2}$.

We can convert different units of energy into Joules, and back given conversion factors. For example, in the US Customary System, we use pounds for mass, feet for length, and seconds for time, it is true that a unit of energy measure is called the FOOT POUND and $1 \text{ ft. lb.} = \frac{1 \text{ lb.} \cdot 1 \text{ ft.}}{1 \text{ sec}^2}$.

And the energy of 1 ft. lb.= 1.356 J

Another example is the *BTU*. A BTU is a measure used in heating or cooling (a BTU is calculated as the amount of energy needed to raise the temperature of 1 lb. of water by 1°F)

It is also true that 1 BTU = 1,055 J.

A *CALORIE*, cal, is another energy unit used (not exactly the same used in food measures, but related)

And 1 cal = 4.186 J.

Problem

We have shown that it takes 50,325 J to accelerated Bosco's car to 30 mph in 20 sec.

How many Calories of energy does Bosco's car use?

We now know that 1 cal = 4.186 J.

Now we convert Joules to Calories $50{,}325 \text{ J} \cdot \frac{1 \text{ cal}}{4.186 \text{ J}} \cdot = \frac{50{,}325 \text{ cal}}{4.186} = 12022.22 \text{ cal}$

So it takes 12022.22 cal of energy to reach the 30 mph.

How many foot pounds of energy is that?

As 1 ft. lb. = 1.356 J

$$50{,}325 \text{ J} \cdot \frac{1 \text{ ft.lb.}}{1.356 \text{ J}} = \frac{50{,}325 \text{ ft.lb.}}{1.356} = 37112.83 \text{ ft. lb.}$$

How many BTU's of energy is that? As 1 BTU = 1,055 J

then $50{,}325 \text{ J} \cdot \frac{1 \text{ BTU}}{1{,}055 \text{ J}} = \frac{50{,}325 \text{ BTU}}{1{,}055} = 47.70 \text{ BTUs}$

POWER

Power is the rate that energy is delivered. We calculate power as energy divided by time. If we are calculating the power of delivering 1 J over 1 sec, we have

$$\frac{1 \text{ J}}{1 \text{ sec}} = \frac{\frac{1 \text{kg} \cdot \text{m}^2}{\text{sec}^2}}{1 \text{ sec}} = \frac{1 \text{kg} \cdot \text{m}^2}{\text{sec}^3}$$

This unit of power, $\frac{1 \text{kg} \cdot \text{m}^2}{\text{sec}^3}$, is called a Watt, abbreviated W,

so $1 \text{ W} = \frac{1 \text{ J}}{1 \text{ sec}} = \frac{1 \text{ kg} \cdot \text{m}^2}{\text{sec}^3}$

So, what was the power in Watts generated by accelerating Bosco's 2,205 lb. car to 30 mph in 20 sec over 75 m?

It took 50,325 J of energy, so

$$\frac{50{,}325 \text{ J}}{20 \text{ sec}} = 2516.25 \text{ W of power.}$$

The US Customary unit of measure for power is the *HORSEPOWER*, abbreviated hp

and 1 hp = 745.7 W.

So the power of Bosco's 2,205 lb. car to accelerate to 30 mph in 20 sec over 75 m is

$$2516.25 \text{ W} \cdot \frac{1 \text{ hp}}{745.7 \text{ W}} = \frac{2516.25}{745.7} \text{hp} = 3.37 \text{ hp.}$$

ADDITIONAL NOTES: HORSEPOWER

Horsepower was developed to equal to the power an average horse exerts to lift 550 lb. of weight 1 ft. in 1 sec.

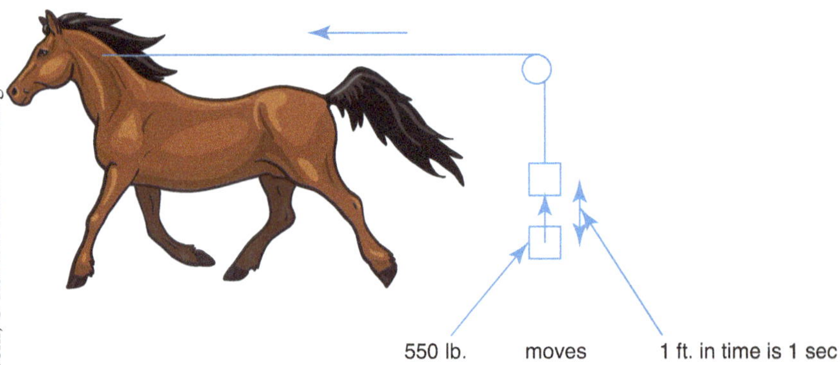

And it looks like this in a formula $1 \text{ hp} = \dfrac{550 \text{ lb.} \times 1 \text{ ft.}}{1 \text{ sec}} = \dfrac{550 \text{ foot pound}}{1 \text{ sec}} = \dfrac{550 \text{ ft. lb.}}{1 \text{ sec}}$

Remember we said 1 hp = 745.7 W.

This means then $745.7 \text{ W} = \dfrac{550 \text{ ft. lb.}}{1 \text{ sec}}$

We also know that $1 \text{ W} = \dfrac{1 \text{ kg m}^2}{\text{sec}^3}$. So $745.7 \text{ W} = \dfrac{745.7 \text{ kg m}^2}{\text{sec}^3}$

We also know that $1 \text{ ft. lb.} = 1.356 \text{ J} = \dfrac{1.356 \text{ kg m}^2}{\text{sec}^2}$

So $550 \text{ ft. lb.} = 550\,(1.356 \text{ J}) = 550\left(\dfrac{1.356 \text{ kg m}^2}{\text{sec}^2}\right) = \dfrac{745.7 \text{ kg m}^2}{\text{sec}^2}$

$$\text{So } 1 \text{ hp} = \frac{550 \text{ ft. lb.}}{1 \text{ sec}} = \frac{\dfrac{745.7 \text{ kg m}^2}{\text{sec}^2}}{1 \text{ sec}} = \frac{745.7 \text{ kg m}^2}{\text{sec}^3} = 745.7 \text{ W}$$

SUMMARY

Units of Force: Newtons (N); Pounds (lb.); Ounces (oz.); Tons (tn.)

Units of Energy: Joules (J); foot pounds (ft.·lb.); BTU; Calories (cal)

Units of Power: Watts (W); Horsepower (hp)

Conversion Factors

FORCE

1 lb. = 4.4482 N

ENERGY

1 ft.·lb. = 1.356 J

1 cal = 4.186 J

1 BTU = 1,055 J

1 BTU = 778 ft.·lb.

POWER

1 hp = 745.7 W

3-6 Exercises

Convert the following units of measure, round to the nearest hundredth when appropriate

1. 350 N = ________lbs
2. 180 lbs = ________N
3. 710 J = ________ft-lbs
4. 185 ft-lbs = ______J
5. 145 J = _____BTUs
6. 3456 BTUs = ______J
7. 65 BTUs = _____ft-lbs
8. 500 J = _______cal.
9. 567 cal. = ______J

10. 34 BTUs = _____cal.

11. 679 cal. = _____ft-lbs

12. 300 hp = _____W

13. 1400 W = _____hp

14. 40 hp = _____W

15. 51W = ______hp

For the problems figure the Acceleration, Force, Energy and Power

16. A 3000 Kilogram vehicle accelerates from 0 kilometers per hour to 64 kilometers per hour in 30 seconds, traveling 80 meters.
    a) Find the acceleration in m/sec^2
    b) Find the force needed in Newtons
    c) Find the energy needed in Joules
    d) Find the power needed in Watts

17. A 3000 pound vehicle accelerates from 0 mile per hour to 40 miles per hour in 40 seconds, traveling 100 feet.
    a) Find the acceleration in ft/sec^2.
    b) Find the force needed in Pounds.
    c) Find the energy needed in Foot-pounds.
    d) Find the power needed in Horsepower.

Answers

1. 78.68
2. 800.68
3. 523.60
4. 250.86
5. .14
6. 3646080
7. 50570
8. 119.45

9. 2373.46

10. 8569.04

11. 2096.09

12. 223710

13. 1.88

14. 29828

15. .07

16. **a)** .59 m/sec^2
 b) 1770 N
 c) 141600 J
 d) 4720 W

17. **a)** 1.47 ft/sec^2
 b) 137.03 lbs
 c) 13703.00 ft-lbs
 d) .62 hp

3-7 TEMPERATURE: CELSIUS, FAHRENHEIT, AND KELVIN

To measure heating and cooling temperature, there are two main measuring systems in use today.

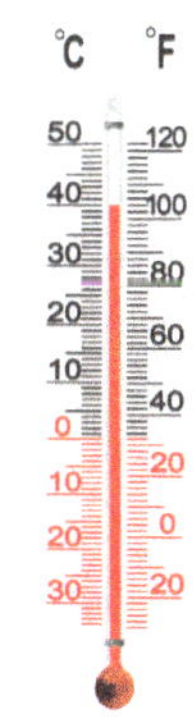

©Tomas Ragina/Shutterstock.com

CELSIUS TEMPERATURE, the most used system, was developed in the 1700s by Andres Celsius, a Swedish astronomer. Celsius determined a scale of degrees (°) based on freezing of water, set at 0° and the boiling of water, set at 100°. It is often called the centigrade scale.

FAHRENHEIT TEMPERATURE, the system used primarily in the United States, was developed about the same time by German physicist Gabriel Fahrenheit, and he also used the freezing and boiling of water. He chose 32° as the freezing point and 212° as the boiling point. (Fahrenheit used the 0 degree point as the freezing of salt water, which corresponded with 32 degrees for freezing of pure water.)

TEMPERATURE CONVERSION

If we look at the range of degrees in Celsius between freezing, 0°, and boiling, 100°, we have 100° units. The range between Fahrenheit freezing, 32°, and boiling, 212°, is 180° units. So for every 180°F, there are 100°C, or there are 1.8°F to 1°C. Or we could use the ratio $\frac{1.8\text{ F}}{1\text{ C}}$.

If both temperature inventors had chosen 0° as the freezing point, as setting F as Fahrenheit degrees and C as Celsius degrees, we could have used 1 F = 1.8 C as our conversion equation. However, as 32° was chosen as the Fahrenheit freezing point and 0° as the Celsius freezing point, we must alter our equation as

$$F = 1.8C + 32.$$

Sometimes this equation shows up as $F = \frac{9}{5}C + 32$, as $\frac{9}{5} = 1.8$

We can use this formula to find corresponding Fahrenheit temperature give Celsius temperature.

We can take the formula F = 1.8 C + 32 and use some algebra to find a formula for Celsius temperature given Fahrenheit temperature.

$$\begin{array}{rr} F = 1.8\,C + 32 & \\ -32 & -32 \\ \hline F - 32 = & 1.8C \\ \div 1.8 & \div 1.8 \\ \hline \multicolumn{2}{c}{\frac{F-32}{1.8} = C} \end{array}$$

This formula sometimes appears as $\frac{5}{9}(F-32) = C$

> So we can use F = 1.8 C + 32 to convert from Celsius to Fahrenheit.
>
> And $C = \frac{F-32}{1.8}$ to convert from Fahrenheit to Celsius.

Example: Convert 38.8°C to Fahrenheit.

$$F = 1.8\,C + 32 = 1.8(38.8) + 32 = 69.84 + 32 = 101.84°F$$

Convert 18°F to Celsius $C = \frac{F-32}{1.8} = \frac{18-32}{1.8} = \frac{-14}{1.8} \approx -7.78°C$

KELVIN TEMPERATURE

The Kelvin scale is used in science calculations when measuring some kind of atomic/molecular/particle motion. Particle motions are combinations of

- Translations, or where particles change location.
- Rotations, where particle stay in one position but spin.
- Vibrations, where particles stay in one position but move side to side.

Absolute zero in Kelvin is the temperature at which all particle motion stops! This has been measured as –273°C. The degree unit itself is the same as the Celsius degree unit.

So Kelvin temperature is K = C + 273

And as $C = \frac{F-32}{1.8}$, this means $K = \frac{F-32}{1.8} + 273$

If it is 23°C what is the temperature in Kelvin? K = C + 273 = 23 + 273 = 296°K

If it is 67°F, what is the temperature in Kelvin?

$$K = \frac{F-32}{1.8} + 273 = \frac{67-32}{1.8} + 273 = 19.44 + 273 = 292.44°K$$

What is Celsius if we know Kelvin? As C + 273 = K, then subtract 273 both sides to get

$$C = K - 273$$

What is Fahrenheit if we know Kelvin?

We can use properties of equality to solve for F in terms of K.

$$\frac{F-32}{1.8} + 273 = K, \text{ subtract 273 both sides}$$

$$\frac{F-32}{1.8} + 273 - 273 = K - 23 \text{ which gives us}$$

$$\frac{F-32}{1.8} = K - 273 \text{ multiply both sides by 1.8}$$

$$1.8\left(\frac{F-32}{1.8}\right) = 1.8\,(K - 273) \text{ simplify, use distributive prop right side}$$

$$F - 32 = 1.8\,K - 491.4 \text{ add 32 both sides}$$

$$\underline{+\,32 \qquad +\,32}$$
$$F = 1.8\,K - 459.4$$

So we have the following formulas you can use for conversions involving Kelvin, Celsius, and Fahrenheit

$K = C + 273$	$K = \frac{F-32}{1.8} + 273$	$C = K - 273$	$F = 1.8\,K - 459.4$

Convert to Celsius and to Kelvin

90°F ______ °C _______ °K 5°F ______ °C _______ °K

−20°F ______ °C _______ °K 0°F _______ °C ______ °K

Convert to Fahrenheit and to Kelvin

60°C ______ °F ______ °K 2°C ______ °F ______ °K

−160°C ______ °F _____ °K 0°C ______ °F ______ °K

Convert to Celsius and to Fahrenheit

40°K ______ °C ______ °F 110°K ______ °C _____ °F

325°K ______ °C ______ °F 0°K ______ °C _____ °F

3-7 Exercises

Convert the following temperatures to Fahrenheit degree

1. 20°C _________
2. 34°C _________
3. 120°C _________
4. –52°C _________
5. –23°C _________
6. 0°C _________
7. 20°K _________
8. 120°K _________
9. 235°K _________
10. 0°K _________

Convert the following temperatures to Celsius degree

11. 20°F _________
12. 34°F _________
13. 120°F _________
14. –52°F _________
15. –2°F _________
16. 0°F _________
17. 20°K _________
18. 122°K _________
19. 635°K _________
20. 0°K _________

Convert the following the Kelvin degree

21. 20°F _________

22. 34°F __________

23. 120°F __________

24. –52°F __________

25. –34°F __________

26. 0°F __________

27. 20°C_________

28. 122°C _________

29. 235°C _________

30. 0 °C _________

31. At what Degree value is Fahrenheit and Celsius the same value?

Answers

1. 68°F
2. 93.2°F
3. 248°F
4. –61.6°F
5. –9.4°F
6. 32°F
7. –423.4°F
8. –243.4°F
9. –36.4°F
10. –459.4 °F
11. –6.67°C
12. 1.11°C
13. 48.89°C
14. –46.67°C

15. −18.88°C

16. −17.78°C

17. −253°C

18. −151°C

19. 362°C

20. −273°C

21. 266.33°K

22. 274.11°K

23. 321.89°K

24. 226.33°K

25. 236.33°K

26. 255.22°K

27. 293°K

28. 395°K

29. 508°K

30. 273°K

31. −40

CHAPTER 4

Two Dimensional Figures: Perimeter and Area

4-1 PLANE GEOMETRIC DEFINITIONS AND SHAPES

The term Geometry evolves from the Ancient Greek, as "geo" - which means earth, and 'metron", which means measurement. Geometry is a branch of mathematics that is concerned with questions of shape, size, relative position of figures, and the properties of space. These questions are dealt with daily in the industrial profession. Lets "define" some common Geometric terms.

<u>Point</u> – This is an "undefined" term and refers to a location in space.

In the world, space around us is made up of infinite "suspended" points. A point cannot really be seen, as it has no thickness or weight or length, etc. This make a point an "undefined" concept. However, we have to accept that points exist and that points are arranged in ways to form figures. We can represent points as dots on flat (or curved) surface, such as on a piece of paper of a whiteboard in a classroom. Points are the "building blocks" of Geometry in order to create figures.

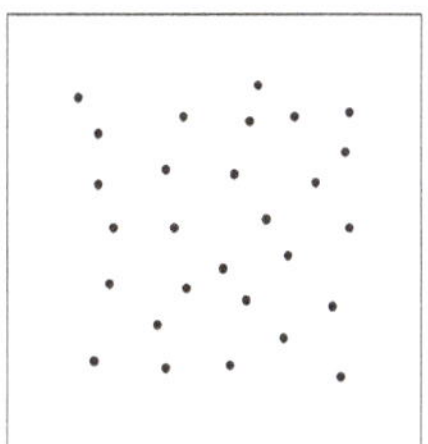

With points, we can create

<u>Line</u> – An infinite collection of points, side by side, that go on forever in both directions.

< . . . . > becomes 

<u>Plane</u> – set of all points making up a flat surface that goes on forever in all direction

<u>Line Segment</u> – This refers to a part of a line that lies between two points, called the endpoints.

<u>Ray</u> – part of a line that has a start point (called an endpoint) and continues in one direction forever

Angle – Made by two rays that share the same endpoint

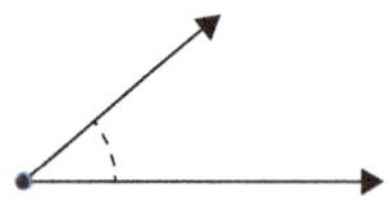

When all of the points making up a figure in a single flat plane of points, we consider the figure as a TWO DIMENSIONAL FIGURE. In other words, figures in two dimensions can be drawn on a surface or plane. We call the study of two dimensional figures PLANE GEOMETRY.

For line segments, we use customary or metric length units of measure, such as inches, feet, meters, centimeters, and so on.

Example: this segment

Is about 2 inches or 5 centimeters

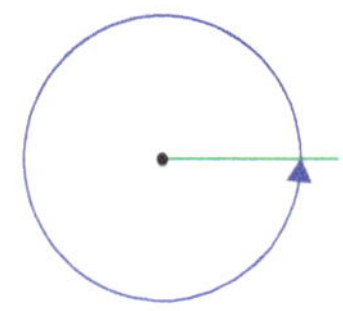

In an angle, we measure the space if we rotate from one ray to the other using a unit called a degree, which uses the symbol ^{c}. In a full angle rotation, or when we complete a circle in degrees, there are 360 degrees. So a rotation of 1/360 of a circle is 1 degree.

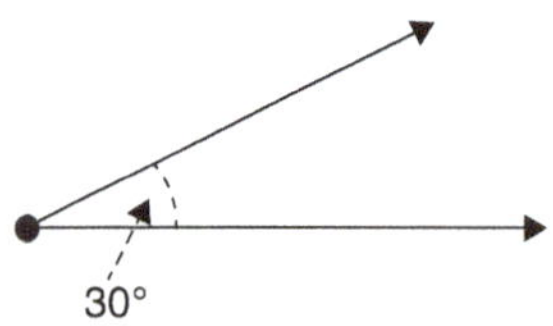

The interior space between these two rays making this angle is about 30^{c}.

Here are some other angle measures

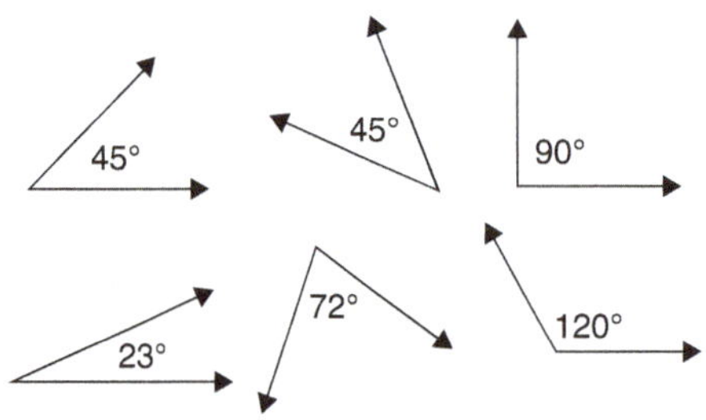

PARALLEL AND PERPENDICULAR LINES AND SEGMENTS

Two (or more) lines that travel in the same direction on a plan are called PARALLEL LINES. Parallel lines never cross or intersect, as they do not share any common points.

Here is an example

Segments on parallel lines are considered parallel segments.

Two lines that cross or intersect such that they form four 90 angles are called PERPENDICULAR LINES.

Here is an example

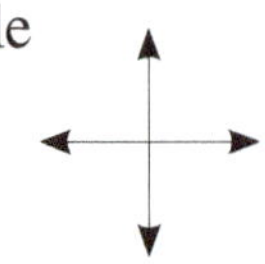

ENCLOSED PLANE FIGURES

When we arrange the points in ways where the figure is enclosed, we can find important measurements called the PERIMETER and the AREA.

Perimeter is the distance if one was to "walk" the boundary of the enclosed figure and is measures in length measure units, such as inches, feet, centimeters, meters, etc. Area makes up the space within the enclosed figure, and is measured by "counting" the number square units that fill the enclosed figure entirely and without overlap. The area measure is in square length units, such as square feet, square miles, square meters, square millimeters, etc. . .

Many plane figures can be challenging to find perimeter and area values. However, there are some plane figures that have characteristics that allow the formation of perimeter and area formulas which can then be used in finding more complicated figure's measures. In the next few sections we will look at some of those figures.

In addition, when two (or more) plane figures have the same shape and size, we consider the figures to be CONGRUENT. So in two congruent figures, and corresponding side lengths are the same measure, all corresponding angle measures are the same degree measure, and so on. Also, any two (or more) segments of the same length measure are considered congruent and as any two (or more) angles of the same degree measure are considered congruent.

Examples of congruent figures

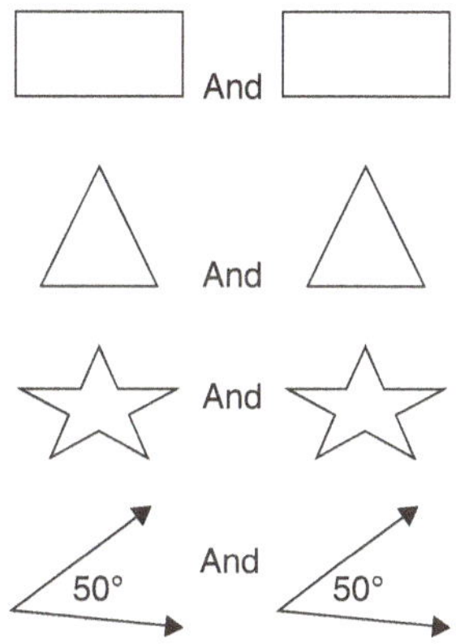

POLYGONAL CHARACTERISTICS

Polygons are plain figures made up of line segments, with each segment endpoint meeting one and only one other segment endpoint. Here are some polygon examples

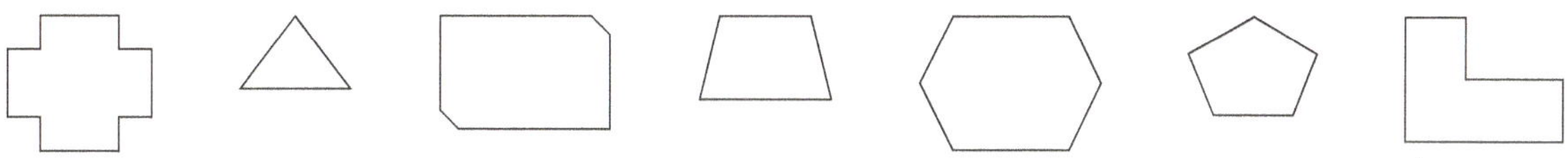

Polygons are figures that are totally enclosed, so we can figure perimeter and area.

First, let's look at some definitions involving polygons.

The segments making up a polygon is called SIDES of the polygon. For instance, this is a four sided polygon.

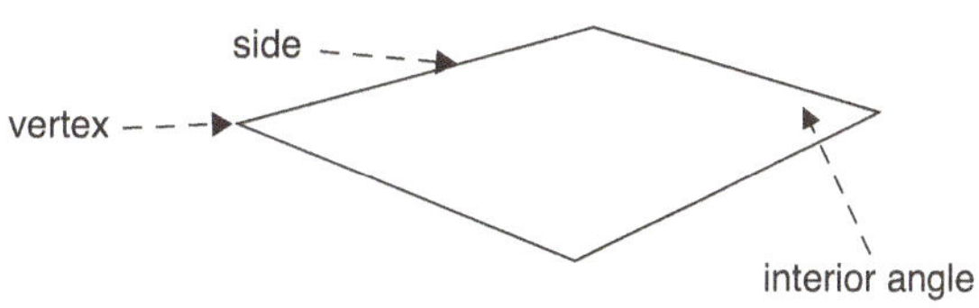

The corners where the sides meet are called VERTICES of the polygon. One of the vertices is called a VERTEX.

And two of the sides that meet at a vertex forms an INTERIOR ANGLE of the polygon.

A segment that goes from one vertex to another vertex across the interior of a polygon is called the DIAGONAL.

Here are some diagonal examples

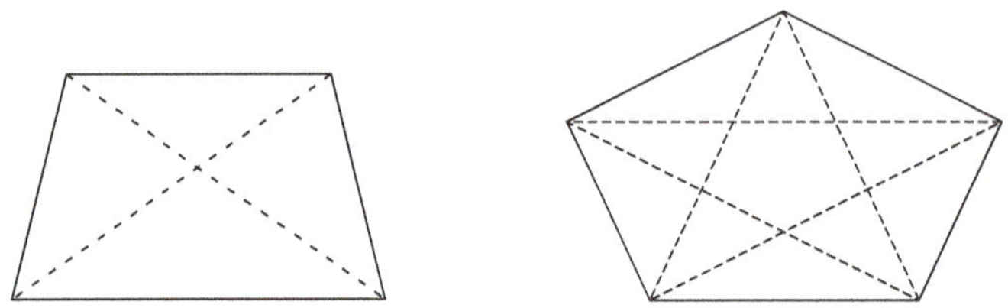

Polygons are named for the number of sides (and angles) in the figure.

Here is a list of the first 10 sided polygons by example and name.

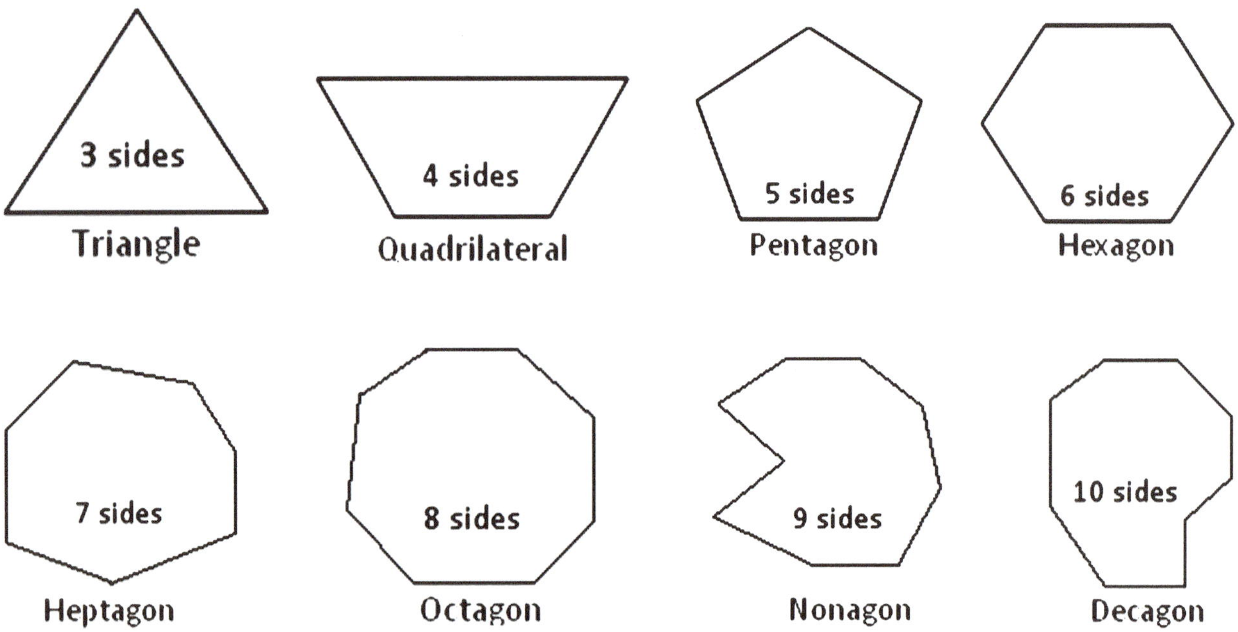

4-1 Exercises

Identify the polygons.

1. 

2. 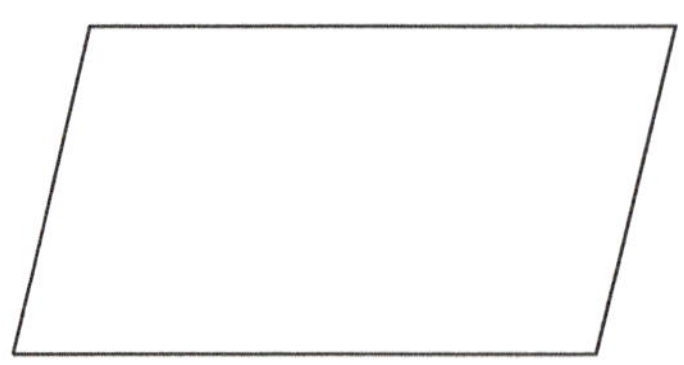

3. 

4. 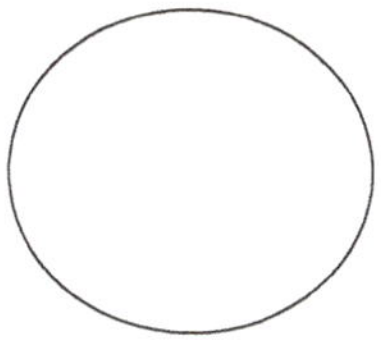

5. 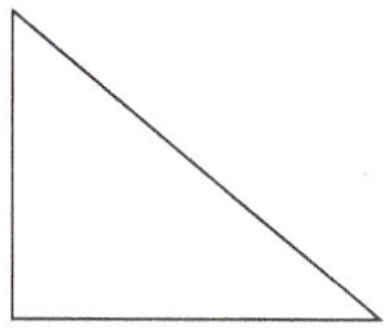

6. 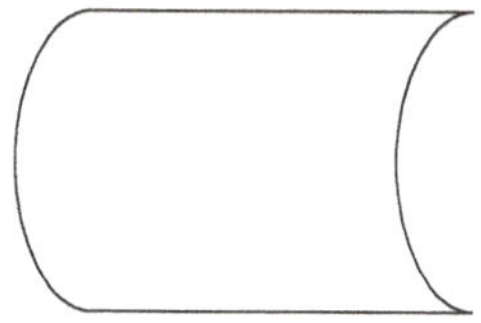

7. 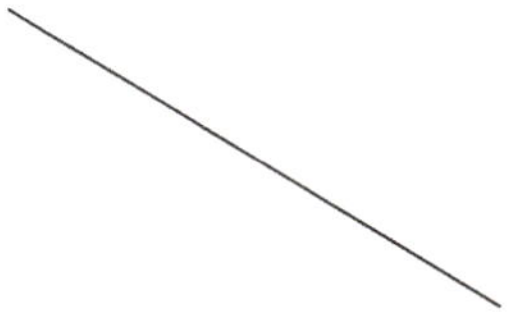

8. 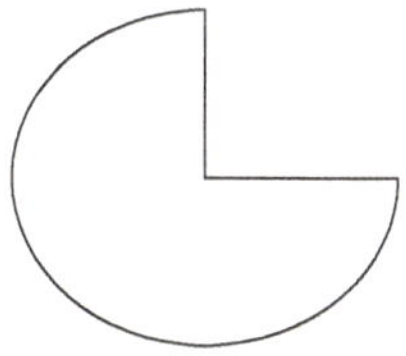

9. 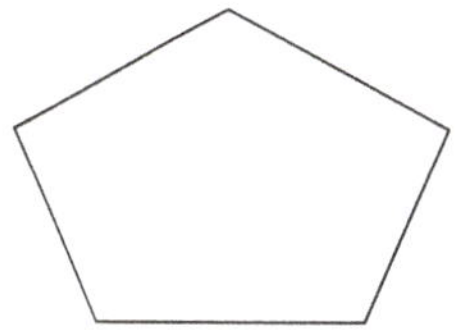

10. 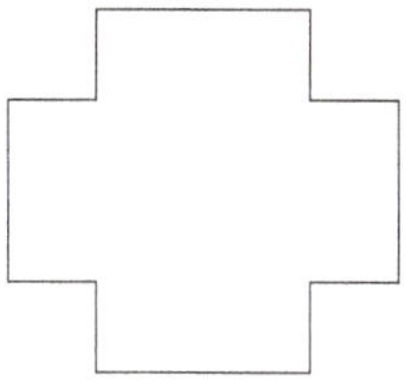

11. 

12. 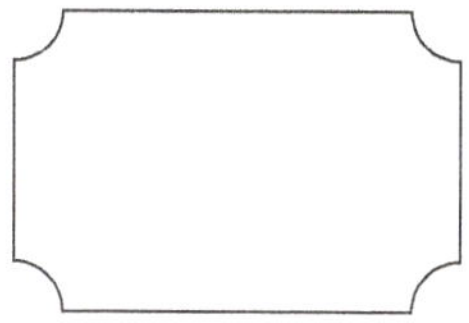

13. For this polygon, identify the side, vertex, and diagonal

a) ____________

b) ____________

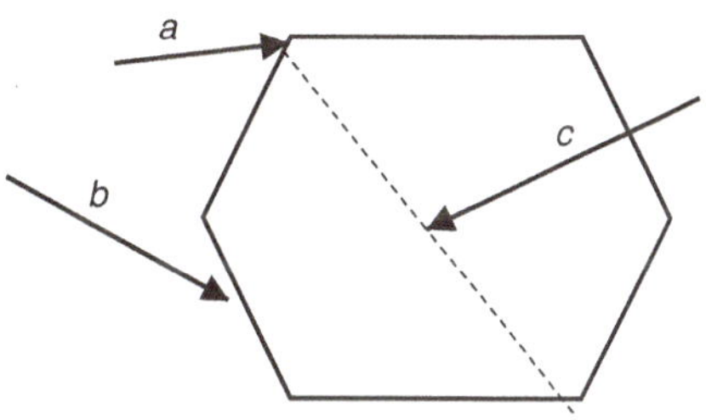

c) ____________

Answers

1. Polygon
2. Polygon
3. Not a polygon
4. Not a polygon
5. Polygon
6. Not a polygon
7. Not a polygon
8. Not a polygon
9. Polygon
10. Polygon
11. Not a polygon
12. Not a polygon
13. a) Vertex b) side c) diagonal

4-2 RECTANGLES

Four-sided polygons are called QUADRILATERALS. As they are polygons, and polygons are figures that are totally enclosed, quadrilaterals have a perimeter and an area.

There are an infinite number of quadrilaterals, many of which require complicated and involved measuring of perimeter and area. However, there are some quadrilaterals, due to their unique arrangement of points, that can be measured using simple perimeter and area formulas.

RECTANGLES

One of the simplest polygonal shape is the rectangle. A rectangle has four line segment sides and has four interior angles with the same degree measure of 90° (called right angles). This means that each side segment is perpendicular to a second side segment.

Here are examples of rectangles:

It is also true in every rectangle.

Two side lengths opposite each other have the same length.

We can indicate the right angle by placing a small □ in the angle position, and the equal side lengths by mark indicators like this

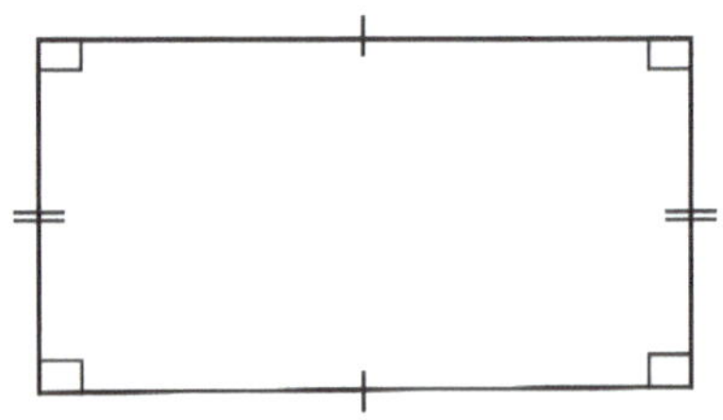

The longest of the pair of side lengths is usually referred to as the length and the shortest of the pair of side lengths is the width.

The PERIMETER of a rectangle is the total length of the four sides.

Find the perimeter of this rectangle:

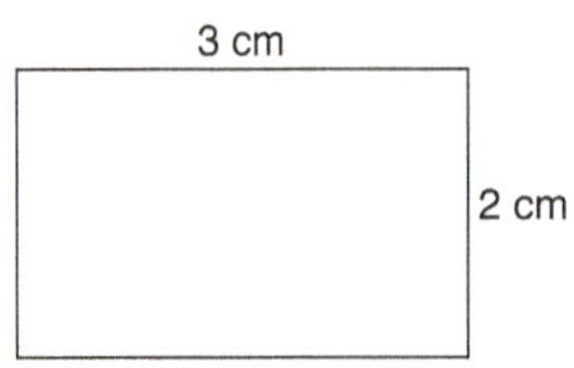

We have a width of 2 cm and a length of 3 cm. So we have two width measures and two length measures in our perimeter.

So, this means the perimeter = 2 cm + 2 cm + 3 cm+ 3 cm = 10 cm

Given W = width and L = length, we make a formula as Perimeter = 2W + 2L

The AREA can be found by multiplying the width times the length.

When we do this, we create (width times length) number of square units inside the rectangle.

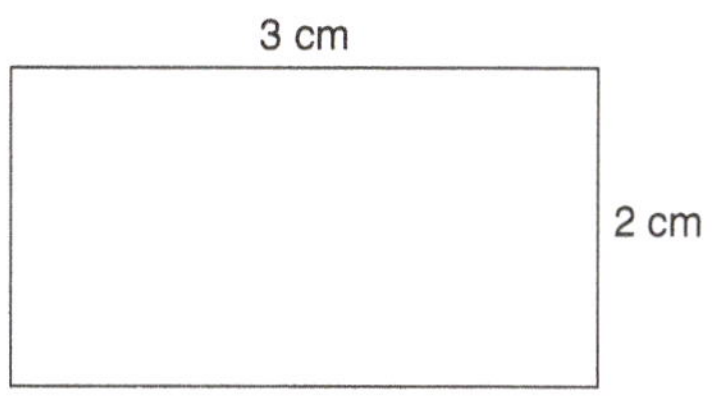

The area is figured as 2 cm × 3 cm = 2 × 3 × cm × cm = 6 cm squared or 6 cm^2.

Let's see what is happening in the rectangle

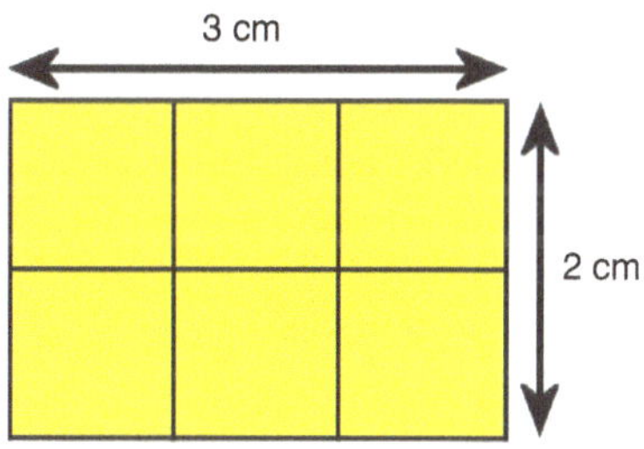

You find six 1 cm × 1 cm squares make up the area of 6 cm^2

Given W = width and L = length, our area formula is Area = W L

Find the perimeter and area for this rectangle

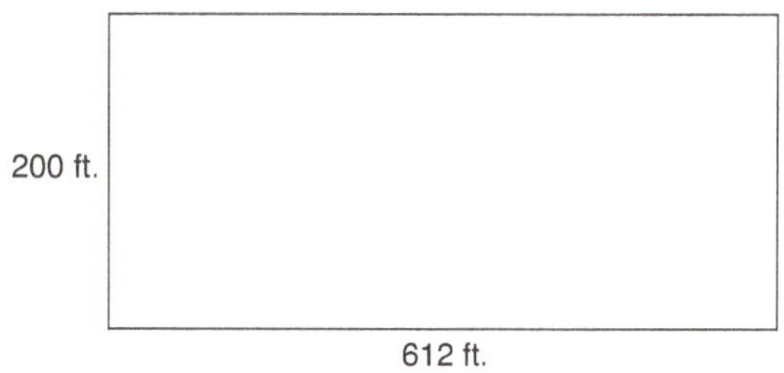

What about this one

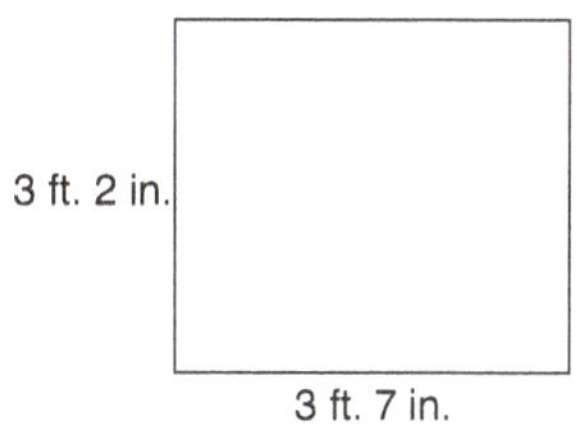

For perimeter, we use Perimeter = 2W + 2L

$$\text{Perimeter} = 2W + 2L = 2\,(3\text{ ft. }2\text{ in.}) + 2\,(3\text{ ft. }7\text{ in.}) =$$

$$2\,(3\text{ ft.} + 2\text{ in.}) + 2\,(3\text{ ft.} + 7\text{ in.}) =$$

$$6\text{ ft.} + 4\text{ in.} + 6\text{ ft.} + 14\text{ in.} =$$

$$12\text{ ft.} + 18\text{ in.}$$

As 18 in. = 1 ft. 6 in., we get

$$12\text{ ft.} + 1\text{ ft.} + 6\text{ in.} = 13\text{ ft. }6\text{ in.}$$

For area, we need to either convert the width and length to feet or to inches.

If we want square feet, we convert to feet

$$\text{Width} = 3\text{ ft. }2\text{ in.} = 3\text{ ft.} + 2\text{ in.} = 3\text{ ft.} + \frac{2}{12}\text{ ft.} = 3\frac{1}{6}\text{ ft.}$$

$$\text{Length} = 3\text{ ft. }7\text{ in.} = 3\text{ ft.} + 7\text{ in.} = 3\text{ ft.} + \frac{7}{12}\text{ ft.} = 3\frac{7}{12}\text{ ft.}$$

Now using Area = W L, we get

$$\text{Area} = W\,L = 3\frac{1}{6}\text{ ft.} \cdot 3\frac{7}{12}\text{ ft.} = (3\frac{1}{6} \cdot 3\frac{7}{12})\text{ ft.}^2 = \frac{19}{6} \cdot \frac{43}{12}\text{ ft.}^2 =$$

$$\frac{817}{72}\text{ft.}^2 = 11\frac{25}{72}\text{ ft.}^2$$

So perimeter is 13 ft. 6 in. and area is $11\frac{25}{72}$ ft.^{2}.

Note: To indicate feet and inches, you will often find single prime to indicate feet and a double prime to indicate inches.

So 3 ft. 2 in. = 3’ 2″ or

200 ft. 11 in. = 200’ 11″

COMPOUND AREAS

How about this problem—find perimeter and area for the polygon shape.

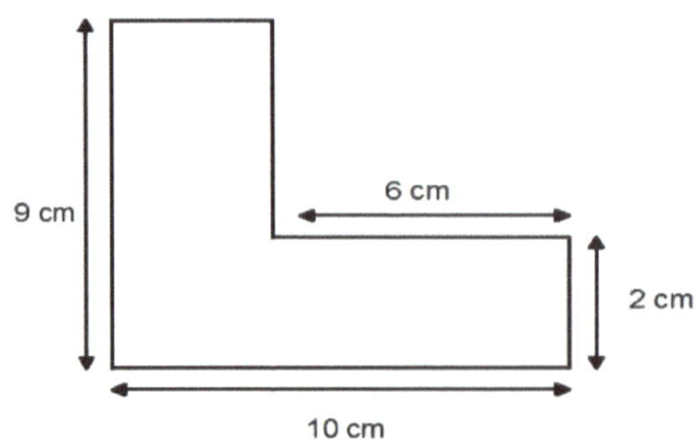

Perimeter: We have side length 9 cm, 6 cm, 2 cm, and 10 cm. We are missing the side length on the top and a right side length as well. If we were to continue the side for the 6 cm side through the figure, it would form two rectangle parts.

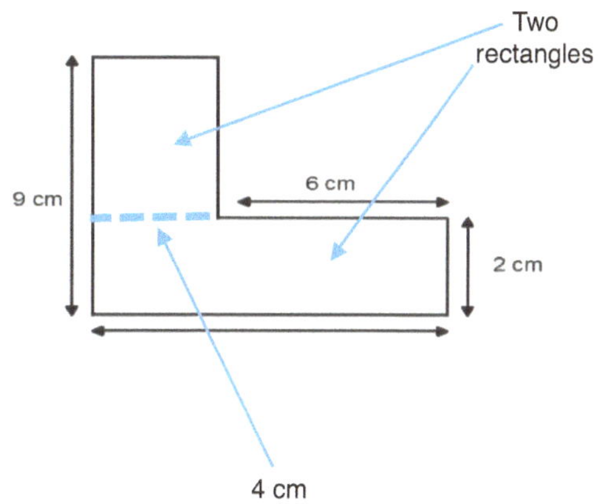

As 6 cm + 4 cm = 10 cm, and opposite sides of rectangles are equal in means the extended "part" must equal 4 cm. This also means the top side length must also equal 4 cm. Also the other missing side is 7 cm, because 9 cm – 2 cm = 7 cm.

So we now have

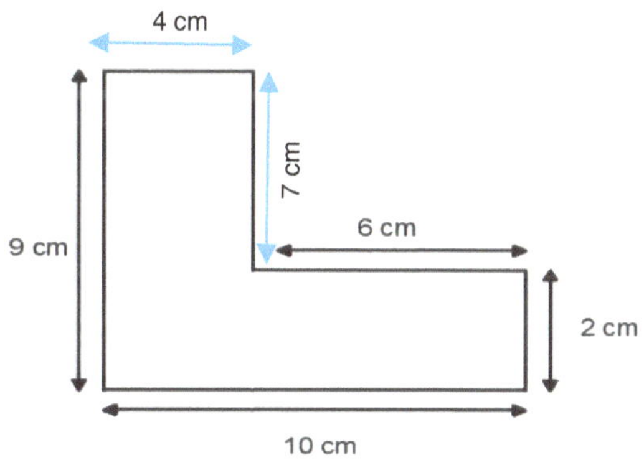

So perimeter = 9 cm + 6 cm + 2 cm + 10 cm + 4 cm +7 cm = 38 cm

Area: Now we have divided the polygon into two rectangles. The bottom rectangle has a width of 2 cm and a length of 10 cm. As for the top rectangle, we know that, the length will be 7 cm and a width of 4 cm.

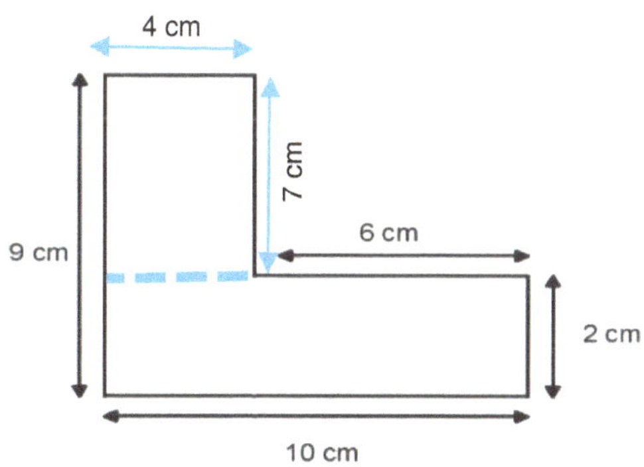

So total area equals bottom rectangle area + top rectangle area =

$2 \text{ cm} \times 10 \text{ cm} + 4 \text{ cm} \times 7 \text{ cm} = 20 \text{ cm}^2 + 28 \text{ cm}^2 = 48 \text{ cm}^2$

Try this one: Perimeter and area

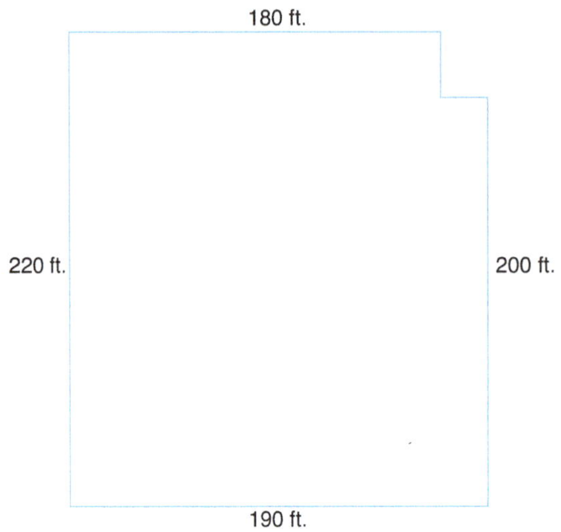

Now, another MEASURE FOR LARGE AREAS IS THE ACRE. 1 acre = 43,500 ft.^2.

What is this area in acres?

Practice

Find perimeter and area

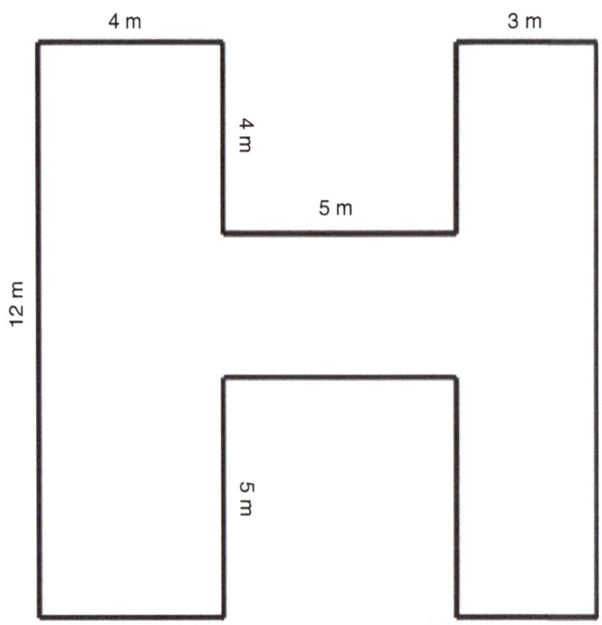

4-2 Exercises

Find the a) perimeter and b) area of these figures. Answer with correct unit of measure.

1.

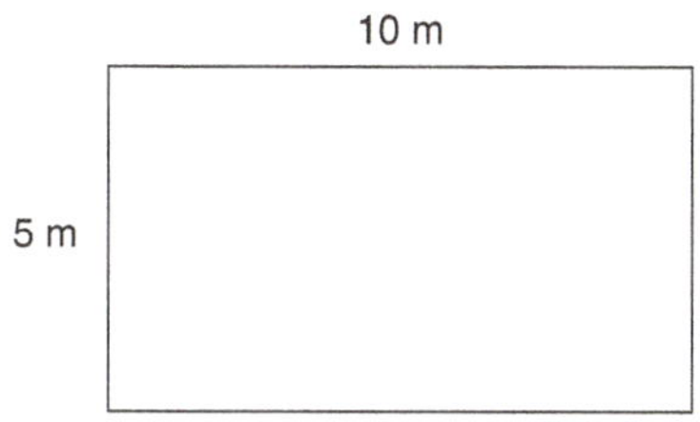

2.

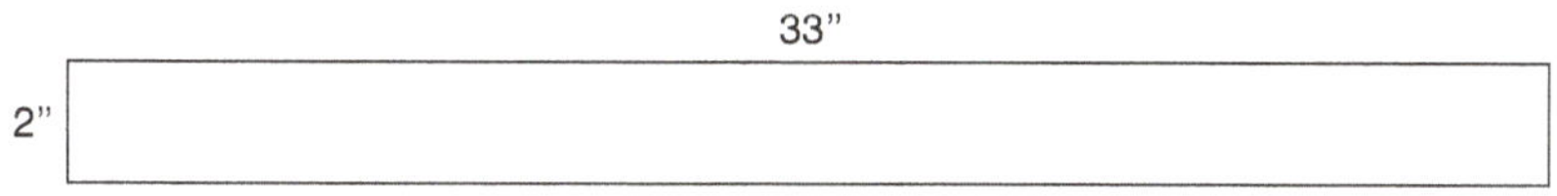

3.

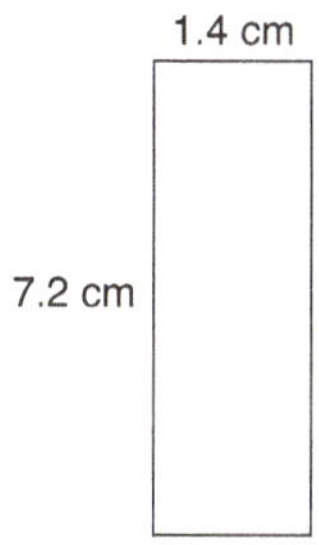

4.

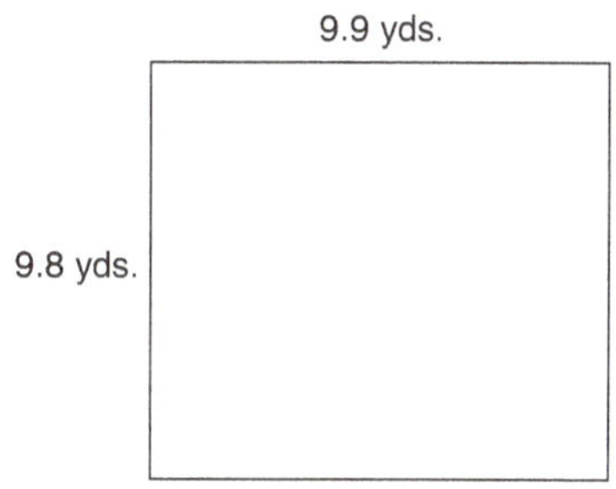

5.

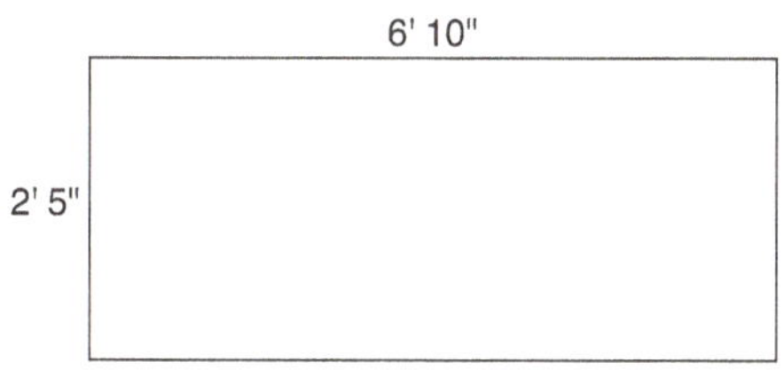

6.

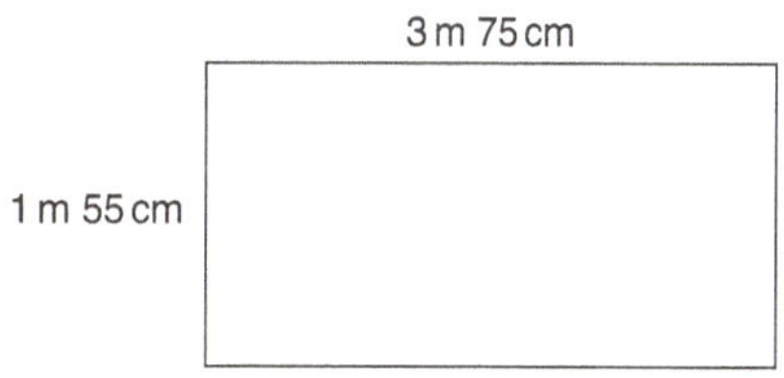

7.

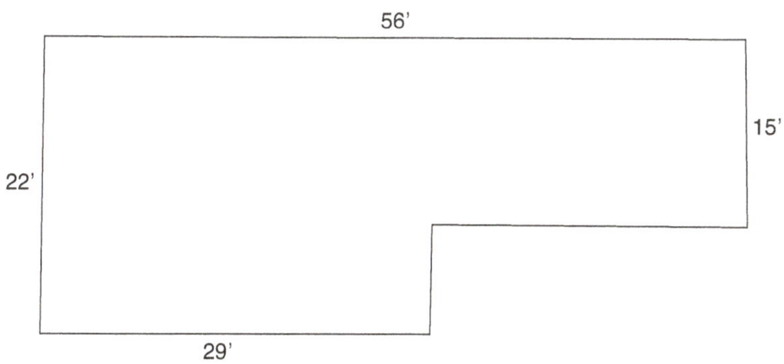

8.

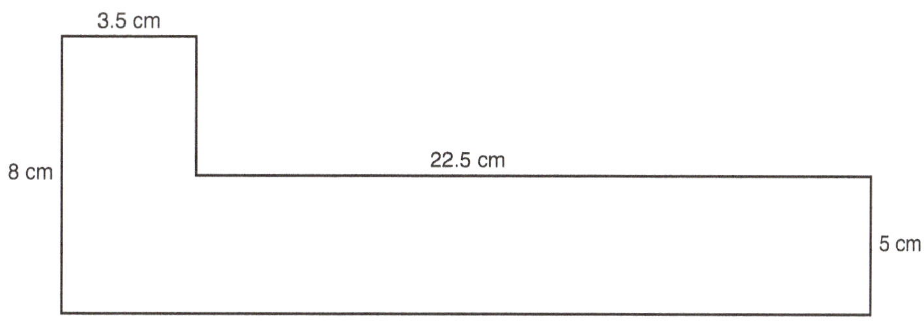

9.

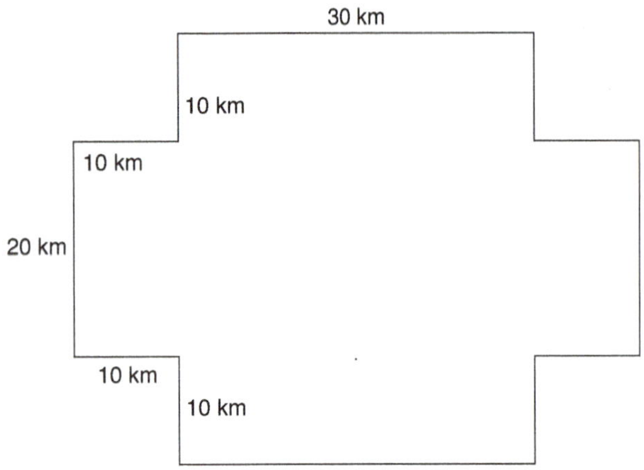

10.

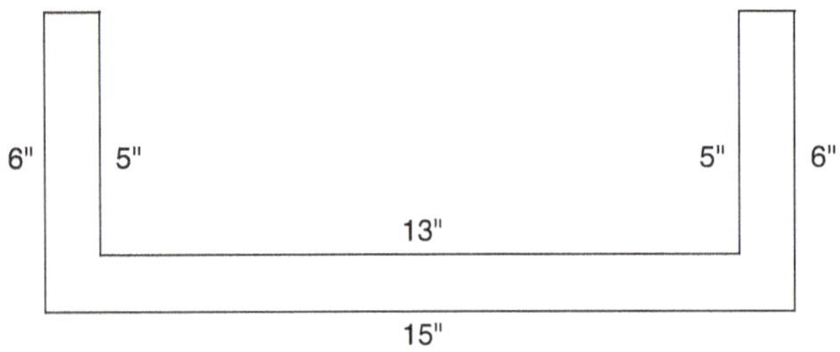

11.

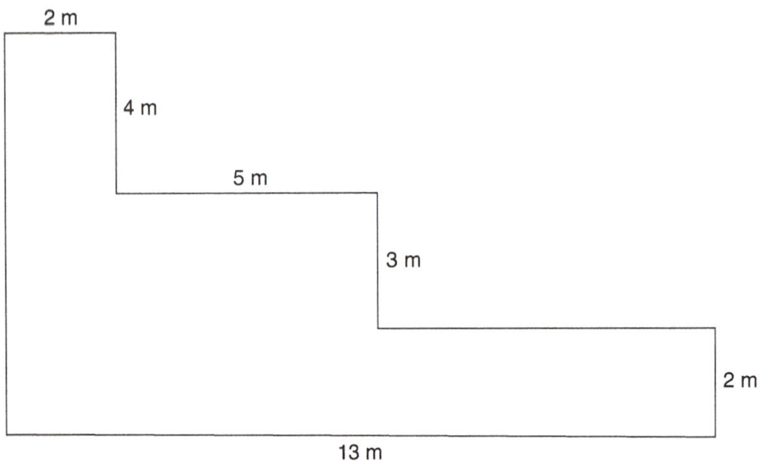

12. Each step is 9" high and 1' wide

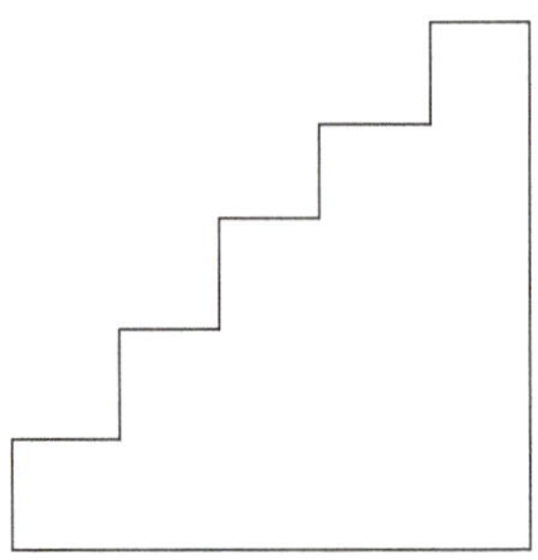

13.

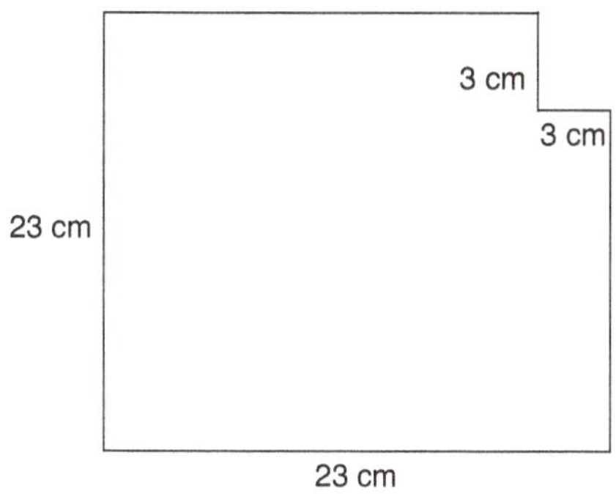

14.

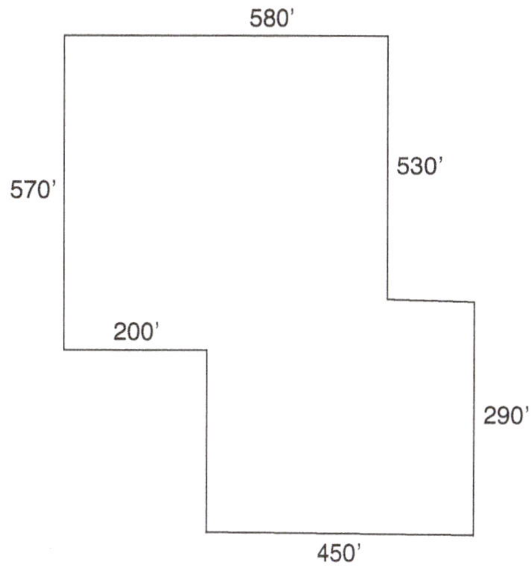

Answers

1.	**a)** 30m	**b)** 50 m^2
2.	**a)** 70″	**b)** 66 in^2
3.	**a)** 17.2 cm	**b)** 10.08 cm^2
4.	**a)** 39.4 yds	**b)** 97.02 yd^2
5.	**a)** 18′ 6″ or 222″	**b)** 16.51 ft^2 or 2378 in^2
6.	**a)** 1060 cm or 10.6 m	**b)** 58125 cm^2 or 5.8125 m^2
7.	**a)** 156′	**b)** 1043 ft^2
8.	**a)** 68 cm	**b)** 140.5 cm
9.	**a)** 180 km	**b)** 1600 km^2
10.	**a)** 52″	**b)** 25 in^2
11.	**a)** 44m	**b)** 55 m^2
12.	**a)** 17′ 6″ or 210″	**b)** 1620 in^2 or 11.25 ft^2
13.	**a)** 92 cm	**b)** 520 cm^2
14.	**a)** 2940′	**b)** 445900 ft^2

4-3 RIGHT TRIANGLE AND THE PYTHAGOREAN THEOREM

A triangle is a polygon with three sides. Triangles also have three interior angles.

A special triangle is the RIGHT TRIANGLE. A right triangle is a triangle that has one interior angle that measures 90°.

Here are some examples of right triangles.

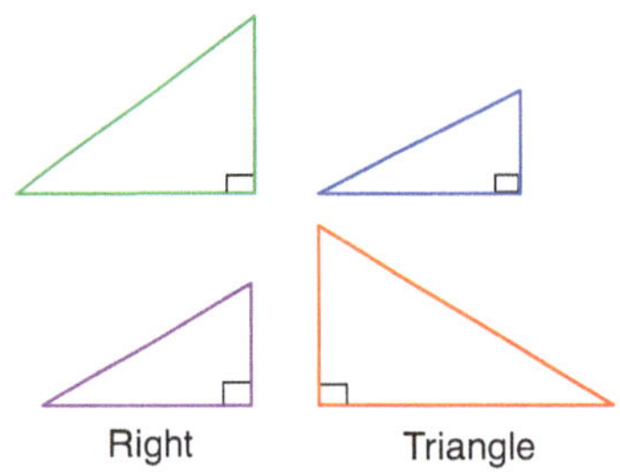

There is a relationship involving the side lengths of every right triangle.

The sides that create the 90° angles are called the LEGS of the right triangle. The third side, which is opposite the 90° angle, is called the HYPOTENUSE of the right triangle.

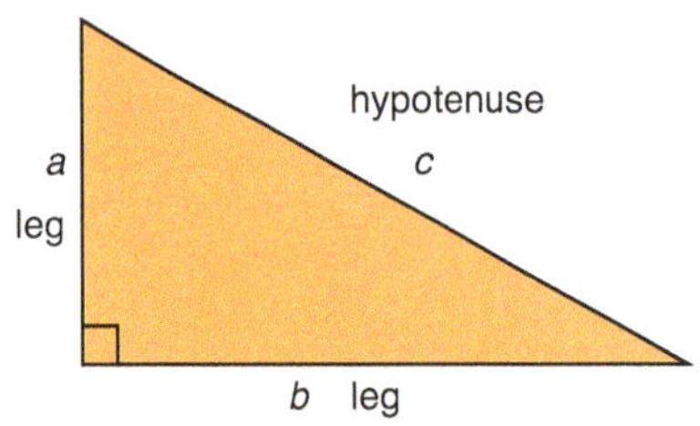

The mathematician Pythagoras, who lived around 600 BC, was given credit for discovering the following:

When the length of each leg is squared then added together, it always equals the square of the hypotenuse length. And this relationship forms what we know today as.

THE PYTHAGOREAN THEOREM

The PYTHAGOREAN THEOREM SAYS: Given leg lengths a and b and hypotenuse length c,

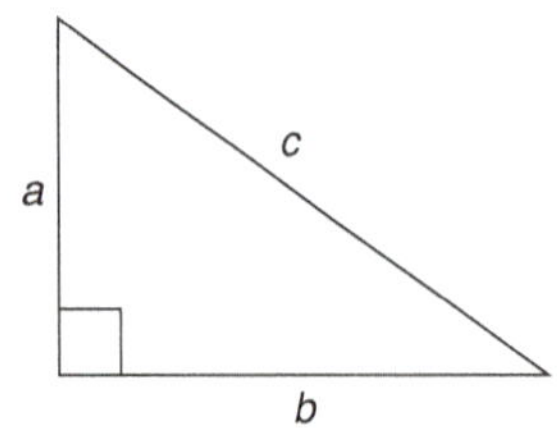

then

$$a^2 + b^2 = c^2$$

We can use the Pythagorean theorem and the principle of square roots to solve problems involving right triangles and their side lengths. The principle of square roots says we can take the square root of both sides of an equation and retain equality.

What if we are given two leg lengths of a right triangle and we want to calculate the hypotenuse length? For example: let $a = 6$ and $b = 8$, let's find c.

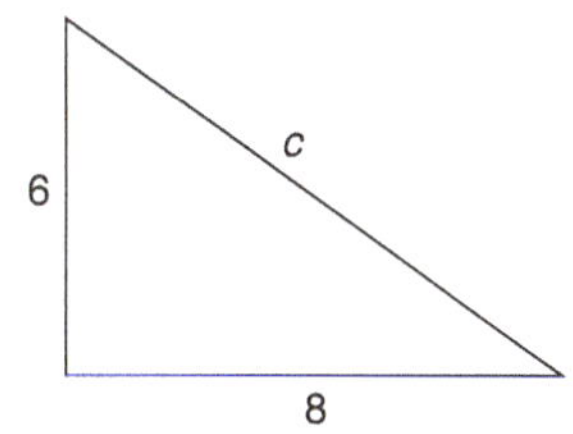

Then $a^2 + b^2 = c^2$ becomes

$$6^2 + 8^2 = c^2 \text{ or}$$

$$36 + 64 = c^2 \text{ or}$$

$$100 = c^2$$

To solve for c we use the principle of square roots, by taking the square root on both sides we get $\sqrt{100} = \sqrt{c^2}$ or $10 = \text{c}$ so the hypotenuse length is 10

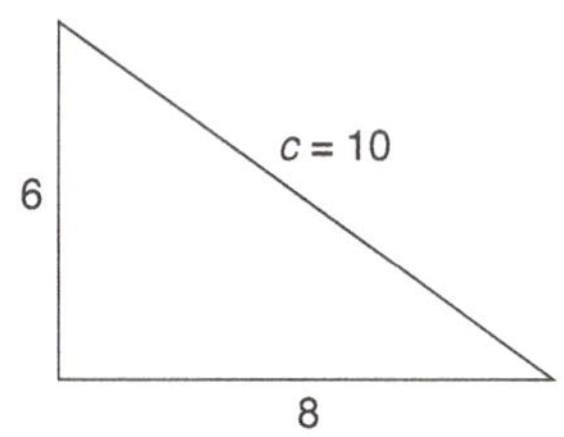

Try this one

The hypotenuse is 33 in. and one leg is 19 in. . . . what is the other leg length?

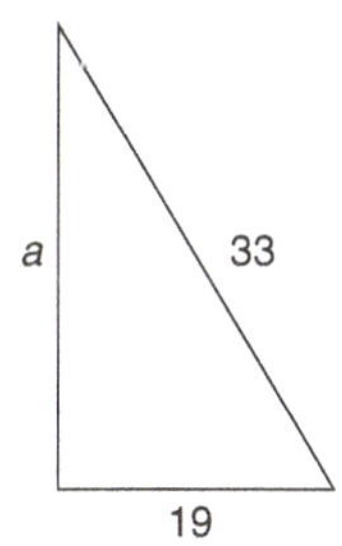

Let's find a. We need to turn in $a^2 + b^2 = c^2$ into an equation solving for a.

If $a^2 + b^2 = c^2$, then $c^2 - b^2 = a^2$. So, let $c = 33$, $b = 19$ and we get $33^2 - 19^2 = a^2$

Using principle of square roots, we take the square root of both sides

$\sqrt{33^2 - 19^2} = \sqrt{a^2}$

$\sqrt{728} = \sqrt{a^2}$

$26.98 \approx a$

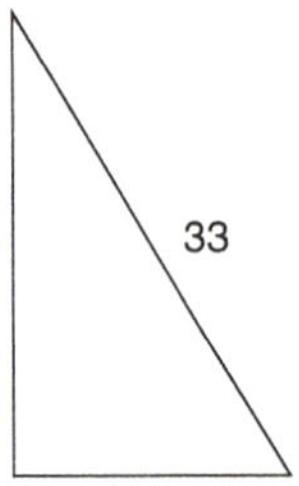

From the Pythagorean theorem $a^2 + b^2 = c^2$, we can create three formulas that you can use to find a missing side when given the other two sides of a right triangle.

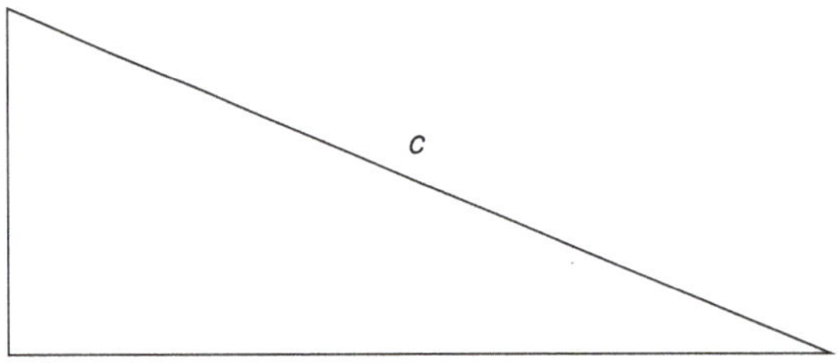

$$c = \sqrt{a^2 + b^2}$$
$$a = \sqrt{c^2 - b^2}$$
$$b = \sqrt{c^2 - a^2}$$

PERIMETER OF A RIGHT TRIANGLE

To find the perimeter of any triangle, we simply add up the three side lengths. Now, given a right triangle, we can find the perimeter if we know two of the side lengths. We use the Pythagorean theorem to find the third side then add the three side lengths to find the perimeter.

Example: Find the perimeter of this right triangle

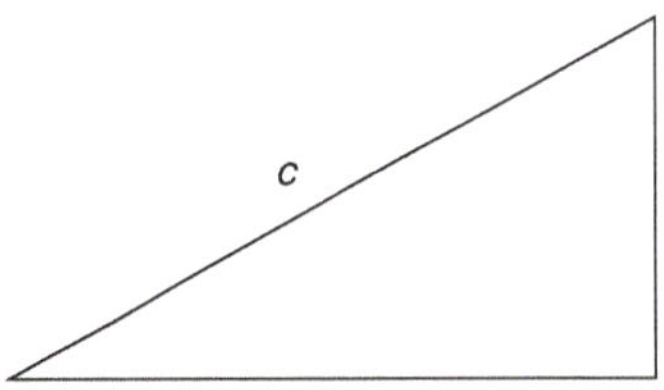

We need to find the hypotenuse length c.

$$c = \sqrt{a^2 + b^2} = \sqrt{(20\text{ cm})^2 + (50\text{ cm})^2} \approx 53.85\text{ cm}.$$

$$\text{Perimeter} = 20\text{ cm} + 50\text{ cm} + 53.85\text{ cm} = 123.85\text{ cm}$$

Practice

Find the perimeter and the area of this right triangle to the nearest hundredth of an inch.

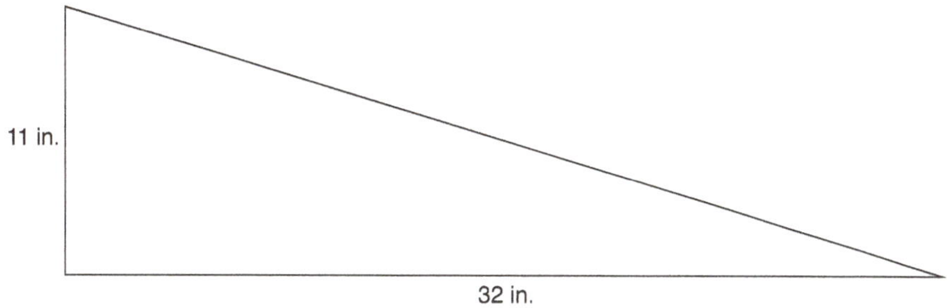

AREA OF A RIGHT TRIANGLE

Look at the rectangle:

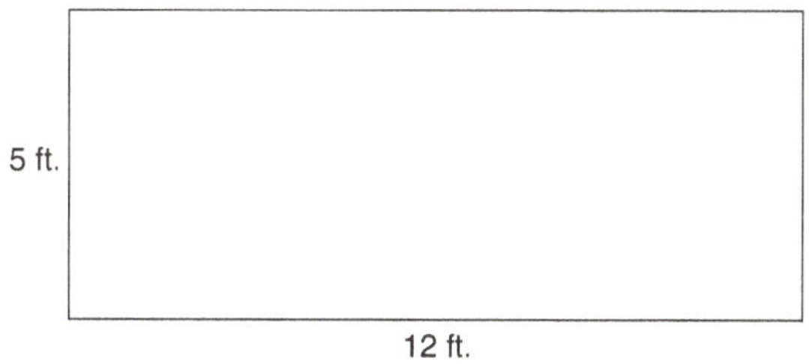

The area is 5 ft. × 12 ft. = 60 ft.^{2}

Now let's cut the triangle in half by drawing a diagonal.

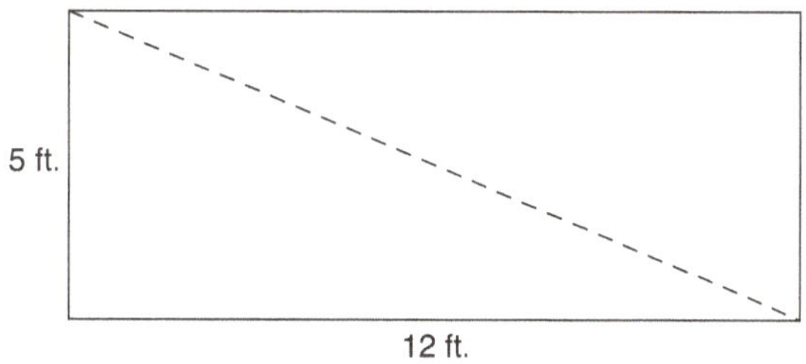

What we have now are two right triangles, each with leg lengths of 5 ft. and 12 ft.

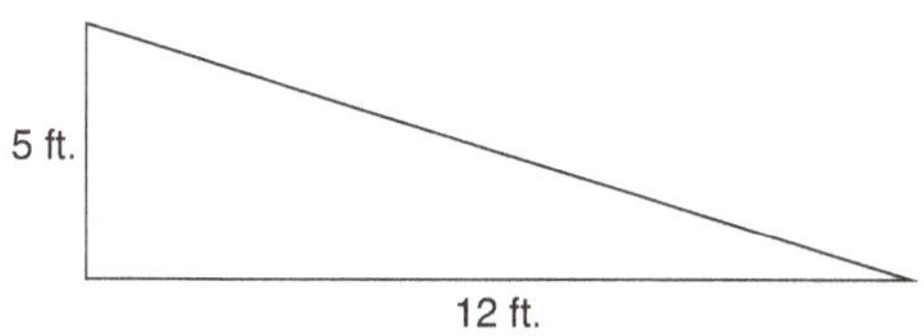

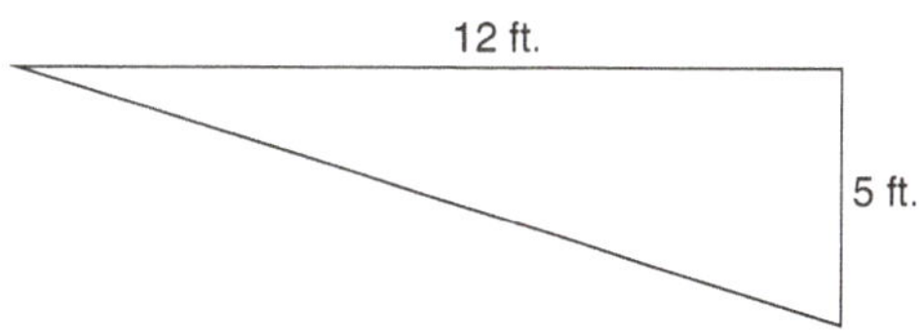

As each right triangle has the same leg lengths (as well as the same hypotenuse length), then each right triangle would have the same area. That area is one half the rectangle area. This means the area of one of the triangles is ½ × 60 ft.^{2} = 30 ft.^{2}.

We can also find this area value by taking ½ (5 ft. × 12 ft.) which is ½ times the product of the two leg lengths. So for a right triangle

given leg lengths h and b,

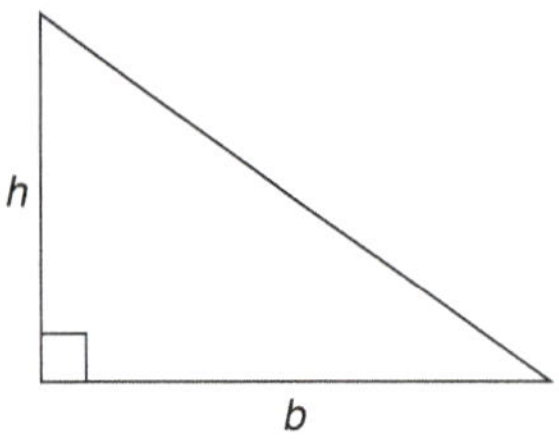

$$\text{Area} = \frac{1}{2}\, hb \text{ or Area} = .5\, hb$$

You will find often this formula using h and b. h stands for the *height* of the right triangle and b stands for the *base* of the right triangle

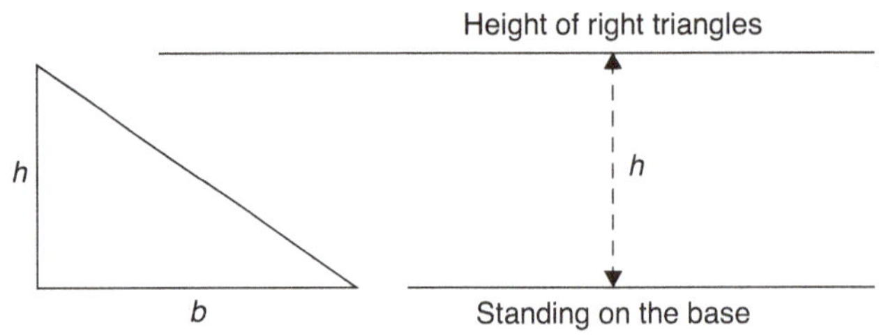

If the triangle is "standing "on the leg that is the base, then the height can be look at as how "tall" the right triangle sits. However, in the area formula, the h and b can be interchanged, that is, changing the height to the base changes the height of the triangle to this new base value.

Now we can use the Pythagorean theorem and the area formula to solve right triangle related problems.

Example: Find the area of this right triangle:

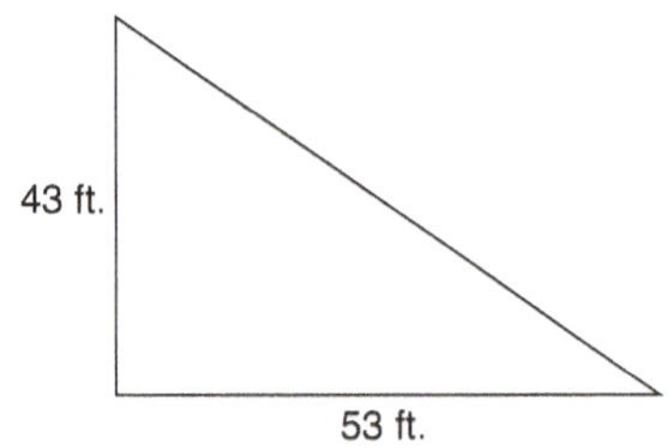

We have the two leg lengths, 43 ft. and 53 ft.

So the

$$\text{Area} = .5\ (43 \text{ ft.})(53 \text{ ft.}) = 1139.5 \text{ ft.}^2$$

Example: Find the area of this right triangle

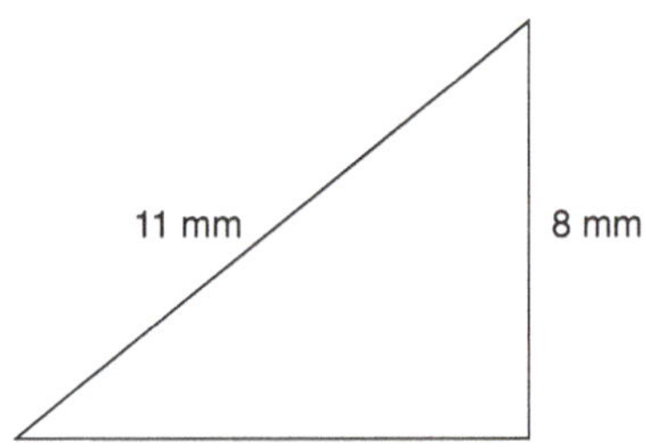

We have the hypotenuse length, 11 mm, and one leg length, 8 mm.

We need the other leg length. Using the Pythagorean theorem for missing leg length, the other leg length $= \sqrt{11^2 - 8^2} \approx 7.55$ mm and the

$$\text{Area} = .5\ (8 \text{ mm})\ (7.55 \text{ mm}) = 30.2 \text{ mm}^2$$

Real life example:

A 2 in. square head bolt is "countersunk" into a circular opening in a plank so it can be attached to a pier. It needs at least ¼ in. of clearance from the circle opening to the bolt so a wrench/pliers can fit to screw it in. What does the diagonal of the circle opening need to be for this work?

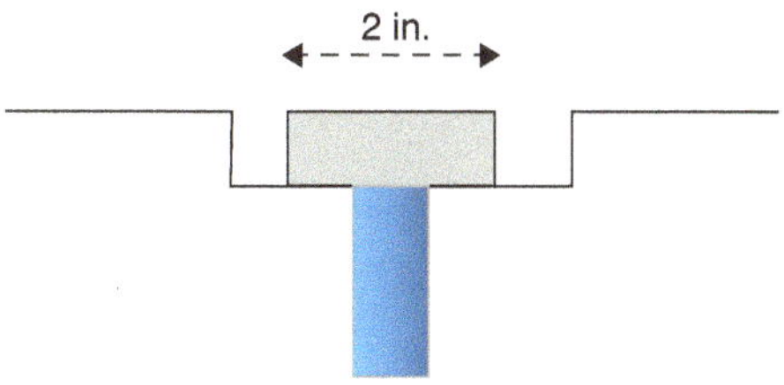

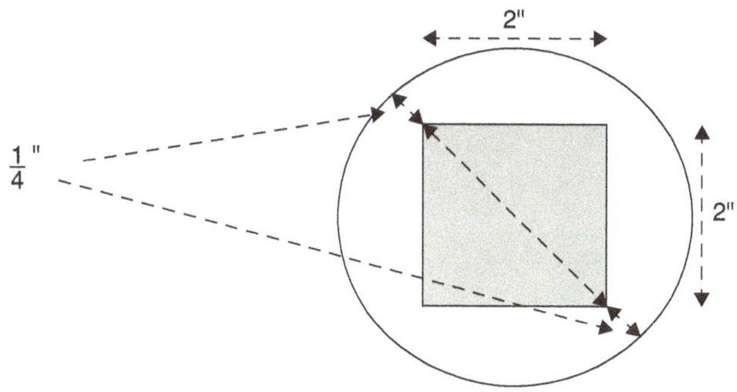

The answer is not to add ¼ in. + ¼ in. + 2 in.!!! This is because the diagonal of the square head of the bolt is longer than the 2 in. side length!

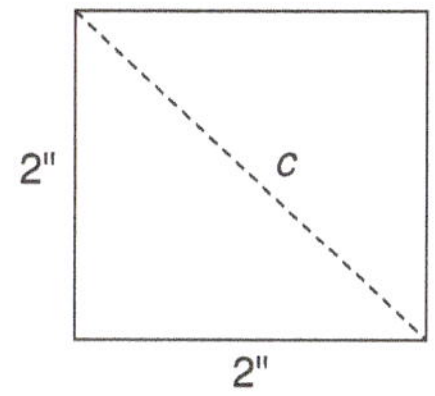

We need to find the diagonal length of the 2 in. bolt head.

The diagonal creates two right triangles, each with leg lengths of 2 in.

So we can use the Pythagorean theorem to find the hypotenuse (which is also the diagonal).

$$a = 2 \text{ and } b = 2 \text{ so the hypotenuse } c = \sqrt{a^2 + b^2} = \sqrt{2^2 + 2^2} = \sqrt{4+4} = \sqrt{8} \approx 2.83 \text{ in.}$$

So the diagonal of the screw head is 2.83 in. As we need clearance of at least ¼ in. = .25 in. on both sides of the diagonal, we get 2.83 in. + .25 in. + .25 in. = 3.33 in. This should be the diameter of the circle opening for countersinking the square head bolt.

Practice Problem:

A Concrete patio was laid out like this

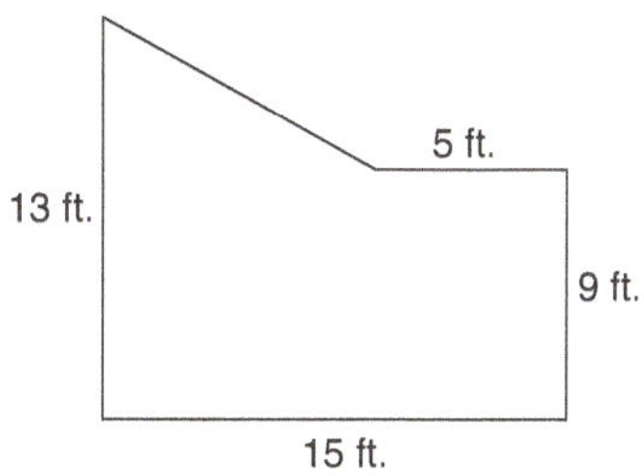

What is the perimeter and what is the area covered for the patio?

4-3 Exercises

Find a) the missing side length of these figures. And b) area of the figure

1.

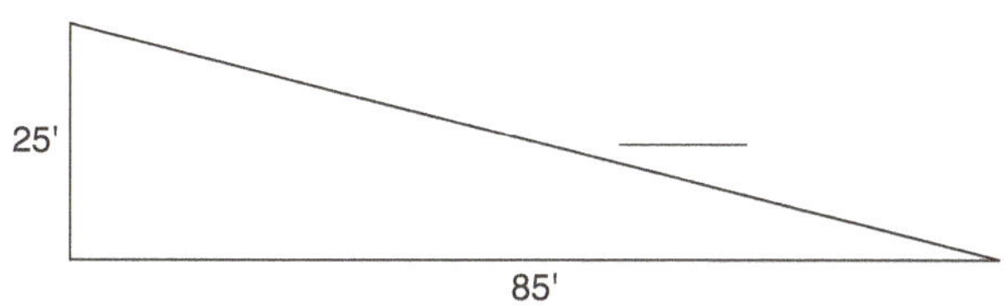

2.

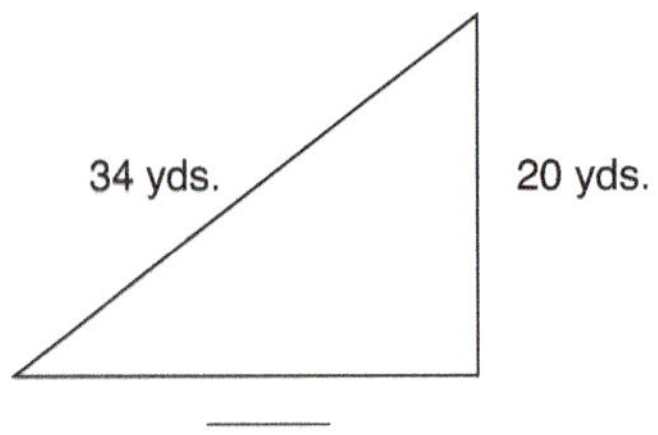

3.

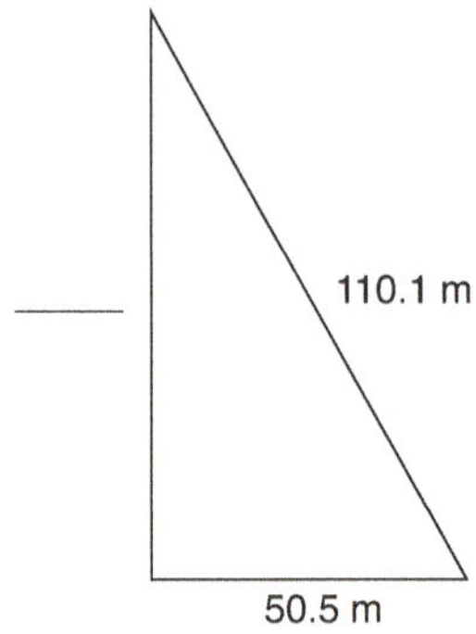

4.

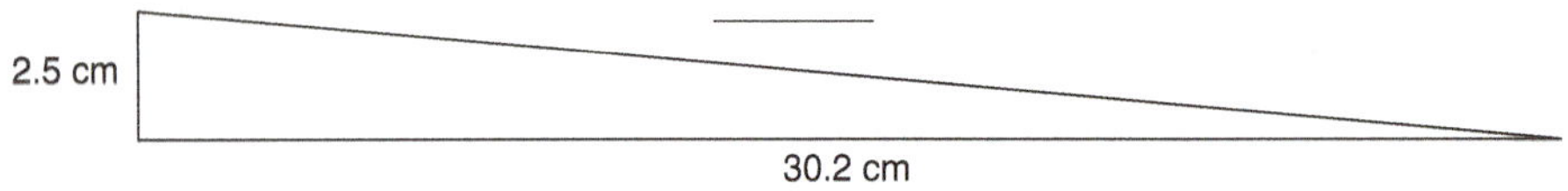

5.

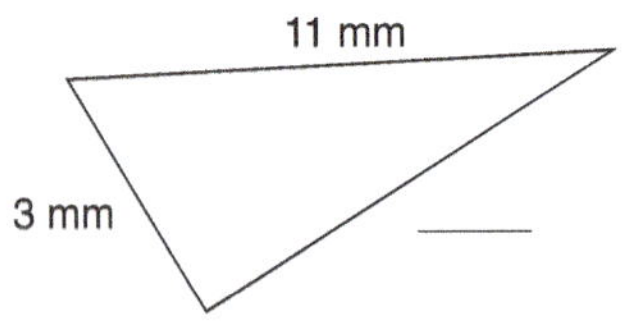

6.

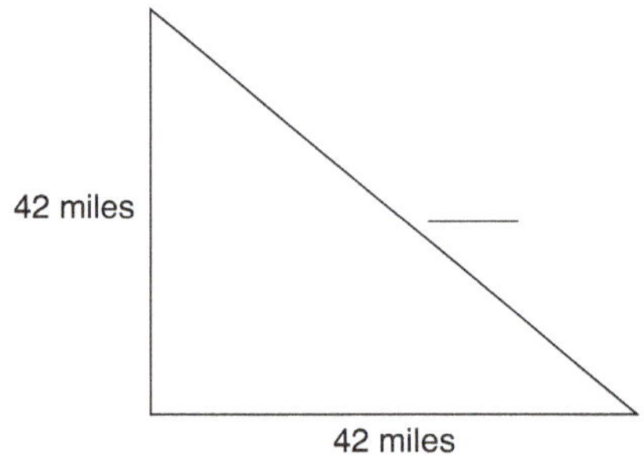

7. Find a) the perimeter b) the area.

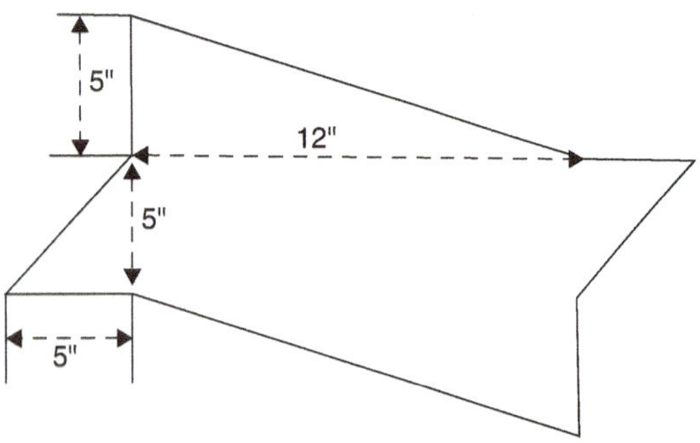

8. Find a) the perimeter b) the area.

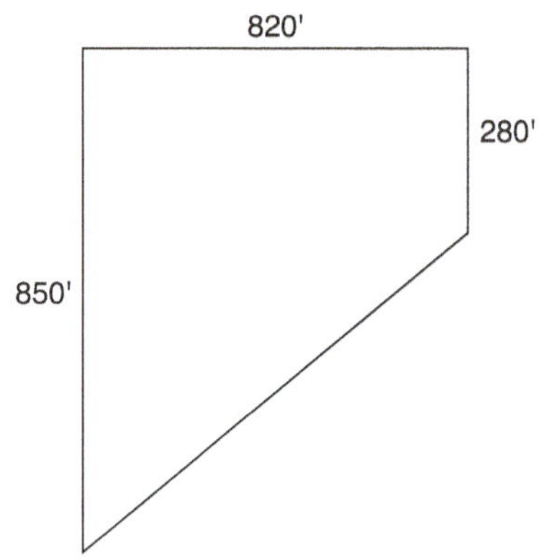

9. A rectangular garden was built in Tacoma. The dimensions of the garden is 1800 yards by 2600 yards. A pathway was built along a diagonal of the park. What is the length of the pathway?

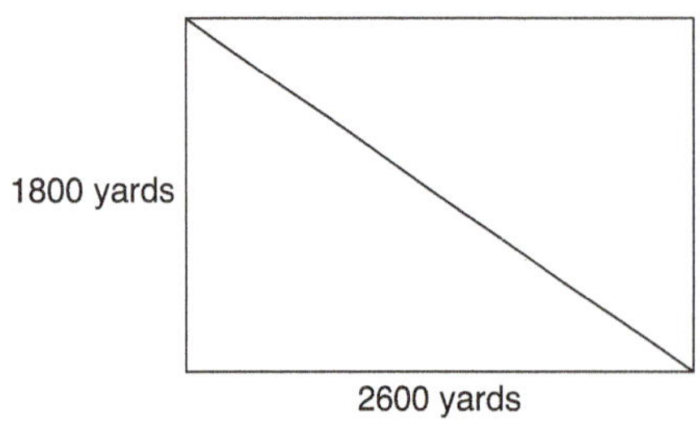

10. 

A flat screen television has a 48" diagonal and a length of 40". What is the height of the screen?

11. To get around Mayberry Lake from point A to point B, one needed to walk along a south bound pathway for 300 feet then an eastbound path for 260 feet.

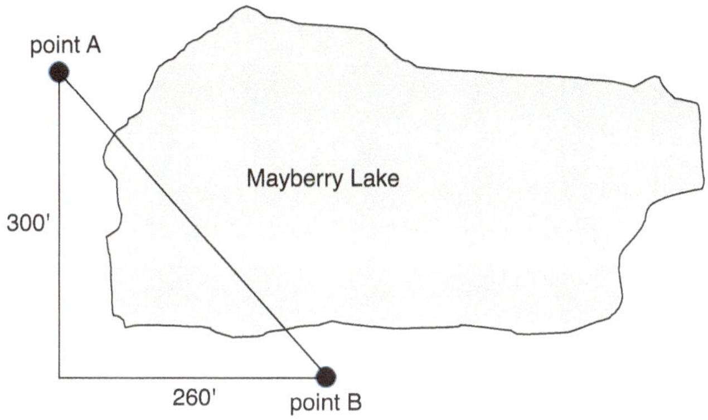

A Bridge pathway was built directly from Point A to Point B across the Lake to save walking distance. How much walking length is saved by the new bridge pathway?

12. 

A 35-foot ladder is leaning against the side of a building and is positioned such that the base of the ladder is 21 feet from the base of the building. How far above the ground is the point where the ladder touches the building?

13. 

The main mast of a fishing boat is supported by a sturdy rope that extends from the top of the mast to the deck. If the mast is 20 feet tall and the rope attached to the deck 15 feet away from the base of the mast, how long is the rope?

14. Acme Cleaners created a company logo in the shape of a right triangle that sits on a rectangular base that is 5 inches by 11 inches. The longest side of the right triangle is 14 inches. Find the perimeter and the area of the logo.

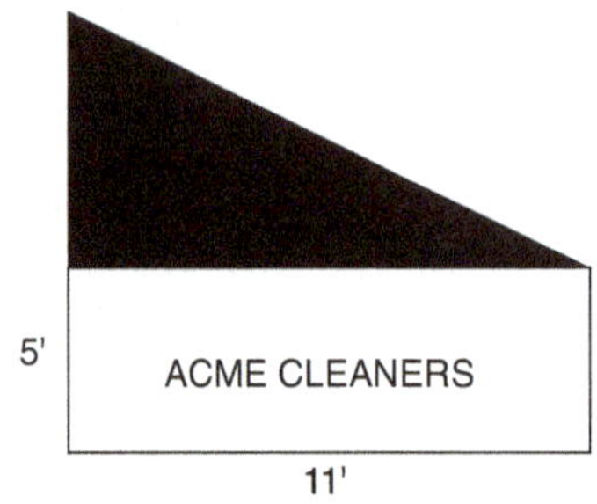

Answers

1. **a)** 88.60′ **b)** 1062.50 ft^2

2. **a)** 27.50 yds **b)** 275 yd^2

3. **a)** 97.84 m **b)** 2470.46 m^2

4. **a)** 30.3 cm **b)** 37.75 cm^2

5. **a)** 10.58 mm **b)** 15.87 mm^2

6. **a)** 59.40 miles **b)** 882 $miles^2$

7. **a)** 60.14″ **b)** 145 in^2

8. **a)** 2948.65′ **b)** 463300 ft^2

9. 3162.28 yds

10. 26.53″

11. Old pathway distance 560′.

New bridge pathway distance 396.99′

560′ – 396.99′ = 163.01′ saved.

12. 28′

13. 25′

14. **a)** 43.67″ **b)** 102.63 in^2

4-4 ISOSCELES TRIANGLES

An ISOSCELES TRIANGLE is a triangle that has two sides of equal length. We say that the two side lengths are congruent.

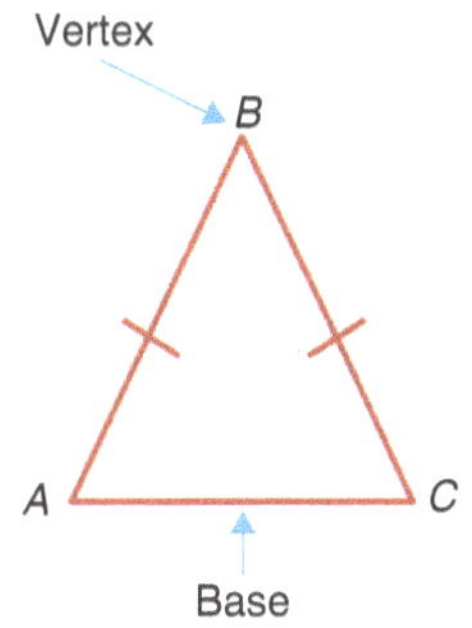

The angle made by the two equal length sides is called the VERTEX ANGLE and the point of the angle is the VERTEX of the isosceles triangle. The side opposite the vertex is called the BASE of the isosceles triangle. The two angles made with the base side are called BASE ANGLES. In an iscosceles triangle, base angles have the same degree measure, so are congruent.

When the vertex angle is cut in half with a segment that goes from the vertex to the base, the isosceles triangle is divided into two congruent right triangles (equal shape and size).

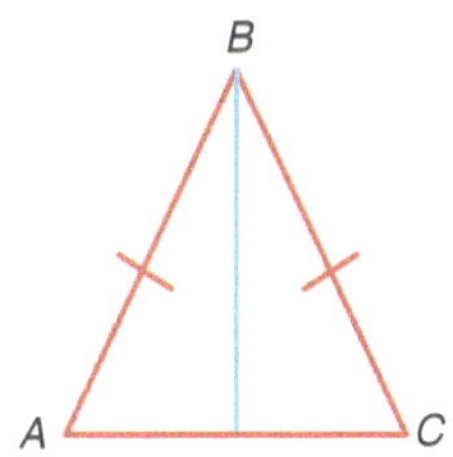

This segment touches the base segment side at its MIDPOINT, and divides the base in half.

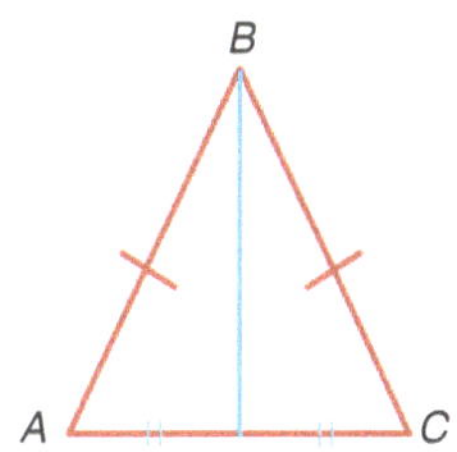

The length from *A* to the midpoint is equal to the length from the midpoint to *C*

Now we can use this to find other measures.

Here is an example: Find the perimeter and area of this triangle.

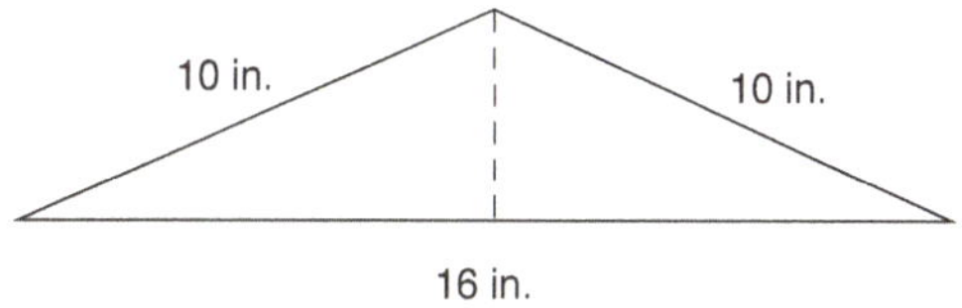

An isosceles triangle has side length of 10 in., 10 in., and 16 in.

The perimeter is easy, 10 in. + 10 in. + 16 in. = 36 in.

WHAT ABOUT AREA

If this isosceles triangle is cut in half, we have two right triangles with a leg of 8 in. and hypotenuse of 10 in.

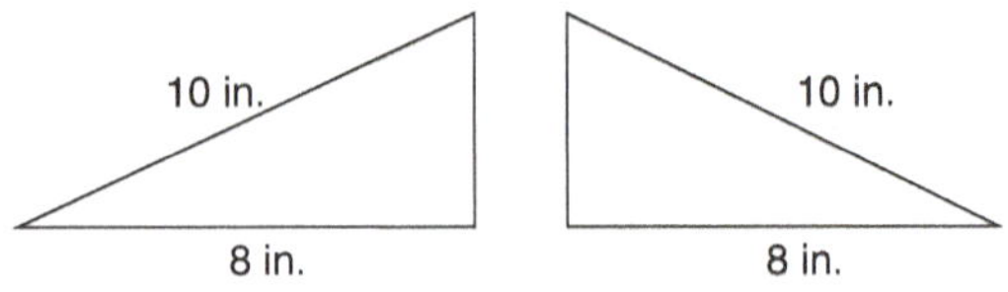

To find length of the other leg of the right triangle(s), we use the Pythagorean theorem.

$$\text{Leg length} = \sqrt{10^2 - 8^2} = \sqrt{100 - 64} = \sqrt{36} = 6 \text{ in.}$$

So we have two right triangles with leg lengths of 8 in. and 6 in. and hypotenuse of 10 in.

The area of each right triangle is $A = .5\ (B \times H) = .5\ (8 \text{ in.} \times 6 \text{ in.}) = .5\ (48) \text{ in.}^2 = 24 \text{ in.}^2$

Now to find the area of the isosceles triangle, as each right triangle has the same area, we double the area of one of the right triangles to get $2 \times 24 \text{ in.}^2 = 48 \text{ in.}^2$

Notice, the 6 in. leg is the height of the isosceles triangle given the 16 in. base!! And notice that area $= 48 \text{ in.}^2 = \frac{1}{2}(8 \times 6) + \frac{1}{2}(8 \times 6) = \frac{1}{2}[(8 \times 6) + (8 \times 6)] = \frac{1}{2}(6\,[8 + 8]\,) = \frac{1}{2}(6 \times 16)$ and that equals $=$ $\frac{1}{2}$(height of isosceles triangle) (base of isosceles triangle).

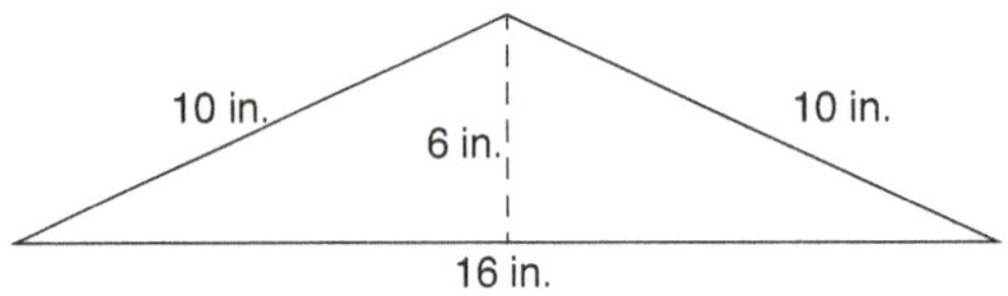

The area formula for an isosceles triangle is Area = .5 times height times base

$$\text{Or } A = .5\ hb$$

If we are given an iscosceles triangle with equal side lengths of k and base length of m, we can use the Pythagorean theorem and square root principle to create formulas for the height of the triangle and the area of the triangle.

So for the isosceles triangle,

we divide it into two right traingles

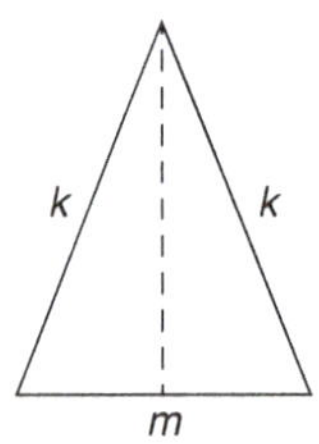

each with hypotenuse k and leg length $\frac{m}{2}$, so the other leg is the height of the isosceles traingle,

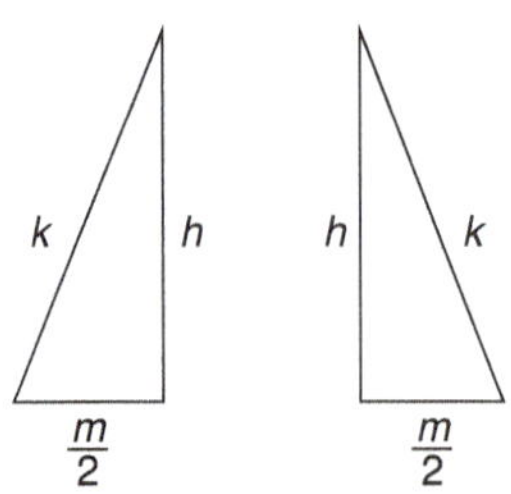

and the height of the isosceles triangle is $h = \sqrt{k^2 - (\frac{m}{2})^2}$

And area of the isosceles triangle is $A = .5\ (h)\ (m) = \ .5\ \ (\mathrm{m})\ (\sqrt{k^2 - (\frac{m}{2})^2})$

Find the height and area of this isosceles triangle

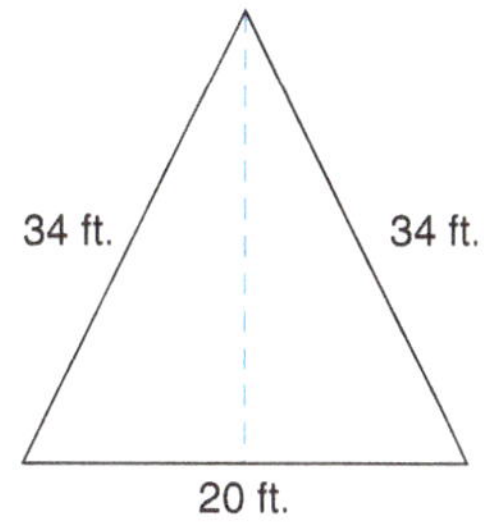

$$k = 34,\ m = 20,$$

$$\text{height} = h = \sqrt{k^2 - (\frac{m}{2})^2} = \sqrt{34^2 - (\frac{20}{2})^2} = \sqrt{1056} = 32.50 \text{ ft.}$$

$$\text{Area} = \frac{1}{2}(h)\ (m) = \frac{1}{2}(32.50)\ (20) = 325 \text{ ft.}^2$$

Practice

Find the height and area of

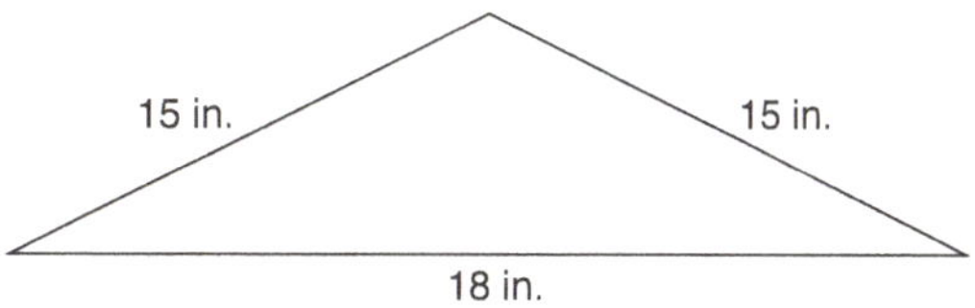

Find the height and area of

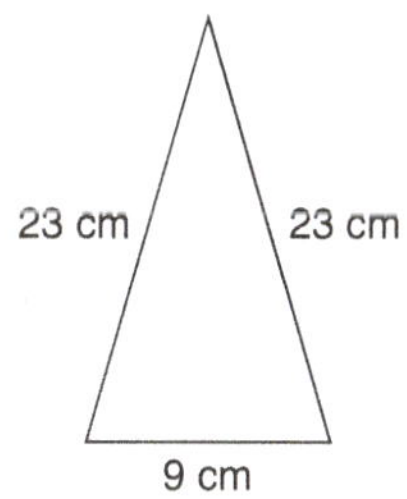

EQUILATERAL TRIANGLES

An equilateral triangle is one that has three equal length sides, so all the sides are congruent.

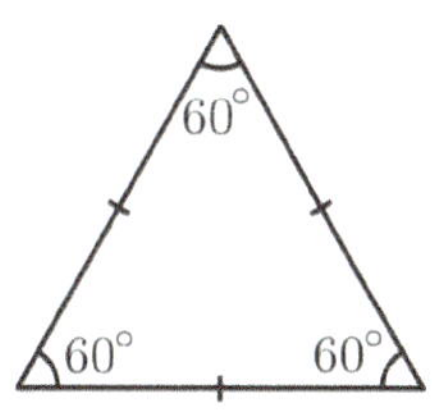

It is also true that all three of the interior angles are equal in measure (so the three angles are congruent). Each angle measure of an equilateral triangle is $60°$.

When a polygon had congruent sides and congruent interior angles, we call them REGULAR POLYGONS. Therefore, an equilateral triangle is a regular triangle.

To find the PERIMETER of an equilateral triangle, we simply multiply the side length by 3.

Given an equilateral triangle of side length x

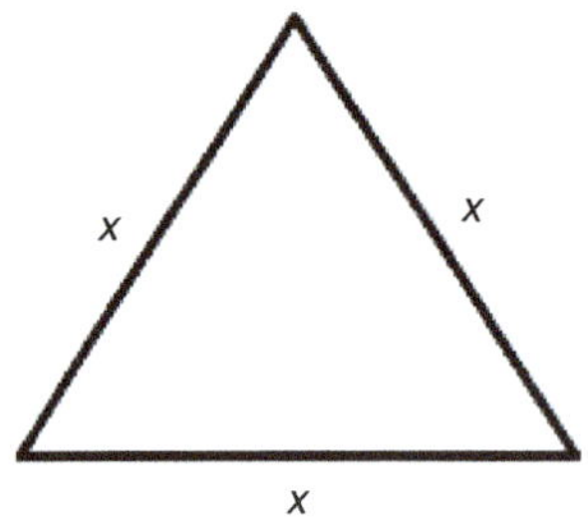

the PERIMETER $= x + x + x = 3x$

To find the AREA, we need to find the height of the equilateral triangle.

Given an equilateral triangle of side length x,

Just like the isosceles triangle, we can divide an equilateral triangle into two congruent right triangles. (Why?) The right triangles will have hypotenuse length of x and one leg length of $\frac{x}{2}$.

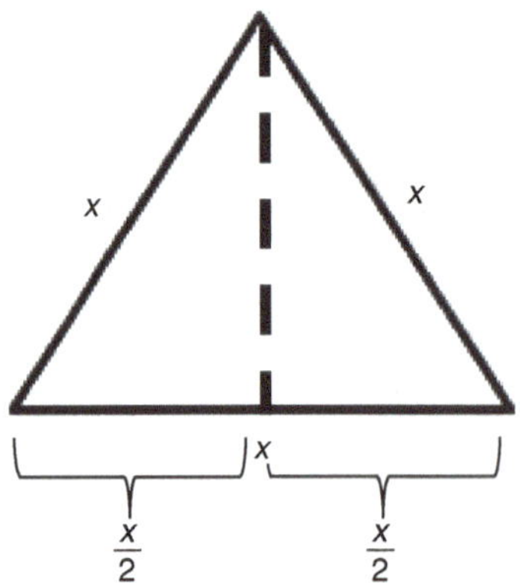

We can use Pythagorean theorem to find the other leg, which will be the height of the equilateral triangle!

So the other leg of the right triangle $= \sqrt{x^2 - (\frac{x}{2})^2} = \sqrt{x^2 - \frac{x^2}{2^2}} = \sqrt{x^2 - \frac{x^2}{4}}$

$$= \sqrt{\frac{4x^2}{4} - \frac{x^2}{4}} = \sqrt{\frac{4x^2 - x^2}{4}} = \sqrt{\frac{3x^2}{4}} = \frac{\sqrt{3}}{2}x$$

This means, given an equilateral triangle with side a, the height = $h = \frac{\sqrt{3}}{2}x$

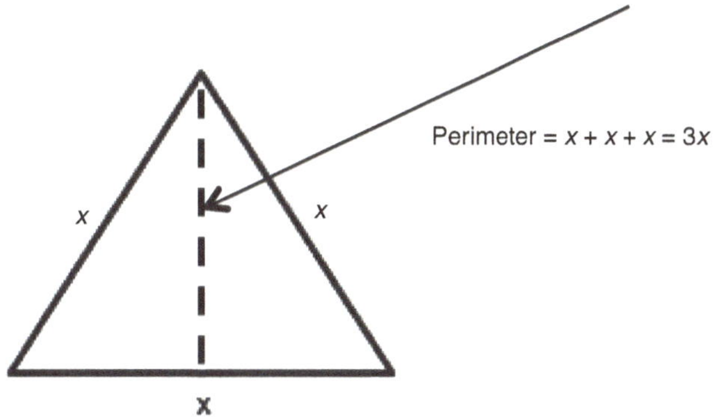

$\frac{\sqrt{3}}{2}$ can be rounded to .866, so an estimated value for the height is $h = .866x$.

Conversely, as $h = .866x$, then dividing both sides by .866, we get

$\frac{h}{.866} = \frac{.866\boldsymbol{x}}{.866}$ or $\frac{h}{.866} = x$ and since $\frac{1}{.866} = 1.155$

$\frac{h}{.866} = x$ becomes $1.155\ h = x$

And we can find the AREA of an equilateral triangle with side length x

by taking Area = .5 (×) h

Or Area = .5 (×) (.866x) = $.433x^2$

We can also find perimeter and area, given just the height, h.

As perimeter of an equilateral triangle with side length a is $3a$, and we have shown that $x = 1.155\ h$, then the perimeter, given the triangle height, h, can be figured as

Perimeter = $3x = 3\ (1.155h) = 3.465h$

And as area of an equilateral triangle with side length a is $.433\boldsymbol{x}^2$, then given height h, area can be figured as

Area = $.433\boldsymbol{x}^2 = .433\ (1.155h)^2 = .433\ (1.334h^2) = .577h^2$

So in summary, for an equilateral triangle with side length x and height h

$h = .866x$	$x = 1.155h$ and
Perimeter = $3x$	Area = $.433x^2$
Perimeter = $3.465h$	Area = $.577h^2$

Example:

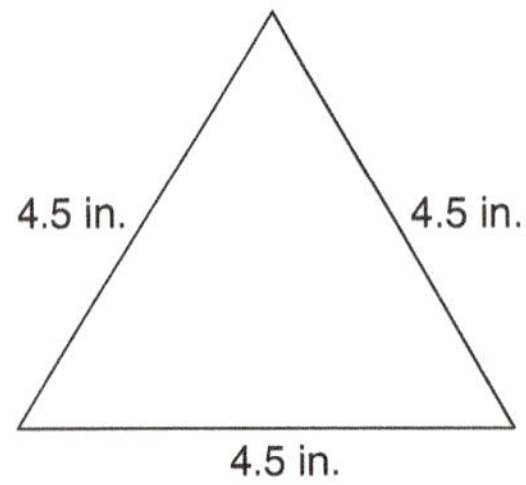

Given equilateral triangle with side length 4.5 in., find the height, perimeter, and area.

Let $x = 4.5$ in., so height $= .866x = .866\ (4.5 \text{ in.}) = 3.897$ in.

$$\text{Perimeter} = 3x = 3(4.5 \text{ in.}) = 13.5 \text{ in.}$$

$$\text{Area} = .433x^2 = .433\ (4.5 \text{ in.})^2 = 8.768 \text{ in.}^2$$

Example:

Given equilateral triangle with height of 67 cm, find side length, perimeter, and area.

Let $h = 67$ cm, so side length $= 1.155h = 1.155\ (67 \text{ cm}) = 77.385$ cm

$$\text{Perimeter} = 3.465h = 3.465\ (67 \text{ cm}) = 232.155 \text{ cm}$$

$$\text{Area} = .577h^2 = .577\ (67 \text{ cm})^2 = 2590.153 \text{ cm}^2$$

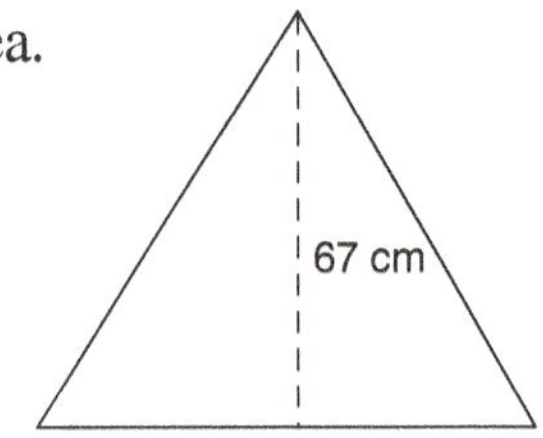

Practice

Find height, perimeter, and area of this equilateral triangle

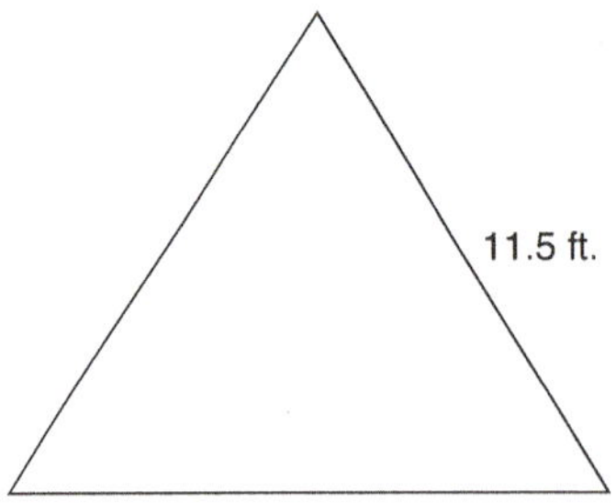

Find side length, perimeter, and area of equilateral triangle with height 20 mm

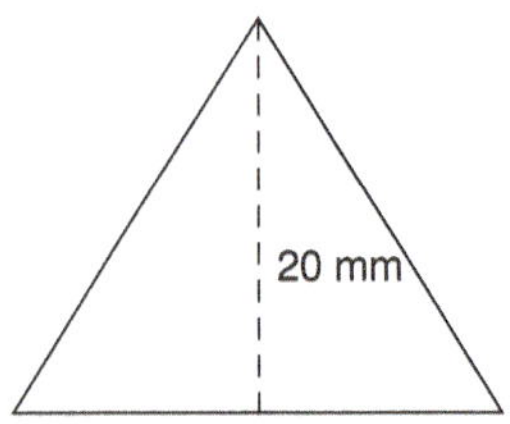

4-4 Exercises

Find the a) perimeter, b) height and c) area of these isosceles triangles, to the nearest hundreth of a unit whe appropriate.

1.

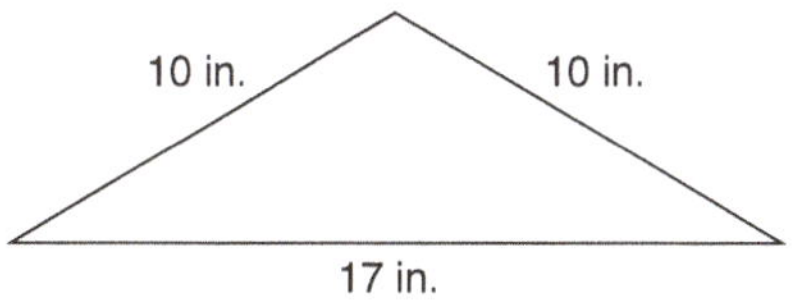

2.

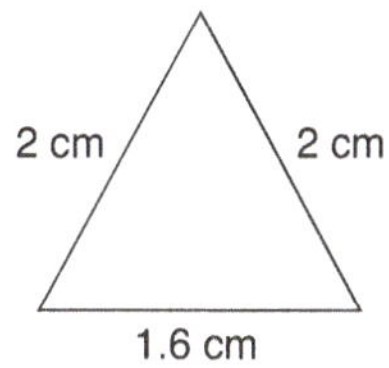

3.

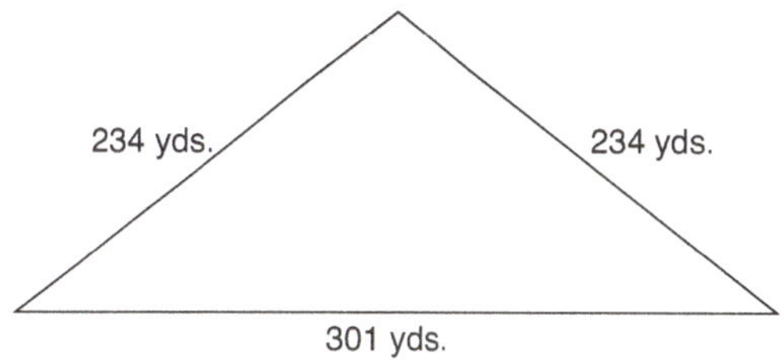

4.

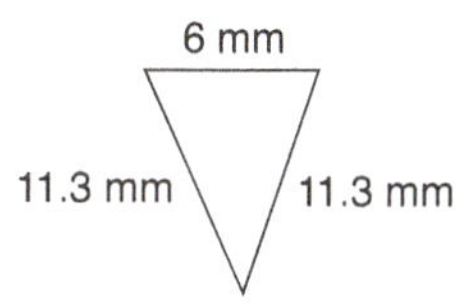

5.

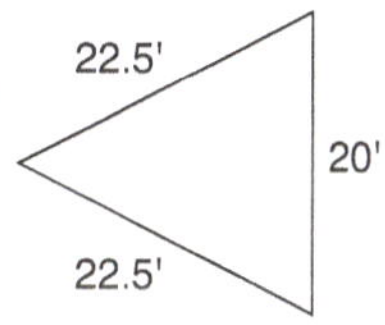

Find the a)perimeter b) height and c) area of these equilateral triangles

6.

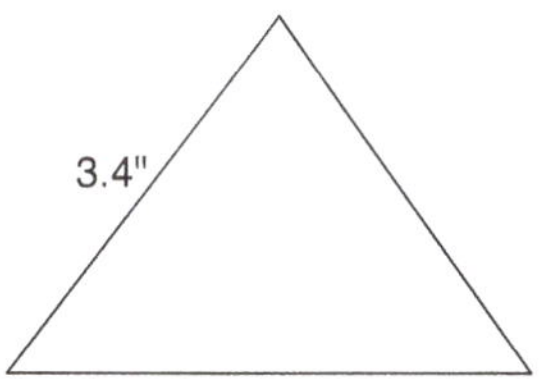

7.

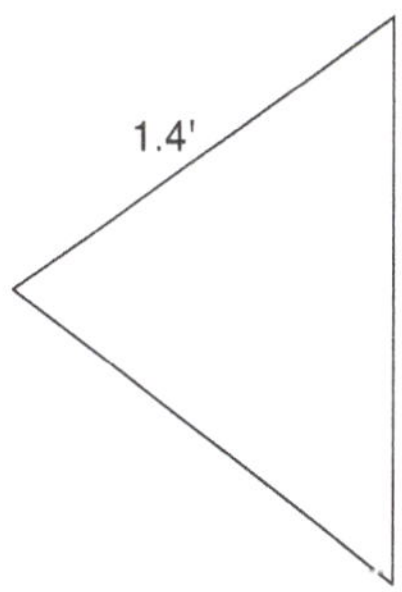

Find the a) side length b) perimeter and c) area of these equliateral triangles

8.

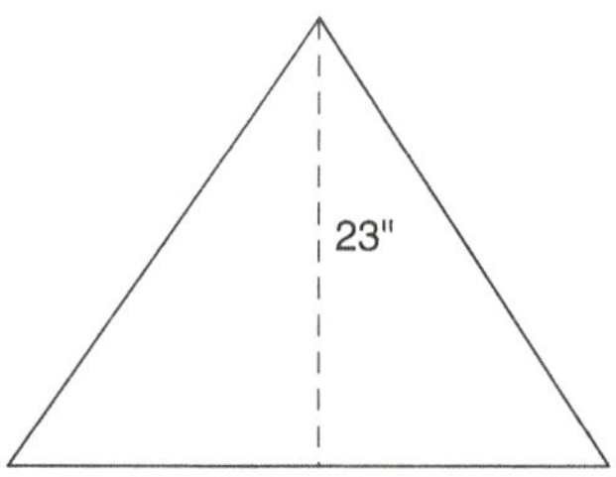

9.

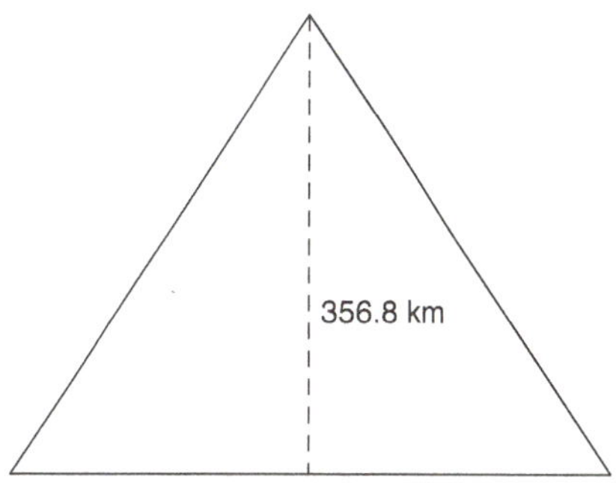

10. An equilateral triangle has an area of 73.18 m^2.

a) What is it's side length? ________

b) perimeter? ________

c) height? ________

11. An equilateral triangle has an area of 100 in^2

a) What is it's side length? ________

b) perimeter? ________

c) height? ________

Find a) perimeter b) area of these figures.

12.

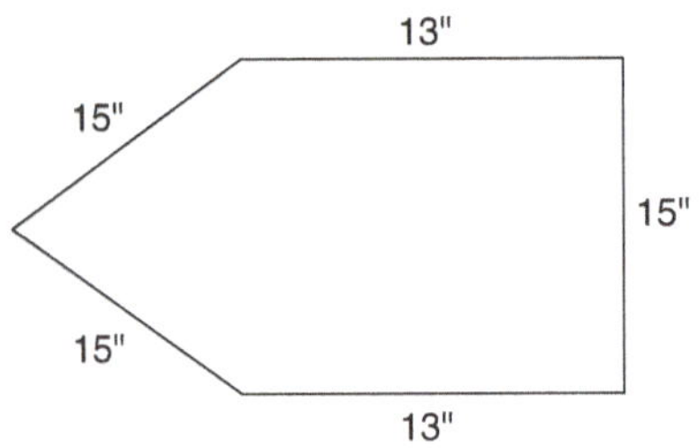

13.

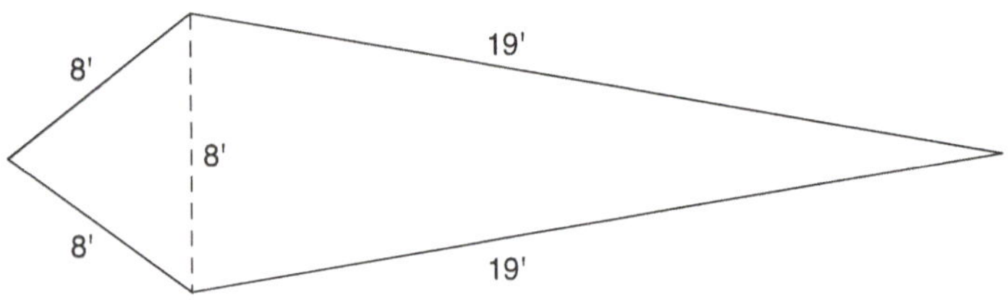

14.

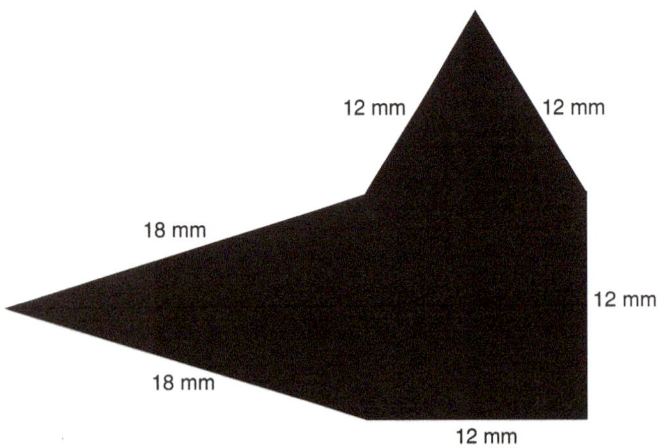

15. Find the area of this saw blade, if each saw "tooth" triangle is an equilateral triangle with side length of 3 inches

Answers

1. a) 37″ b) 5.27″ c) 44.80 in^2
2. a) 5.6 cm b) 1.83 cm c) 1.46 cm^2
3. a) 769 yds. b) 179.18 yds., c) 26966.59 yd^2
4. a) 28.6 mm b) 10.89 mm c) 32.67 mm^2
5. a) 65′, b) 20.16′, c) 201.6 ft^2
6. a) 10.2″, b) 2.94″ c) 5.01 in^2
7. a) 4.2′ b) 1.21′ c) .85 ft^2
8. a) 26.57″ b) 79.71″ c) 305.68 in^2
9. a) 412.1 km b) 1236.3 km c) 73534.84 km^2
10. a) 13 m b) 39 m c) 11.26 m^2
11. a) 15.2″ b) 45.6″ c) 13.16 in^2
12. a) 71″, b) 292.43 in^2
13. a) 54′ b) 102.01 ft^2
14. a) 84 mm b) 308.18mm^2
15. Bottom of blade = $\sqrt{33^2 - 10^2}$ = 31.45″; Area = 3(.433)$(3\text{ in})^2$ + .5 (10 in)(31.45 in) =200.12 in^2

4-5 SQUARES, PARALLELOGRAMS, AND TRAPEZOIDS

We learned in Chapter 4.4 that a REGULAR POLYGON is a polygon with all the side lengths the same length and all the interior angles are the same measure. A quadrilateral that is a regular polygon is called a SQUARE.

A SQUARE has four side lengths that are equal in length and four interior angles are equal in measure, each being 90°. And because all four angles are equal in measure, that makes a square a special kind of rectangle.

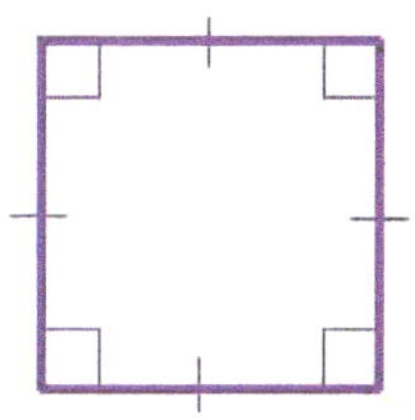

The Perimeter of a square is easy. If the side length of a square is s units, The PERIMETER, *P*, of *a* square is four times the side length or

$P = 4s$ And if we know the perimeter of the square, then the side length is $s = \frac{P}{4}$.

As we learned in Chapter 4.2 that the area of a rectangle is $A = L\,W$, and if the length and width of a square are both the same length, then the AREA, *A*, of a square with side length *s* is

$$A = s{\cdot}s = s^2$$

And if we know the area, *A*, of a square with side length s, the side length is $s = \sqrt{A}$

EXAMPLES

Find the perimeter of this square 14.3 mm  $P = 4(14.3\text{ mm}) = 57.2\text{ mm}$

Find the area $A = (14.3\text{ mm})^2 = 204.49\text{ mm}^2$

If a square has an area of 521 yd.2, find its side length: $s = \sqrt{521\text{yd.}^2} = 22.83\text{ yd.}$

Find its perimeter $P = 4(22.83\text{ yd.}) = 91.32\text{ yd.}$

If a square has a perimeter of 67 in., what is its side length? $s = \frac{67\text{ in.}}{4} = 16.75\text{ in.}$

We learned in Chapter 4.1 that a DIAGONAL of a polygon is a segment that goes from one vertex to another vertex across the polygon. There are two diagonals in a square.

For a square with side length *s*, a diagonal cuts the square into two congruent isosceles right triangle, each with legs of length *s*.

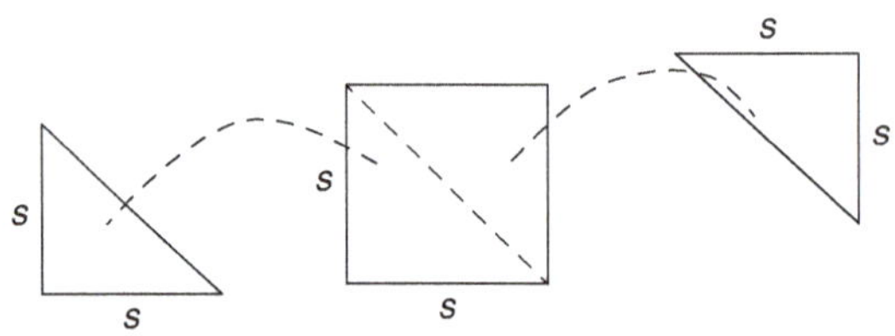

Using the Pythagorean theorem, the hypotenuse length of the right triangle is equal to $\sqrt{s^2 + s^2} = \sqrt{2s^2} = s$ times $\sqrt{2}$. As this hypotenuse length is the length of the square diagonal, then the length of the diagonal of any square of side length *s* is $s\sqrt{2}$. And since $\sqrt{2}$ is ≈ 1.414, then we can use the formula diagonal $d = 1.414s$

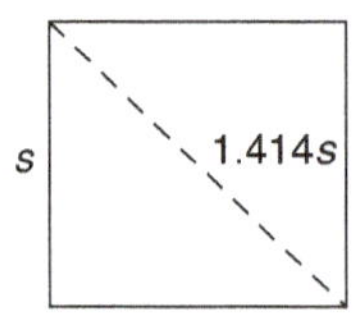

And if we are given the diagonal, *d*, of a square we can find the side length, *s*.

If $d = 1.414s$, then $s = \frac{d}{1.414}$. And since $\frac{1}{1.414} = .707$, then $s = .707d$

Find the diagonal length for this square with side length 13.5 in.

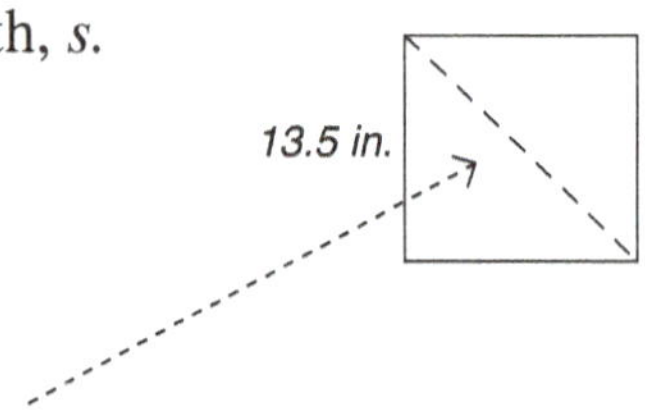

$$d = 1.414\text{s} = 1.414\ (13.5\text{ in.}) = 19.09\text{ in.}$$

Find the side length of this square with diagonal length 2.5 mm.

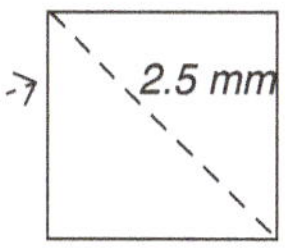

$s = .707d = .707\ (2.5\text{ mm}) = 1.77\text{ mm}$

PARALLELOGRAMS

A PARALLELOGRAM is a quadrilateral that has two pairs of parallel sides.

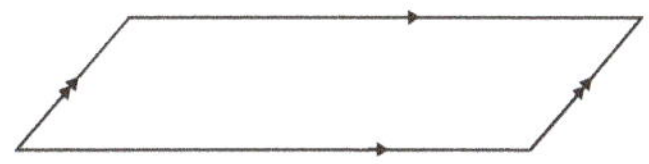

It is also true that in a parallelogram, opposite sides are equal in length (or are congruent).

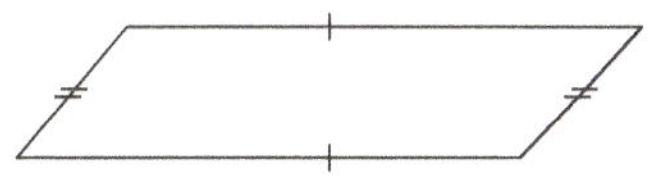

PERIMETER of a parallelogram is the sum of the four side lengths.

Example: Find the perimeter of this parallelogram.

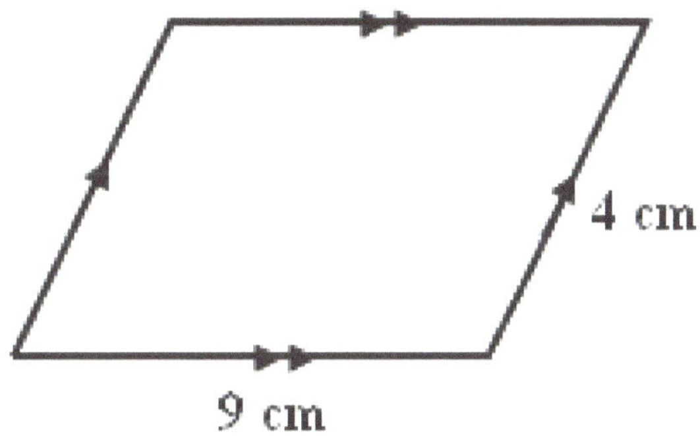

As opposite side lengths are congruent, the Perimeter = 9 cm + 4 cm + 9 cm + 4 cm = 26 cm

AREA of a parallelogram is Area = Base time Height, or $A = bh$. In a parallelogram, given one side as the base, b, there is a corresponding height, h. Give a base side, the height is the perpendicular distance between two parallel sides.

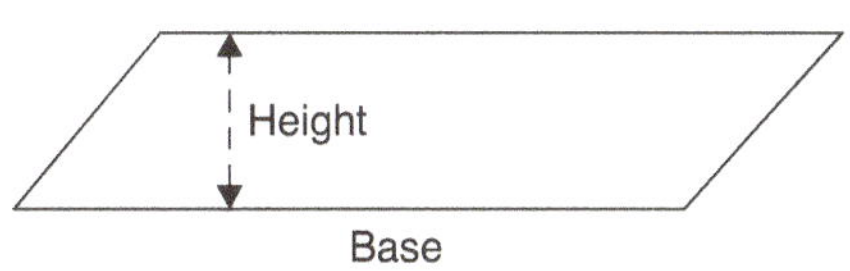

And here is why the Area = base time height. If we draw a line from a vertex perpendicular to an opposite side, we create a right triangle. (This line length is the height of the parallelogram.)

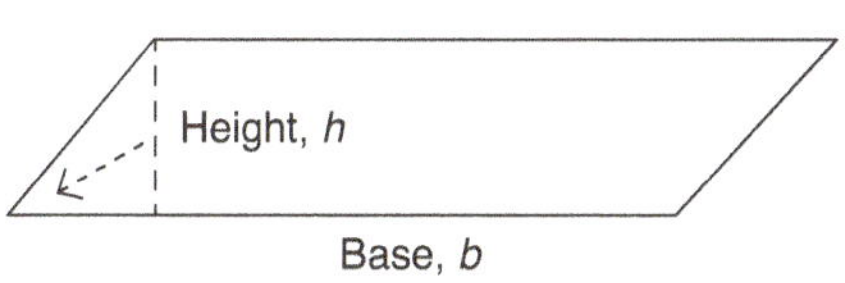

We can then "move" this right triangle to the other end of the parallelogram to fit exactly like this

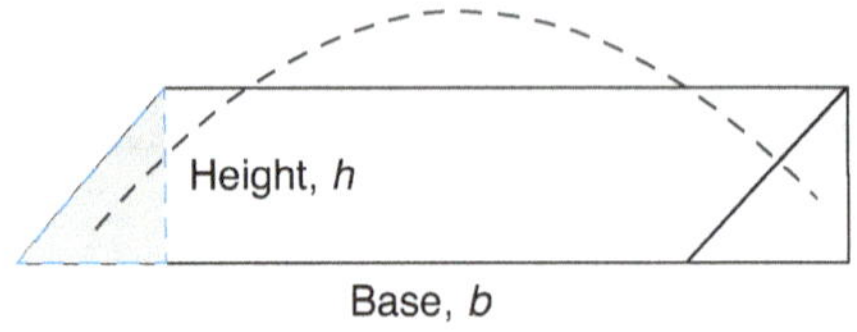

We have "changed" the parallelogram base b and height h into a rectangle with base b and height h. The area of the parallelogram equals the area of the rectangle.

As the area of the rectangle is $A = bh$, then the area of the parallelogram is also $A = bh$.

Example, Find the perimeter and the area of this parallelogram.

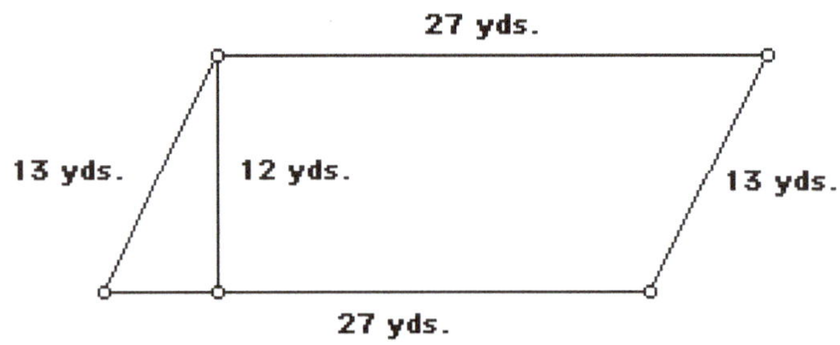

The perimeter = 27 yd. + 13 yd. + 27 yd. + 13 yd. = 80 yd.

The base is 27 yd., the height is 12 yd. $A = bh$, so $A = (27 \text{ yd.})(12 \text{ yd.}) = 324 \text{ yd.}^2$

TRAPEZOIDS

A TRAPEZOID is a special quadrilateral (four-sided polygon) where two of the sides are parallel.

Examples of trapezoids:

For a trapezoid, the two parallel sides are called the BASES of the trapezoid and are referred to as b_1 and b_2. Also the distance between the two parallel sides is the HEIGHT of the trapezoid, referred to as h. The height is the perpendicular distance between the two parallel sides.

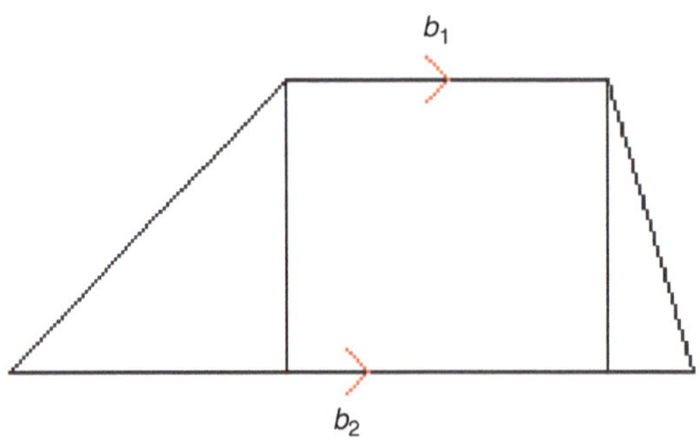

Perimeter is the sum of the four side lengths of the trapezoid.

Example: Find the perimeter of this trapezoid.

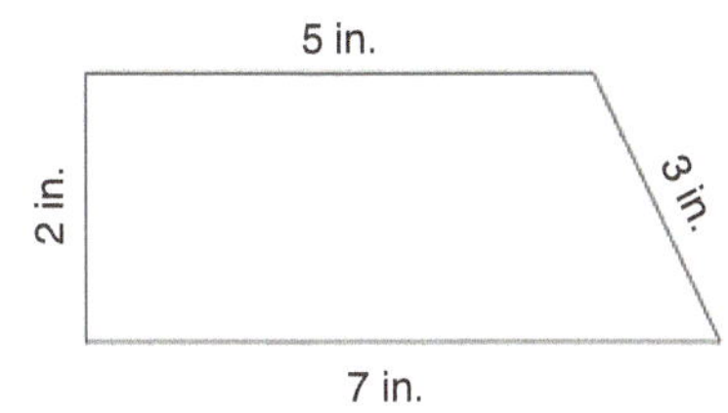

Perimeter = 7 in. + 3 in. + 5 in. + 2 in. = 17 in.

AREA

Area of a trapezoid is found by taking ½ the product of the sum of the two bases length and the height length

$$\text{Or } A = .5\,(b_1 + b_2)\,(h)$$

And here is why the area of a trapezoid is $A = .5\,(b_1 + b_2)\,(h)$ Start with a trapezoid:

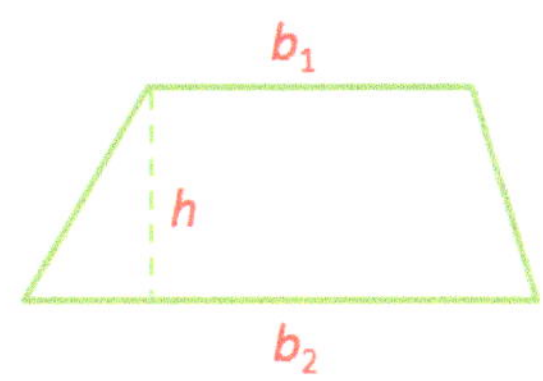

Duplicate it then rotate the copy around to this position

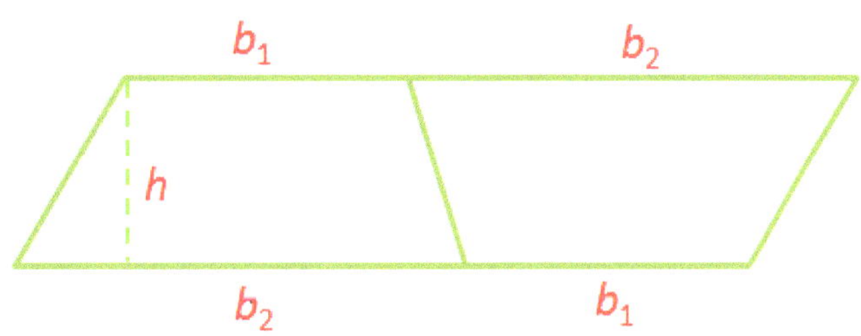

What we have now is a parallelogram with base of length $(b_1 + b_2)$ and height of h. The area of this parallelogram is $A = (b_1 + b_2)\,h$. Since the trapezoid area is one half the area of the parallelogram, then the area of the trapezoid is $A = .5\,(b_1 + b_2)\,(h)$

Example: This trapezoid has bases of 6 units and 8 units. The height of the trapezoid is 4 units. Find the perimeter and area of the trapezoid.

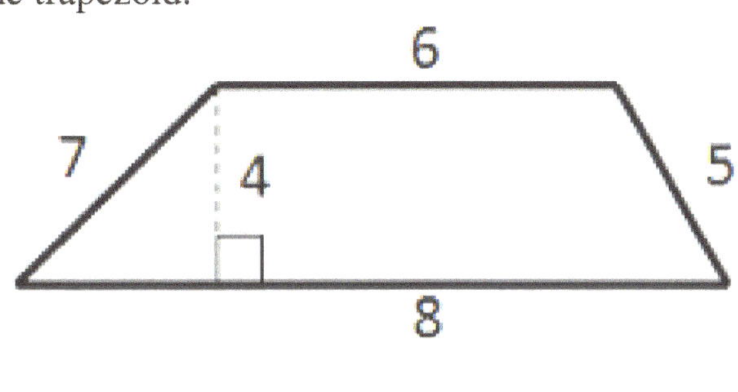

Perimeter is the sum of the side lengths. So, for example, the perimeter of this trapezoid above is 7 + 6 + 5 + 8 = 26 units.

Area is $A = .5\ (b_1 + b_2)\ (h) = .5\ (6 + 8)(4) = 28$ square units.

Find the perimeter and area of this trapezoid

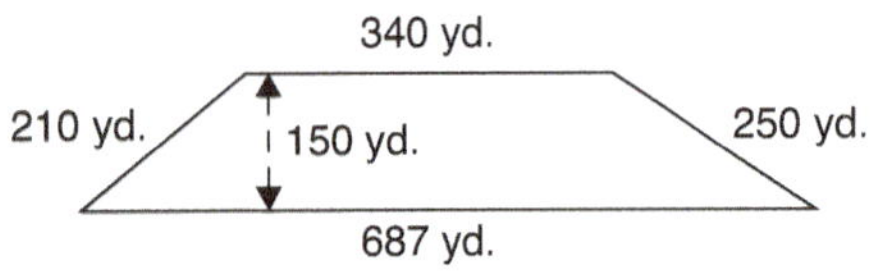

Perimeter = 210 yd. + 340 yd. + 250 yd. + 687 yd. = 1487 yd.

Area = $.5\ (b_1 + b_2)\ (h) = .5$ (340 yd. + 687 yd.) (150 yd.) = 77025 yd.^2

ISOSCELES TRAPEZOID

A special kind of trapezoid is the ISOCELES TRAPEZOID. When the two nonparallel sides of a trapezoid are equal in measure (or are congruent), then we call the trapezoid an ISOSCELES TRAPEZOID.

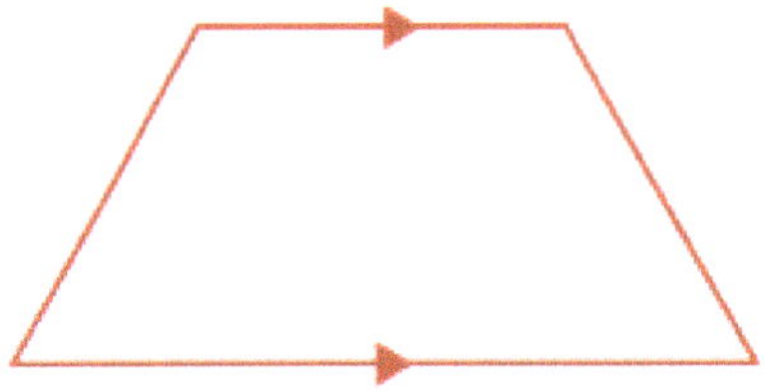

Isosceles Trapezoid

When we have an isosceles triangle where we know the base lengths and one of the nonparallel side length, let's say *k*,

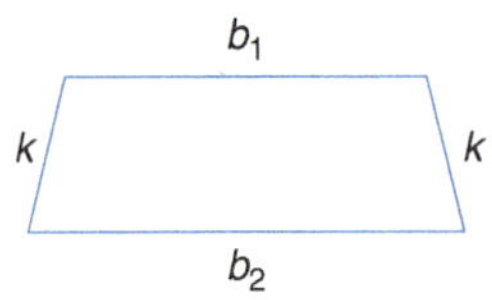

The perimeter $P = b_1 + b_2 + 2k$

The area formula $A = = .5\ (b_1 + b_2)\ (h)$.

With an isosceles trapezoid, we can find the height given the bases and the congruent nonparalle side length.

For example, GIVEN an isosceles trapezoid with base lengths of 15 units and 27 units and nonparallel side length of 10 units.

Draw the two heights such that we create a rectangle inside the trapezoid. This rectangle should have two sides that are 15 units. Also we now have two congruent right triangles with hypotenuse of 10. Let's call the "base leg" of the right triangles *x*. Now, to find *x*, we can see that twice the *x* length equals 27 − 15,

$$\text{or } 2x = 27 - 15 \text{ or } 2x = 12 \text{ or } x = 6.$$

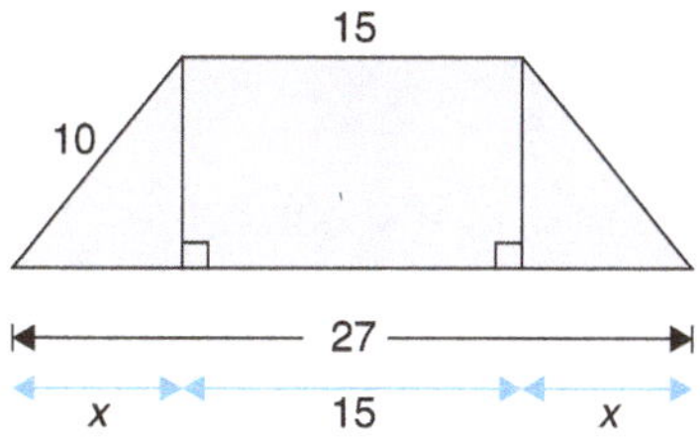

We now can change this to

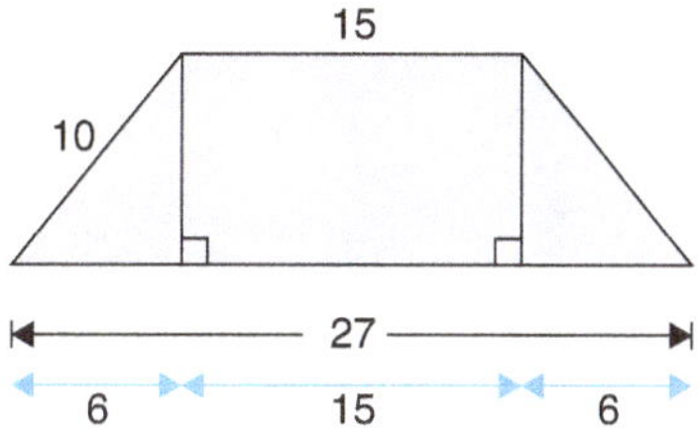

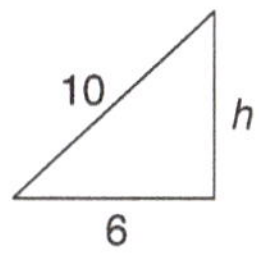

We have two congruent right triangles with leg lengths of 6 and h and a hypotenuse length 10. And the h length is the height of the trapezoid. using the Pythagorean Theorem. We can find the height, h, using the Pythagorean Theorem. Of the trapezoid by finding the other leg of the right triangle the height = $\sqrt{10^2 - 6^2} = \sqrt{64} = 8$

The height of the isosceles trapezoid is 8 units.

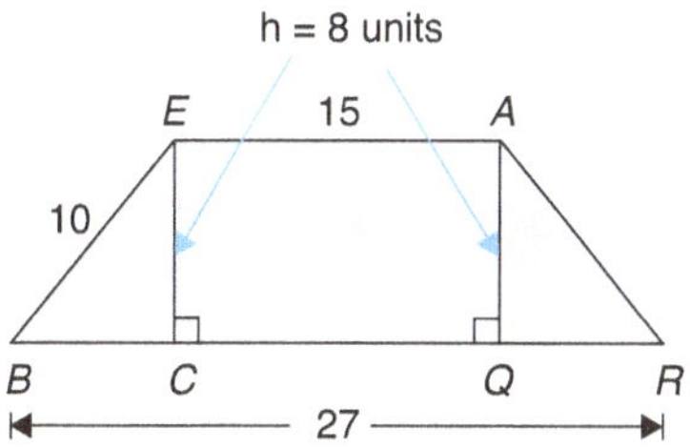

$$\text{Perimeter is } P = b_1 + b_2 + 2k = 15 + 27 + 2(10) = 62 \text{ units}$$

$$\text{Area is } A = .5\ (b_1 + b_2)\ (h) = A = .5\ (15 + 27)\ (8) = 168 \text{ units}^2$$

Practice: Find the height, perimeter, and area of this iscosceles trapezoid.

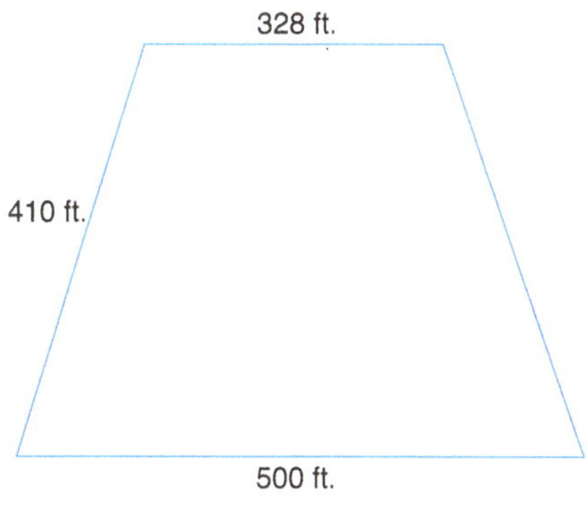

4-5 Exercises

Find answers to the nearest hundredth of a unit when appropriate.

1. A square has side length of 45 cm, find
   a) the perimeter ____________
   b) the diagonal length ________
   c) the area __________

2. A square has a perimeter of 76 inches , find
   a) the side length ________
   b) the diagonal length ________
   c) the area ______

3. A square has a diagonal length of 6 mm, find
   a) the side length _________
   b) the Perimeter _________
   c) the are _______

4. A square has an area of 196 square feet, find
   a) the side length _________
   b) The perimeter _______
   c) The diagonal length _________

5. For this parallelogram

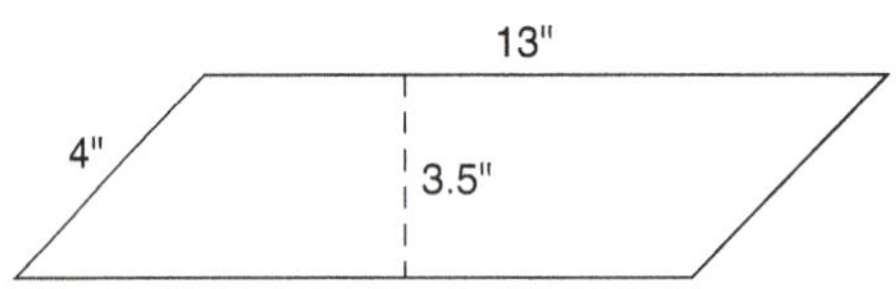

Find a) the perimeter _________ b) the area __________

6. For the parallelogram

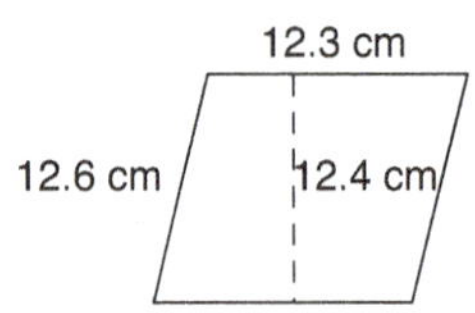

Find a) the perimeter _________ b) the area __________

7. For this trapezoid,

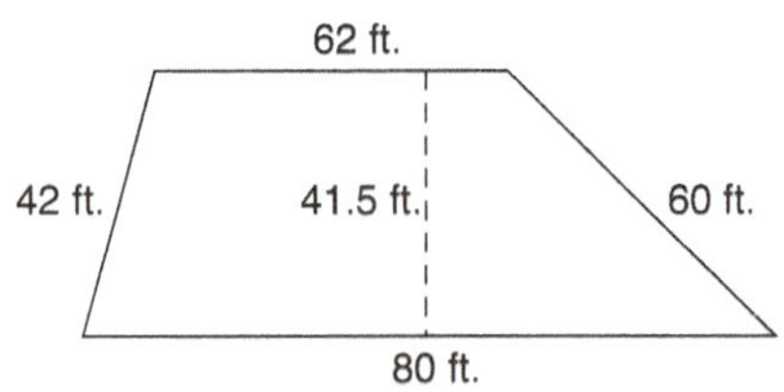

Find a) the perimeter _________ b) the area __________

8. For this trapezoid,

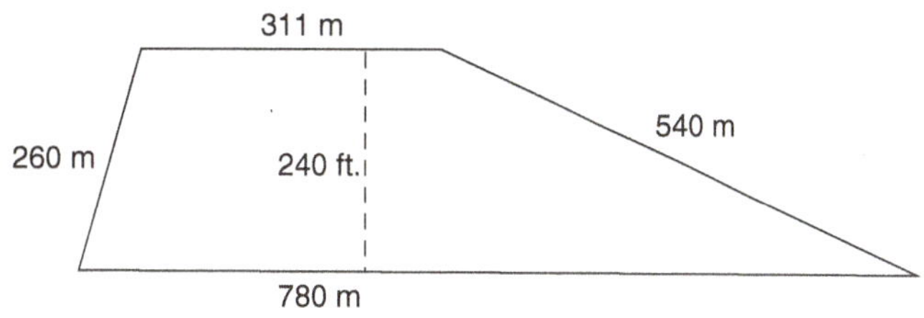

Find a) the perimeter __________ b) the area __________

9. For this isosceles trapezoid,

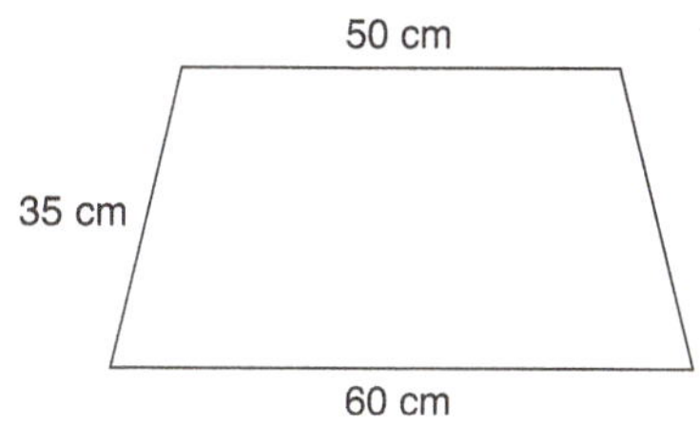

Find a) the perimeter __________ b) The height ______________ c) the area ____________

10. For this isosceles trapezoid,

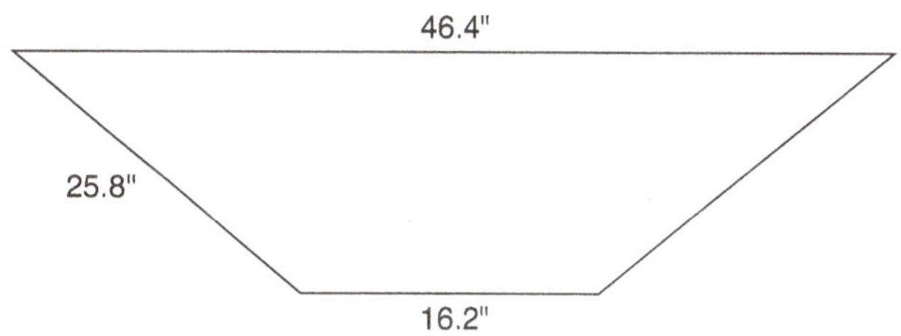

Find a) the perimeter _________ b) The height ______________ c) the area ____________

Answers

1. a) 180 cm **b)** 63.63 cm **c)** 2025 cm^2

2. a) 19″ **b)** 26.87″ **c)** 361 in^2

3. a) 4.24 mm **b)** 16.97 mm **c)** 17.98 mm^2

4. a) 14′ **b)** 56′ **c)** 19.80 ft^2

5. a) 34″ **b)** 45.5 in^2

6. a) 49.8 cm **b)** 152.52 cm^2

7. a) 244′ **b)** 2946.5 ft^2

8. a) 1891 m **b)** 130920 m^2

9. a) 180 cm **b)** $\sqrt{35^2 - 5^2} = 34.64$ cm **c)** 1905.20 cm^2

10. a) 114.20″ **b)** $\sqrt{25.8^2 - 15.1^2} = 20.92''$ **c)** 654.80 in^2

4-6 REGULAR HEXAGON

A HEXAGON is a six-sided polygon. When a hexagon has all six sides equal in measure and all six interior angles equal in measure, it is called a REGULAR HEXAGON. Each angle measure of a regular hexagon is 120°.

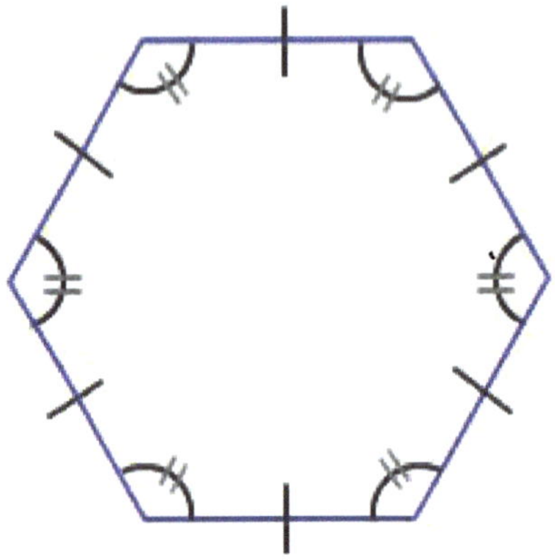

PERIMETER of a regular hexagon with side length s is six times the side length or $P = 6s$. And given the perimeter of a regular hexagon, a side length is $s = \frac{P}{6}$

What is the perimeter of this regular hexagon? 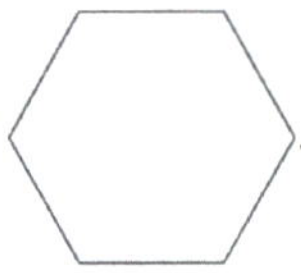 . $P = 6(5 \text{ in.}) = 30 \text{ in.}$

What is the side length of a regular hexagon with perimeter of 600 mi.?, $s = \frac{600 \text{ mi.}}{6} = 100 \text{ mi.}$

AREA

As for area, we need to look at some of the characteristics of the regular hexagon to figure this.

Diagonals: IN A REGULAR HEXAGON, there are nine diagonals

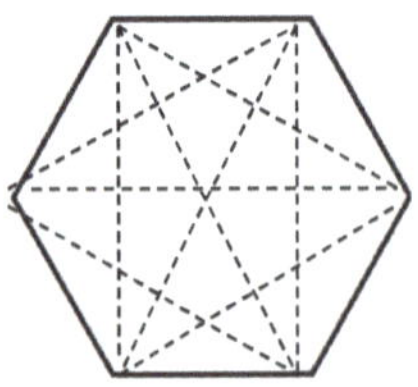

Let's focus on the diagonals that are the longest, and crosses the polygon through the true center of the regular polygon. There are three diagonals that fit this description.

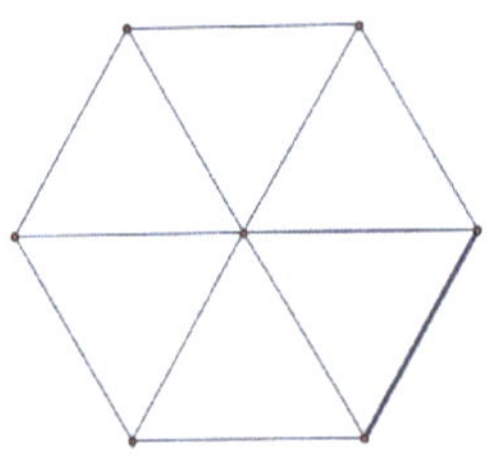

Given side length s, let's develop a formula for this diagonal length using equilateral triangles.

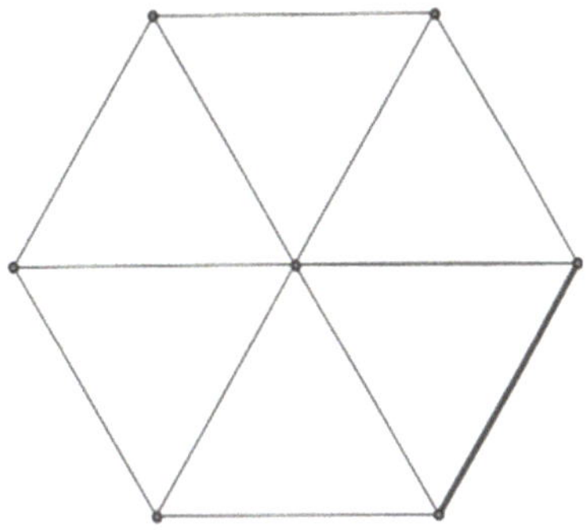

Drawing these three diagonals divides the regular hexagon into six congruent triangles. Better yet, each triangle is an equilateral triangle with side length of s. So, if each side of all the equilateral triangles is the same, then this diagonal length, called d, is simply twice s or $d = 2s$

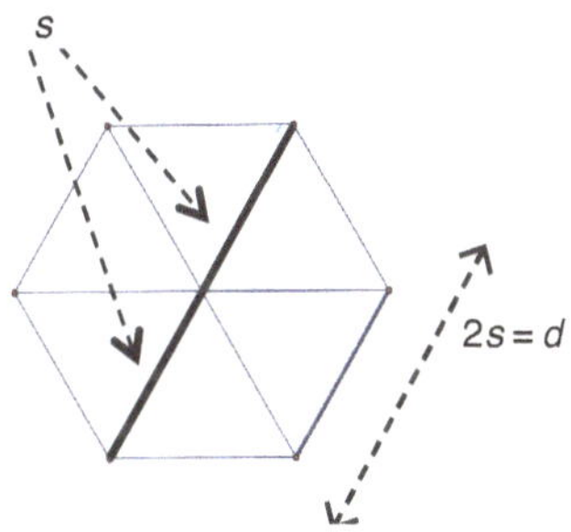

What is the length across the regular hexagon if we measure from one side to its respective opposite side? This length represents the height of the regular hexagon. We can see this length is twice the height of one of the equilateral triangle, and we have shown that the height of an equilateral triangle with side length is $h = .866s$

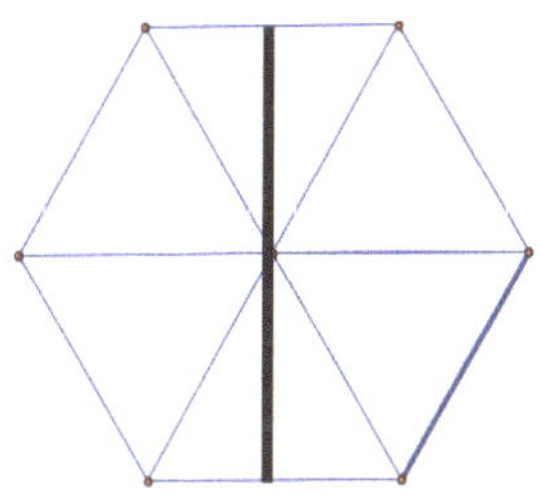

The length of this hexagon height measure, we will call f, is twice $.866s$ so $f = 2(.866s)$ or $f = 1.732s$

We can also manipulate formulas to find d and f in the following way.

As $d = 2s$, then $s = \frac{d}{2} = .5d$ and $f = 1.732s$, then $f = 1.732(\ .5d\ ) = .866d$

As $f = 1.732\text{s}$, then $s = \frac{f}{1.732} = .577f$ so as $d = 2s$, then $d = 2(.577f) = 1.155f$

Now, what about AREA of a regular hexagon with side length of s?

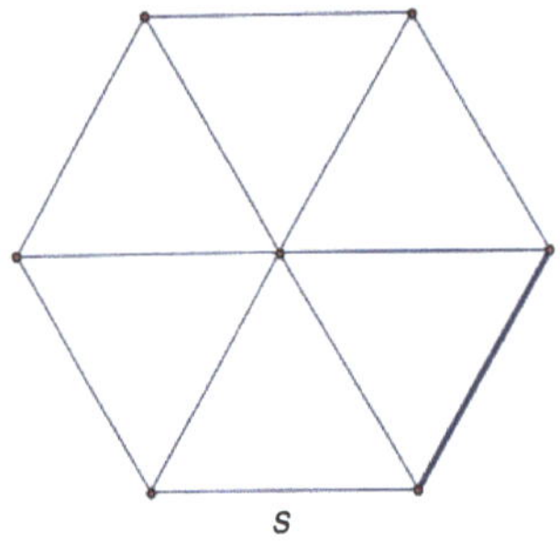

As we have six equilateral triangles, all with side length of s, in the regular hexagon, then the total area is six times the area of one of the equilateral triangle. Area of one equilateral triangle is $.433s^2$

So the area of a regular hexagon with side length of s is

$$A = 6\,(.433s^2) = 2.598s^2$$

Given the area A, we can find the side length.

As $A = 2.598s^2$ then $\frac{A}{2.598} = s^2$ Taking the square root of both sides,

$$\sqrt{\frac{A}{2.598}} = \sqrt{s^2} = s.$$

As $\frac{1}{2.598} = .385$, then $\sqrt{.385\,A} = s$.

We can also find area in terms of the diagonal, d, and the height of the hexagon, f.

Given $d = 2s$, then $s = .5d$ so $A = 2.598s^2 = 2.598\,(.5d)^2 = .650d^2$

And

Given $f = 1.732s$ then $s = .577f$ so $A = 2.598s^2 = 2.598\,(.577f)^2 = .866f^2$

So in summary

GIVEN A REGULAR HEXAGON

WITH SIDE LENGTH OF s.

And d = diagonal length, f = height of hexagon, and A = area

$s = .50d$ $\quad$ $s = .577f$

$d = 2s$ $\quad$ $d = 1.155f$

$f = 1.732s$ $\quad$ $f = .866d$

$A = 2.598s^2$ $\quad$ $A = .650d^2$ $\quad$ $A = .866f^2$

And $s = \sqrt{.385\,A}$

For this regular hexagon, find the diagonal length, then height of the hexagon and the area.

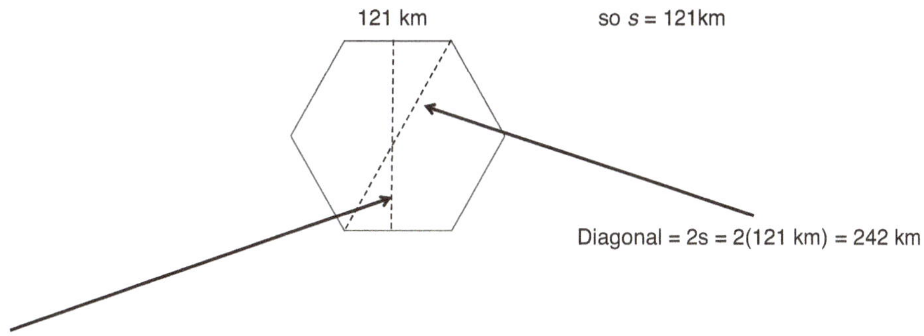

Height $f = 1.732s = 1.732\ (121\text{ km}) = 209.572\text{ km}$

Area $= 2.598s^2 = 2.598\ (121\text{ km})^2 = 38037.318\text{ km}^2$

Problem: Given the area of 300 sq. in., what is the side length, diagonal, and length from side to side of a regular hexagon?

TRICK OF THE TRADE

Another way to find the area of a regular hexagon if we know the side length, s, and the height of the hexagon, f, is the simple formula

$$A = 1.5\ sf$$

Why does this work?

Here is the regular hexagon with side s, diagonals, d, and segments for f.

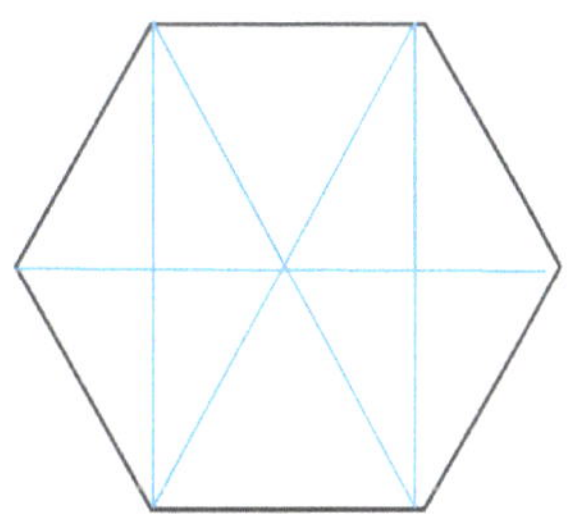

So we can break this into a rectangle and two congruent isosceles triangles.

The rectangle has dimensions of s and f, and the two triangles have a base of f and a height of $.5s$

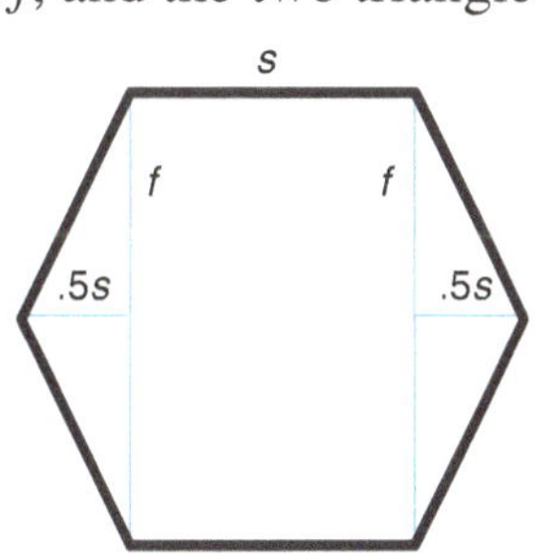

The rectangle has area of sf. One of the triangles has area of $.5\,(.5s)(f) = .25\,sf$

$$\text{So the total area} = sf + .25\,sf + .25\,sf = 1.5\,sf.$$

Example: A regular hexagon has a side length of 6 ft. and a height of 10.4 ft. What is the area?

$$\text{Area} = 1.5\,sf = 1.5\,(6\text{ ft.})\,(10.4\text{ ft.}) = 93.6\text{ ft.}^2$$

4-6 Exercises

1. For this regular hexagon,

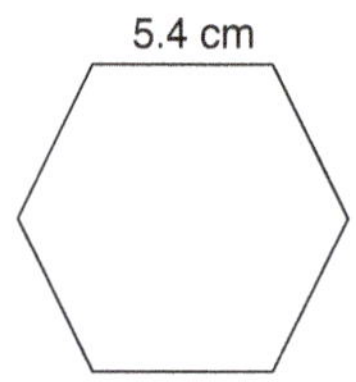

Find a) the perimeter ______ b) diagonal length _____ c) the height length __________ d) area _____

2. For this regular hexagon,

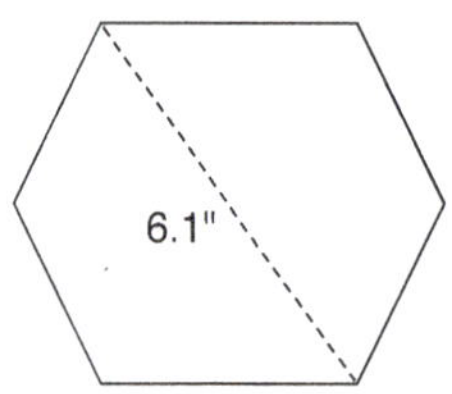

Find a) the side length ______ b) perimeter _____ c) the height length __________ d) area _____

3. For this regular hexagon,

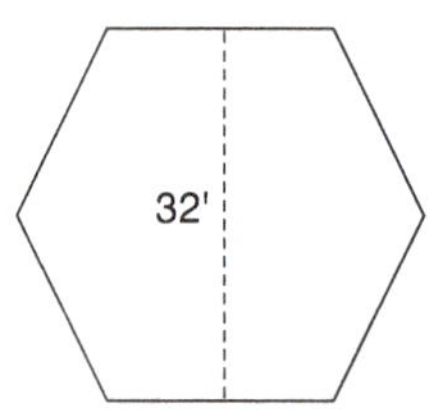

Find a) the side length ______ b) perimeter _____ c) the diagonal length __________ d) area _____

4. A Regular hexagon has an area of 25980 square feet. Find

   a) the side length ____ b) perimeter ____ c) the diagonal length ____ d) Height length____

5. Find the a) perimeter and b) area of this figure.

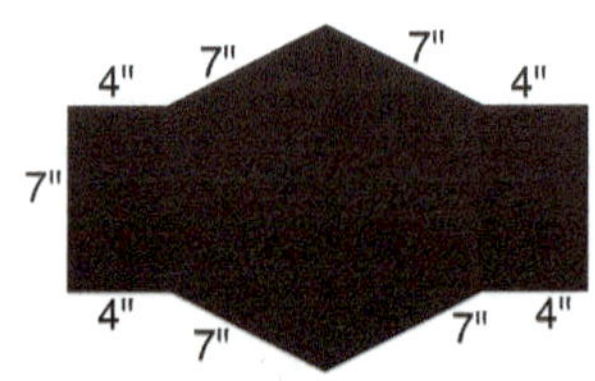

6. Find the a) perimeter and b) area of this figure. Assume the figure is symmetrical.

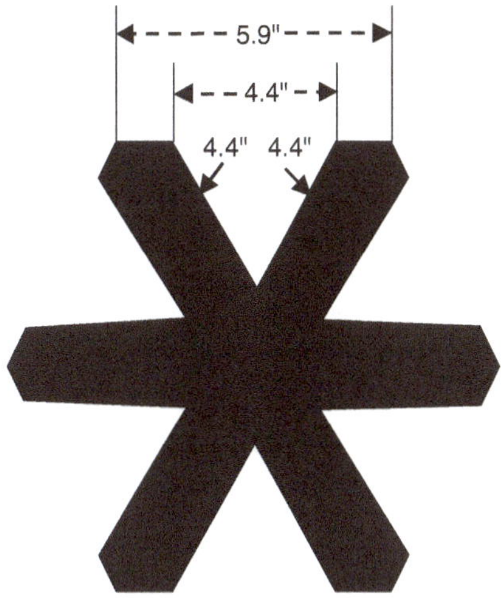

7. Find the a) perimeter and b) area of this figure

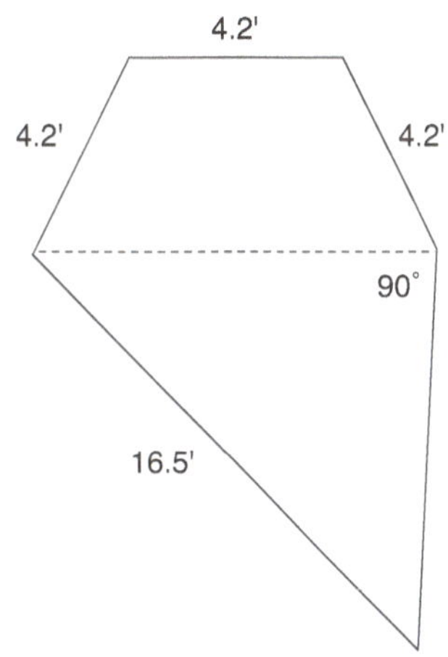

Answers

1. a) 32.4 cm b) 10.8 cm c) 9.35 cm d) 75.76 cm^2
2. a) 3.05″ b) 18.3″ c) 5.28″ d) 24.17 in^2
3. a) 18.46′ b) 110.76′ c) 36.96′ d) 886.78 ft^2
4. a) 100′ b) 600′ c) 200′ d) 173.2 ft
5. a) 58″ b) 183.3 in^2
6. a) 61.8″, b) 40.14 in^2
7. (The sides of the bottom right triangle are 8.4′ and 14.2′)
   a) 43.3′ b) 82.55 ft^2

4-7 REGULAR OCTAGONS

An octagon is a polygon with eight sides. It also has eight interior angles. A regular octagon has all side lengths equal in measure and all interior angles equal in measure. Each of the angle measure is always 135°.

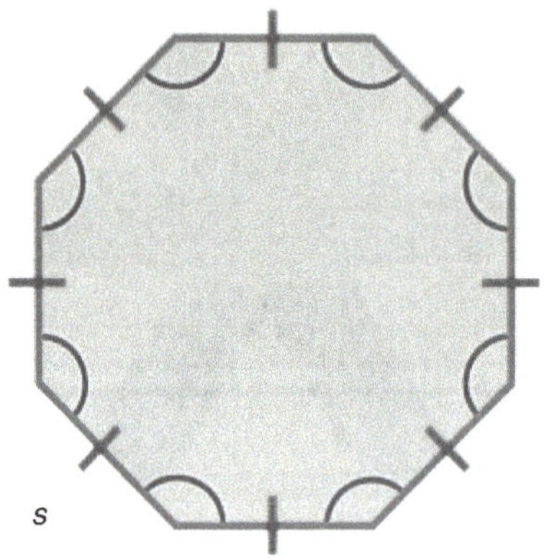

Perimeter of a regular octagon with side lengths is $P = 8s$

And given the perimeter, the side length is $s = \frac{P}{8}$

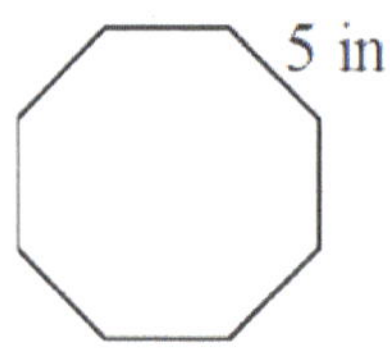

Find the perimeter of this octagon: $P = 8(5 \text{ in.}) = 40 \text{ in.}$

What is the side length of a regular octagon is its perimeter is 300 cm? $s = \frac{300 \text{ cm}}{8} = 37.5 \text{ cm}$

DIAGONAL

If we draw the longest diagonal segments across the regular octagon, there are four of them. These diagonals divide the regular octagon into eight congruent triangles. In fact, each of the triangles is an isosceles triangle.

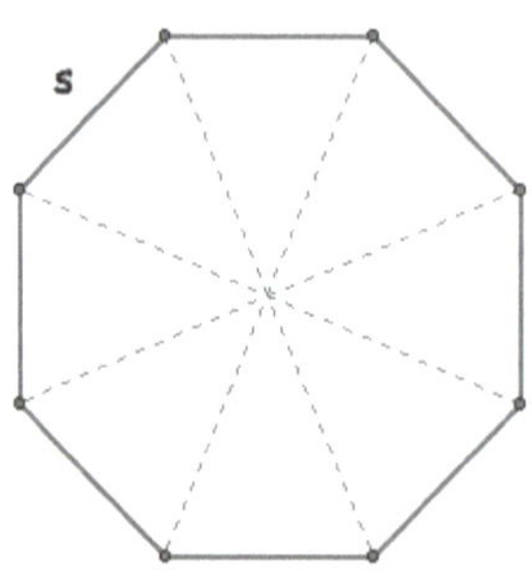

Each isosceles triangle has a vertex angle of 45^{c} and two base angles of 67.5^{c}.

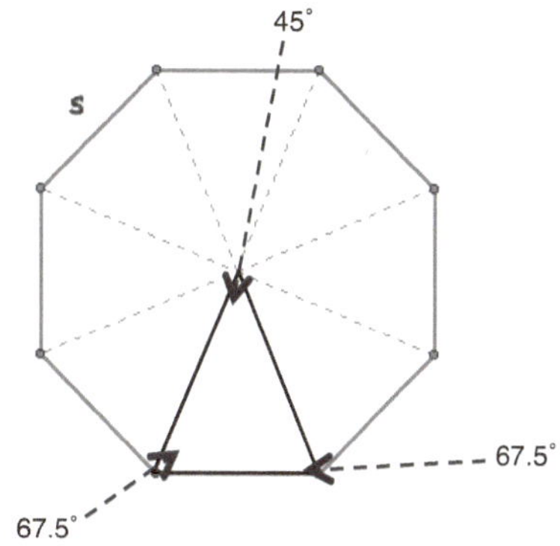

Given the side length of s for the regular octagon, the length of one of these diagonals, d, is $d = 2.613s$ And if given the diagonal, d, the side length is $s = .383\ d$

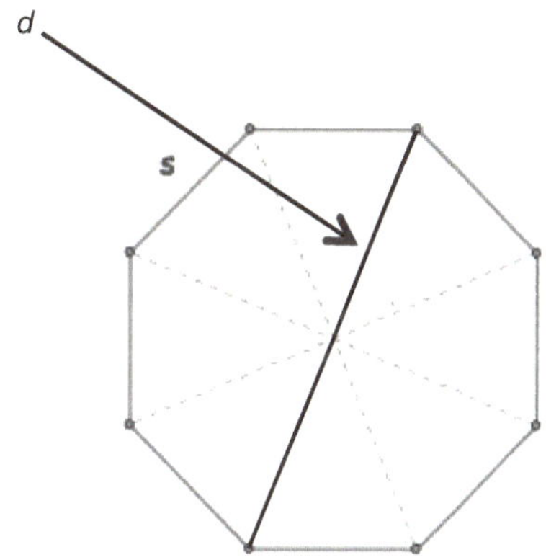

Given side length s of the regular octagon, The height of the triangle, f, is figured as $f = 2.414$ s, and if given the height length, f, the side length is $s = .414f$

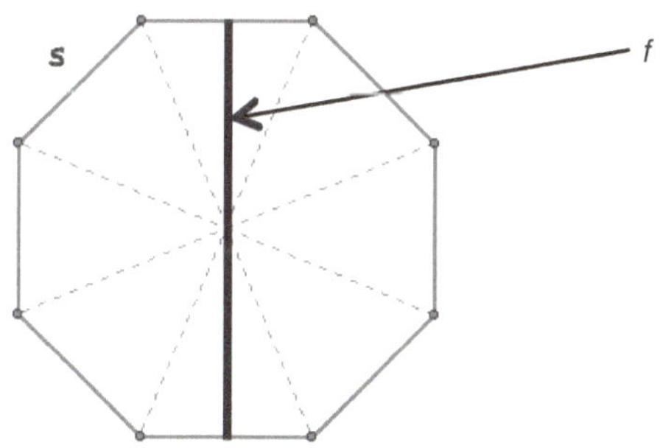

And manipulating the formulas, we find $f = .924d$ and $d = 1.083f$

Finally, AREA of a regular octagon with side length of s is

$$A = 4.828s^2 \text{ and } s = \sqrt{.207A}$$

And given d and or f, $A = .707d^2$ and $A = .828f^2$

In summary, for a regular octagon with side length s, p = perimeter, d = diagonal, f = distance from side to side, and A = area

$p = 8s$	$s = \frac{p}{8}$
$d = 2.613s$	$s = .383d$
$f = 2.414s$	$s = .414f$
$f = .924d$	$d = 1.083f$
$A = 4.828s^2$	$s = \sqrt{.207A}$
$A = .707\,d^2$	$A = .828\,f^2$ $\quad A = 2fs$

Practice: Find the perimeter, diagonal, distance from side to side and area of this regular octagon, with $s = 210$ in.

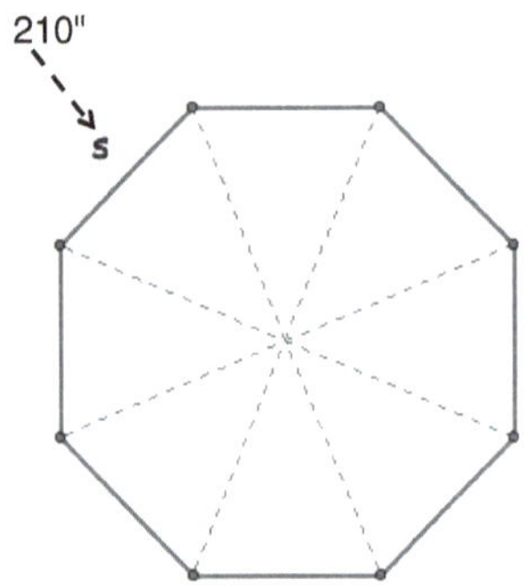

Practice: Find side length, perimeter, distance from side to side and area, for a regular octagon with diagonal of 45 mm.

TRICK OF THE TRADE

In a regular octagon, Area = $2sf$

Why does Area = $2sf$ work in a regular octagon.

We know the diagonal divide the regular octagon into eight congruent isosceles triangles.

Now, given side length s and height f

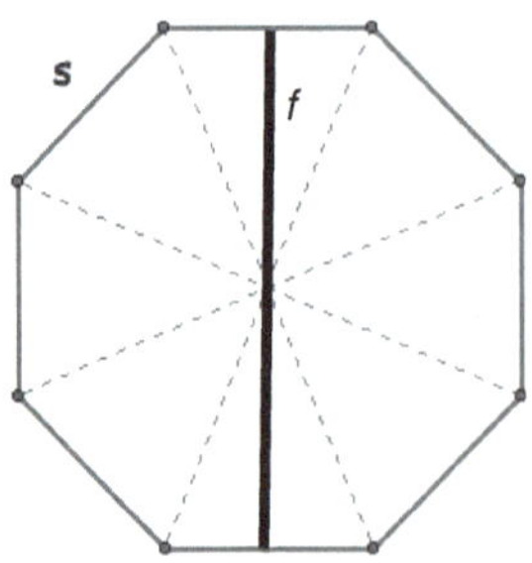

Notice, each of the isosceles triangles has a base of s and a height of half of f, or $.5f$.

The area of an isosceles triangle is .5 base times height, then the area of each of these isosceles triangles is $.5\,(s)\,(.5f) = .25sf$

As there are eight of these triangles, the total area of the regular octagon is

$$A = 8(.25sf) = 2sf.$$

Example, in a regular octagon, the side length is 6 in. and the height is 14.5 in.

So $s = 6, f = 14.5$, Area $= 2sf = 2$ (6 in.) (14.5 in.) = 174 in.^{2}

4-7 Exercises

1. For this Regular octagon with side length = 23.9 cm

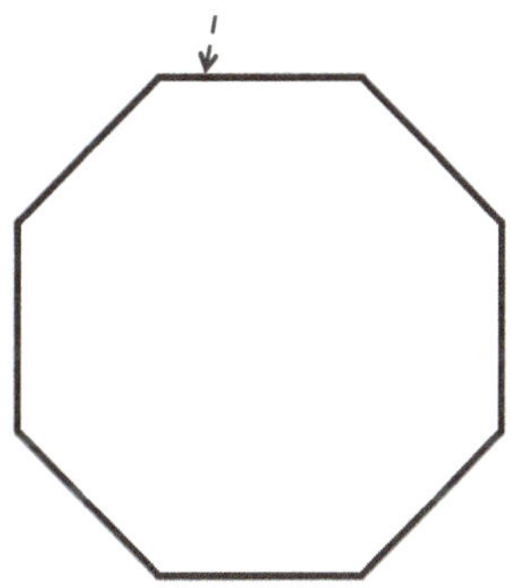

Find a) perimeter, b) diagonal length c) height length and d) area

2. For this regular octagon with diagonal length 5.5 inches

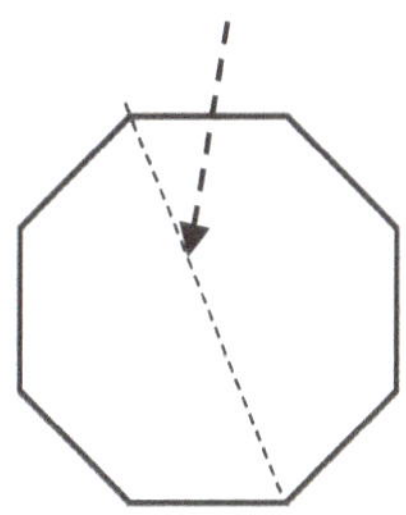

Find a) side length b) perimeter, c) height length and d) area

3. For this regular octagon with height 78 yards

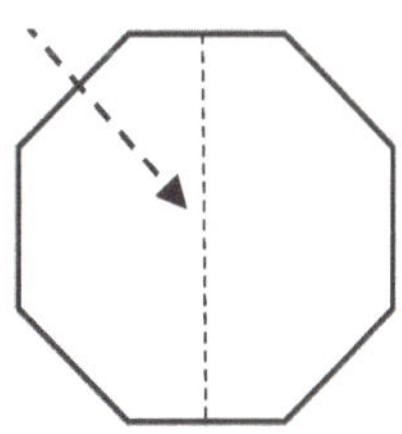

Find a) side length b) perimeter, c) diagonal length and d) area

4. For a regular octagon with side length 212 inches,
   Find a) perimeter, b) diagonal length c) height length and d) area

5. For a regular octagon with diagonal of 2.1 feet
   Find a) side length b) perimeter, c) height length and d) area

6. For a regular octagon with height of 45 mm.
   Find a) side length b) perimeter, c) diagonal length and d) area

7. For a regular octagon with area of 400 square feet
   Find the a) side length b) perimeter, c) diagonal length and d) height length

8. For a regular octagon with area of 2.4 square miles
   Find the a) side length b) perimeter, c) diagonal length and d) height lengt

Answers

1. a) 191.2 cm b) 62.45 cm c) 57.69 cm d) 2757.80 cm^2
2. a) 2.11″ b) 16.88″ c) 5.09″ d) 21.49 in^2
3. a) 32.29 yds b) 258.32 yds c) 84.37 yds d) 5033.89 yd^2
4. a) 1696″ b) 553.96″ c) 511.77″ d) 216989.63 in^2
5. a) .80′ b) 6.4′ c) 1.93′ d) 3.09 ft^2
6. a) 18.63 mm b) 149.04 mm c) 48.68 mm d) 1675.69 mm^2
7. a) 9.10′ b) 72.8′ c) 23.78′ d) 21.97′
8. a) .70 mi b) 5.6 mi. c) 1.83 mi d) 1.69 mi.

4-8 CIRCLES

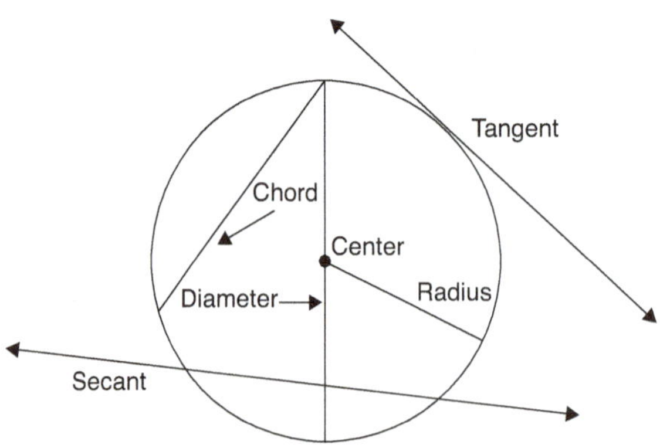

A circle is a common plane figure. By definition, a circle is made up of a set of points that are equal distance from some point on a plane, called the CENTER of the circle. The center "defines" the position of the circle and the equal distance from that point sets the points of the line. Note: The center is not part of that set of points that make up the circle.

Definitions to know:

- The equal distance from the center point out to the circle is called the RADIUS of the circle.
- A segment that has endpoints on the circle is called a CHORD.
- The chord that goes through the center of the circle is called the DIAMETER.
- The perimeter, or the distance around a circle, is called the CIRCUMFERENCE.
- A line on the exterior of a circle except that it touches the circle at one point is called a TANGENT to the circle.
- A line that starts outside the circle and goes completely through the circle is called a SECANT to the circle.

DIAMETER AND RADIUS

As the radius is the distance from the center of the circle to the circle points, and the diameter goes from one point on the circle through the center to another point on the circle, the diameter must travel a length of two radius lengths. So, we say given the radius, r, the diameter, d, is twice the r or $d = 2r$. It also holds that the radius, r, is half the diameter, d, or $r = \frac{d}{2}$

So, a circle with a radius of **6in. has** a diameter of 2(6 in.) = 12 in.

And a circle with diameter of 45 cm has a radius of $\frac{45}{2}$ cm = 22.5 cm

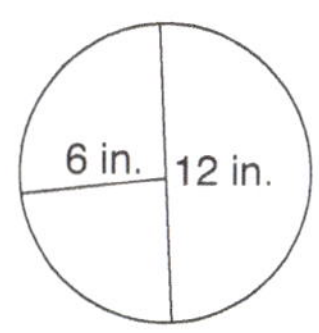

CIRCUMFERENCE

The CIRCUMFERENCE is the "perimeter" of the circle, in other words, the distance if "walking" completely around the circle.

It has been shown that for EVERY CIRCLE, when you divide the circumference value by its diameter value, you always get the same number, a special irrational number that is about 3.141592654…

This special number is called "pi" and uses the special symbol π

So it follows that if the circumference of a circle is C and the diameter is d, then

$$\frac{C}{d} = \pi$$

So the circumference = π times diameter or

$$C = d\pi$$

And, since $d = 2r$, then it is also true that

$$C = 2r \cdot \pi \text{ which is often written } C = 2\pi r$$

So for the circle with diameter 5 in.

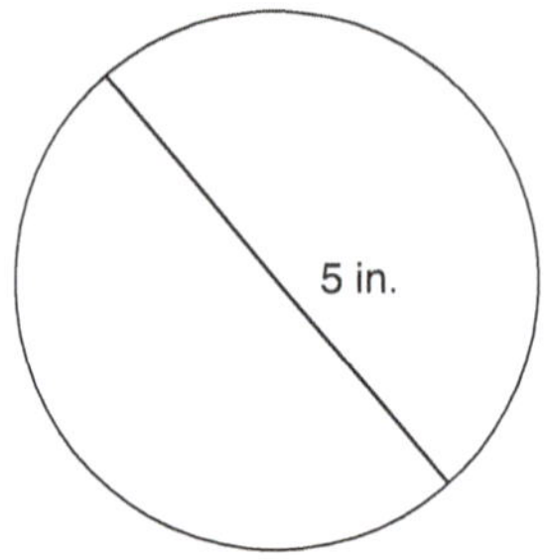

$d = 5$ in. and for $C = d\pi$ then circumference is $C = 5\pi$ in.

EXACTNESS VERSUS APPROXIMATE CIRCUMFERENCE

If we want an EXACT circumference, we need to leave the answer in π as in $C = 5\pi$ in. As π is an irrational number, we can never get an exact decimal answer for the circumference. The best we can do is use π a rounded decimal. Often, π rounded to the nearest hundredth decimal, 3.14, is used in calculations. So $C = 5\pi$ in. $\approx 5\,(3.14)$ in. ≈ 15.7 in. Our answer of 15.7 in. is an APPROXIMATE answer, but is really very close to what the exact answer would be (if we could calculate it!) Of course, if more accuracy is needed, the more decimal places we round π, the more accurate the decimal answer will be for the circumference.

Find the circumference or a circle with radius of 13 mm.

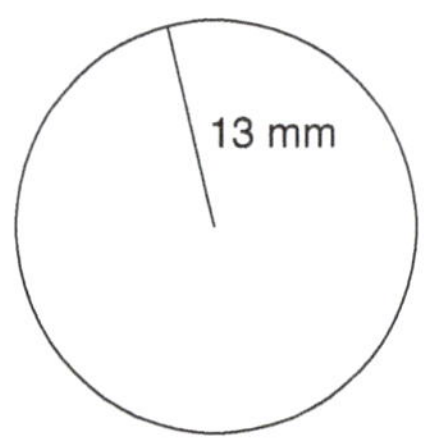

$r = 13$ mm, and $C = 2r\pi$

The exact circumference is $C = 2(13)\pi$ mm $= 26\pi$ mm.

If using 3.14 for, the Circumference $C \approx 2(13)(3.14)$ mm ≈ 81.64 mm.

AREA

The area, A, of any circle give radius, r, is pi times the square of the radius or

$$A = \pi r^2$$

If we start with a circle with radius r, and we create eight "pie slices" from the circle,

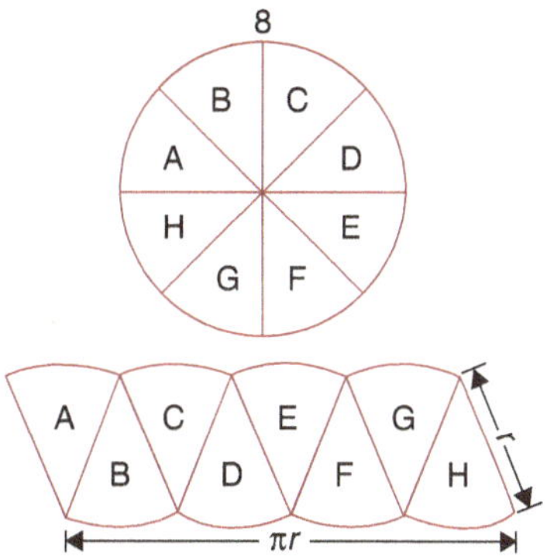

We can arrange the eight "pie slices" like this. The new shape resembles a parallelogram. The "height" is the radius, r. The "base" is half the circumference of the circle,

$$\text{so } \frac{1}{2}C = \frac{1}{2}(2\pi r) = \pi r$$

Imagine that our original circle is divided into more and more pie slices and arranged into the parallelogram shape. Then our "parallelogram" will look more and more like a real parallelogram, one with a height of r and a base of πr.

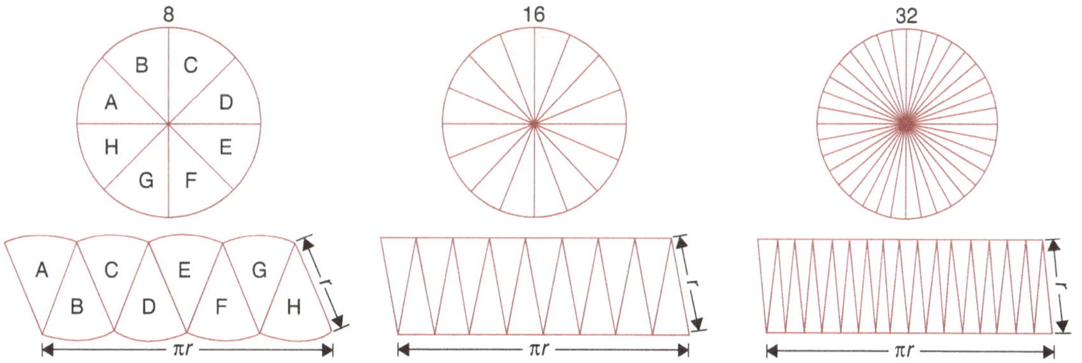

Here is going from 8 pie slices to 16 pie slices to 32 pie slices. The change in our "parallelogram" results in a sharper figure looking more like an actual parallelogram.

What if we could create an infinite number of pie slices. Then our rearranged shape would (if it was possible) become a true parallelogram with height of r and base of πr. In fact, it would become a rectangle with height of r and base of πr.

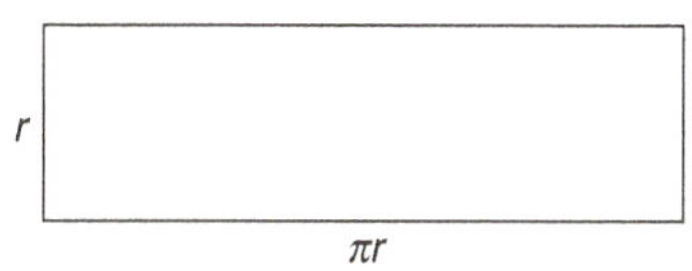

And a rectangle with height of r and base of πr has an area of $a = bh = (r)(r) = \pi r^2$. Therefore, it follows that the area of a circle with radius r is $A = \pi r^2$

Find the area for a circle with radius of 13 mm

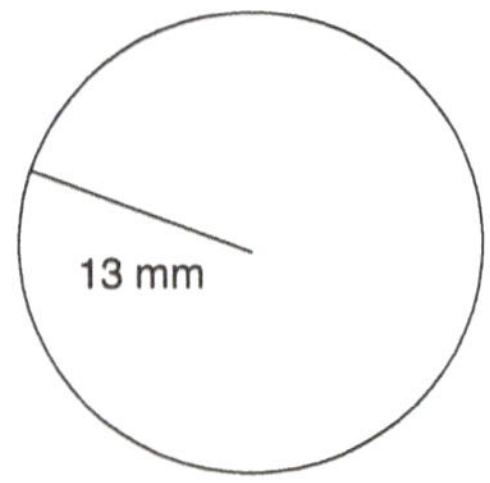

The area is $A = \pi r^2 = \pi\,(13\text{ mm})^2 = 169\pi\text{ mm}^2$, an exact answer. If we want the answer to the nearest hundredth of a square millimeter, we can use 3.14 for π

We get $A \approx (3.14)\,(13\text{ mm})^2 \approx (3.14)(169)\,(\text{mm}^2) \approx 530.66\text{ mm}^2$

The area for the circle with diameter 5 in.

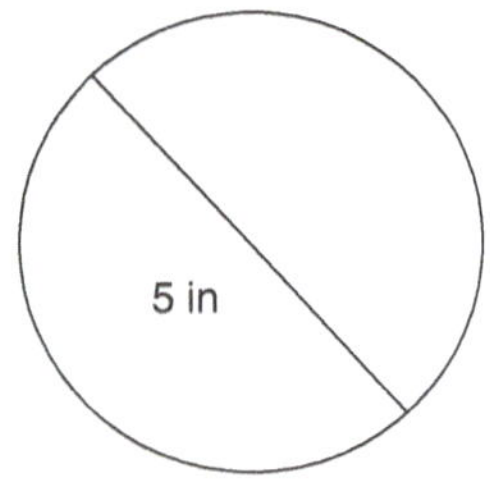

We need to find the radius as $\frac{5}{2}$ in. = 2.5 in.

And area is $A = \pi r^2 = \pi(2.5\text{ in.})^2 = 6.25\pi\text{ in.}^2$

If we want the answer to the nearest hundredth of a square inch, we can use 3.14 for π

We get $A = (3.14)\,(2.5\text{ in.})^2 = (3.14)\,(6.25)\text{ in.}^2 = 19.625\text{ in.}^2$, so nearest hundredths it would be 19.62 in.^2

Now we can use Circumference/Area Formula to work out problems.

A 2 ft. by 5 ft. rectangular tile has a semicircle (half circle) with radius of .75 ft. cut out of it. What is the area of the part of the tile remaining?

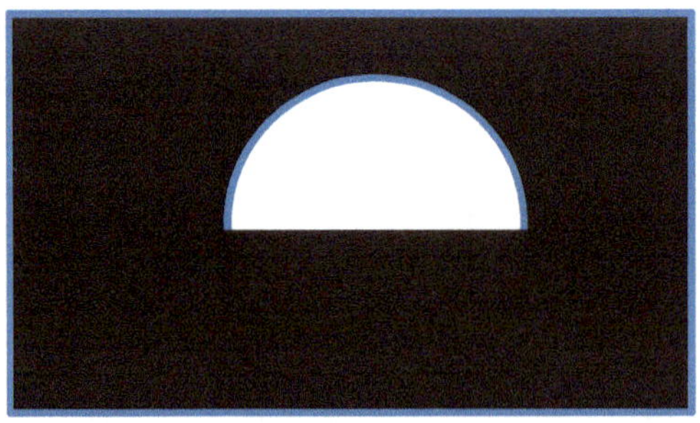

A park design features a semicircle, rectangle, and right triangle in this shape.

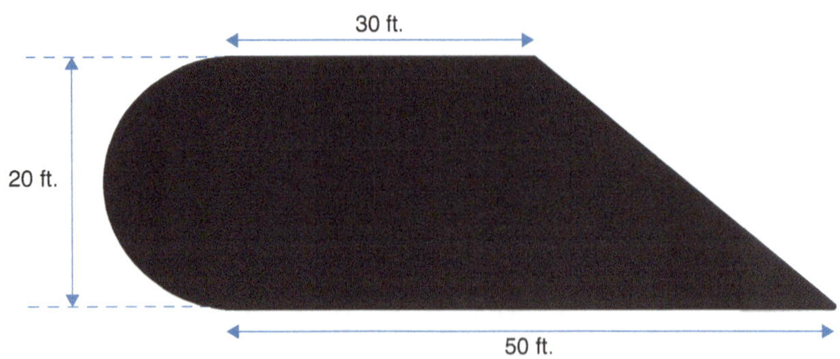

Find the perimeter and the area of the park.

4-8 Exercises

Find a) circumference and b) area of these circles. Give answer to nearest hundredth of a unit when appropriate.

1.
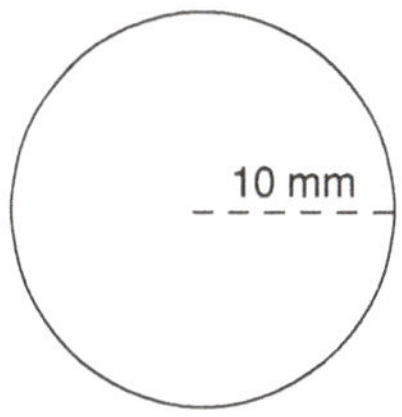

2.
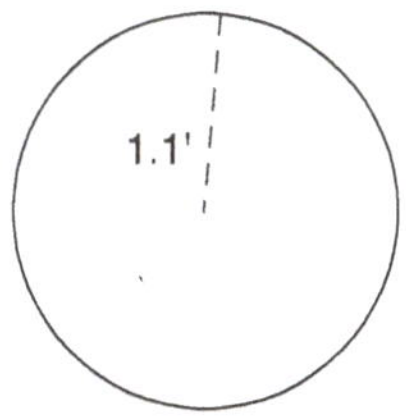

3.
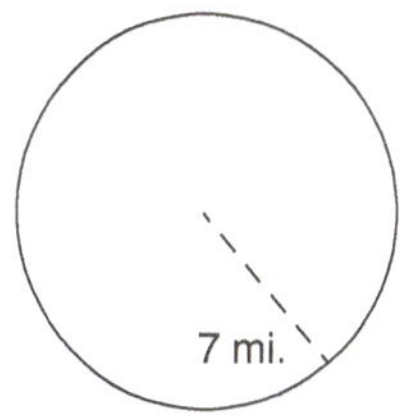

4.
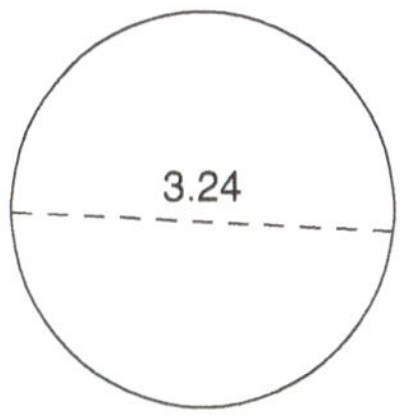

5.
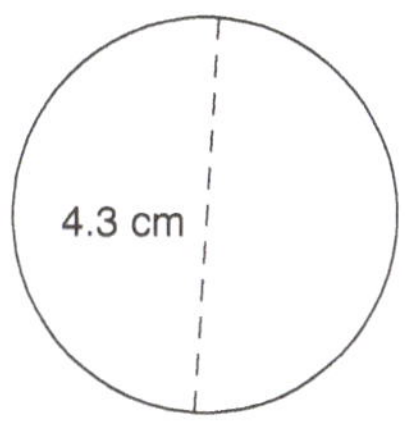

6.
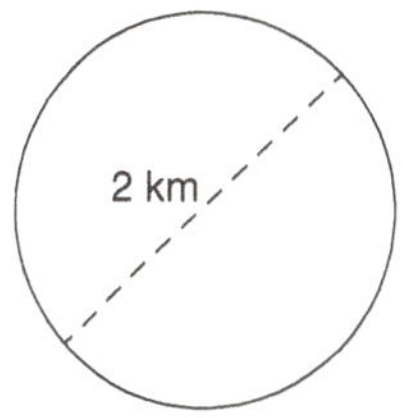

7. A circle has a radius of 345 inches. Find it's a) circumference _________b) area___________

8. A circle has a diameter of 55 m. Find it's a) circumference _______ b) area _____________

9. Circle has area 234 mm^2 find a) radius __________ b) circumference _______

10. Circle has area 1000 cm^2 find a) radius __________ b) circumference _______

11. Circle has area 628.2 in^2 find a) radius __________ b) circumference _______

12. Circle has circumference 345 ft. find a) radius _________ b) area __________

13. Circle has circumference 20 mm. find a) radius _________ b) area __________

14. Circle has circumference 1234 in. find a) radius _________ b) area __________

For 15 and 16, Find a) perimeter b) area of these figures.

15. This shape consists of two semicircles and a rectangle

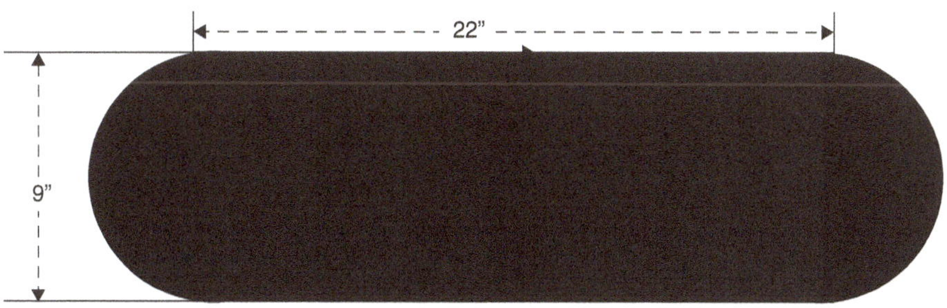

16. This shape consists of a semicircle, rectangle and right triangle

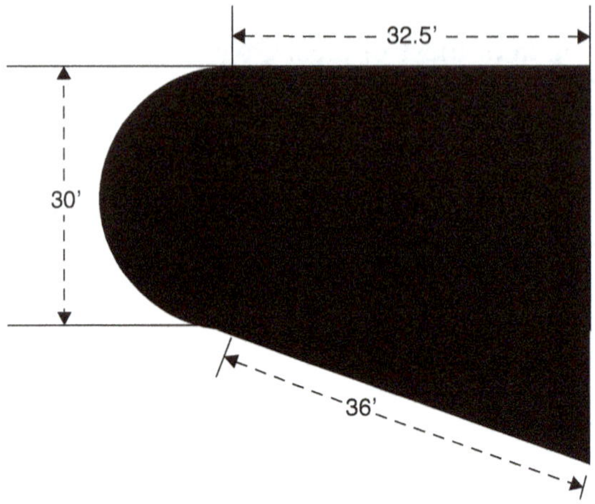

17. Find the shaded area of this rectangular figure with circles. Assume all the circles have the same diameter.

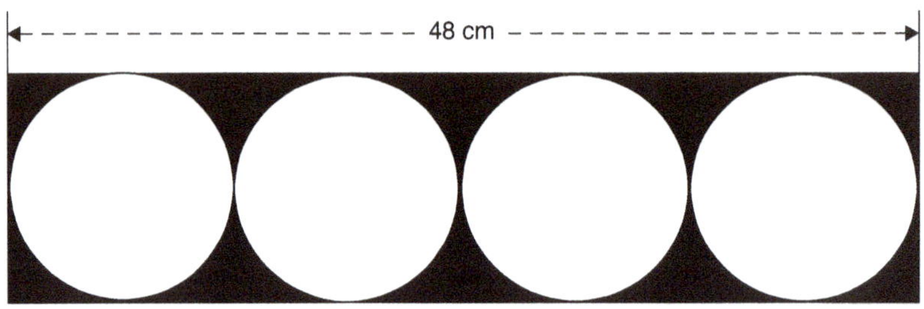

18. A farmer's rotating irrigation hose throws out water a distance of 30 meters How much of an area of land can this hose can cover?

19. A search light from a tower can reach 200 feet. What is the area that the light can cover?

20.

Domino Hut's Pizza offers pizzas in this way:

One Large 18 inch pizza for \$14.50

Or

Two medium 12 inch pizzas for \$14.50.

Which is a better deal and why? (the "inch" for the pizza refers to the diameter.)

Answers

1. a) 62.83 m b) 314.16 mm^2
2. a) 6.91′ b) 3.8 ft^2
3. a) 43.98 mi b) 153.94 mi^2
4. a) 10.18″ b) 8.24 in^2
5. a) 13.51 cm b) 14.52 cm^2
6. a) 6.28 km b) 3.14 km^2
7. a) 2167.70″ b) 373928.07 in^2
8. a) 172.79 m b) 2375.83 m^2
9. a) 8.63 mm b) 54.22 mm
10. a) 17.84 cm b) 112.09 cm
11. a) 14.14″ b) 88.84″
12. a) 54.91′ b) 9472.24 ft^2
13. a) 3.18 mm b) 31.77 mm^2
14. a) 196.40″ b) 121180.53 in^2
15. a) 72.27″ b) 261.62 in^2
16. a) 161.10′ b) 1580.04 ft^2
17. 123.61 cm^2
18. 2827.43 m^2
19. 125663.70 ft^2
20. The 18″ pizza area is 254.47 in^2, two 12″ pizza area is 226.19 in^2 so the 18″ pizza is a better deal for $14.50.

4-9 MORE ON CIRCLES: ARCS AND SECTORS

Circle: As we **learned in Chapter 4.1, if we take a segment** from the center of the circle out to the circle itself, and made a sweep all the way around the circle, it would cover 360 units called degrees, symbolized by °. So there are 360° in a circle.

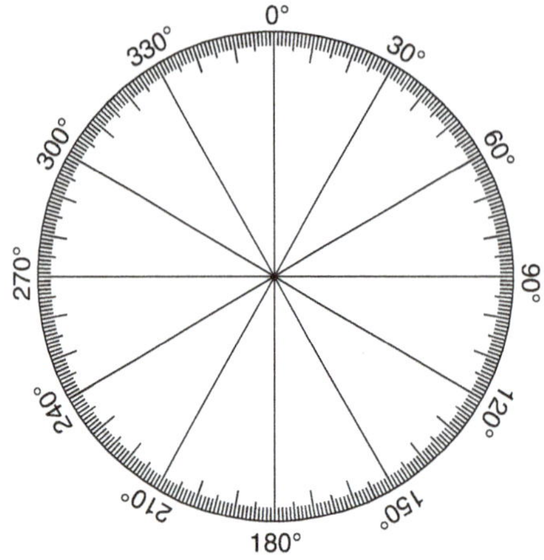

ARCS

An ARC is part of a circle's circumference. It is the path if you were to "walk" part way around the circle. This part of the circle that goes from point A to point B is an arc.

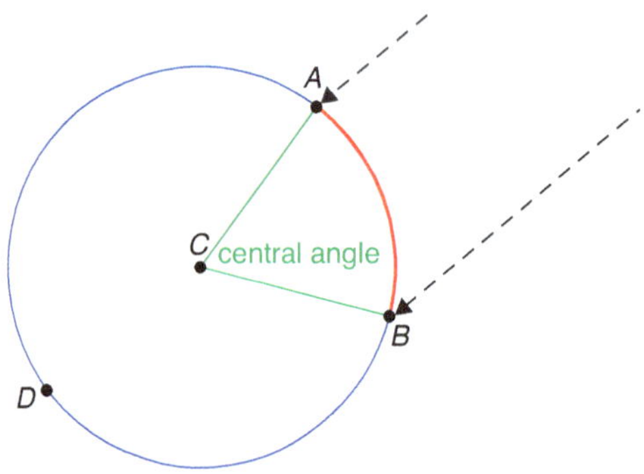

The angle made by drawing segments from the center of the circle out to the "endpoints" of the arc is called the CENTRAL ANGLE of the arc. We say that the central angle "intercepts" the arc. The central arc of this arc above is about 80°.

When the arc travels less than halfway around the circle so it's central angle is less than 180°, then it is called a MINOR ARC. So the arc from A to B above is a minor arc. However, given any two points on a circle, there are actually two arcs. The arc that goes from point A to B through point D is more than 180° and is called a MAJOR ARC.

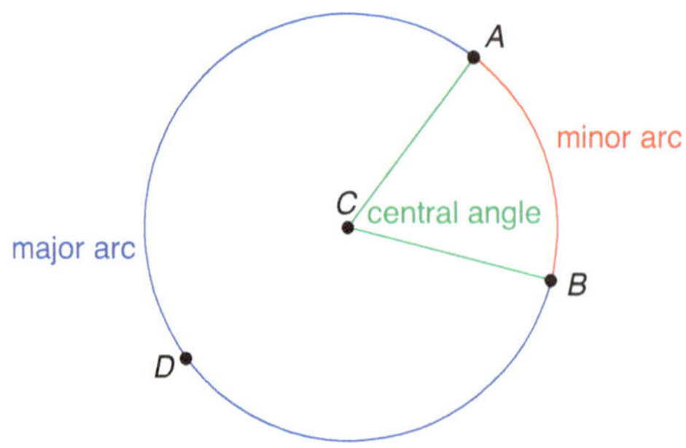

When the arc is halfway around the circle, the central angle of both angle is 180° and the arcs are called SEMICIRCLE ARCs.

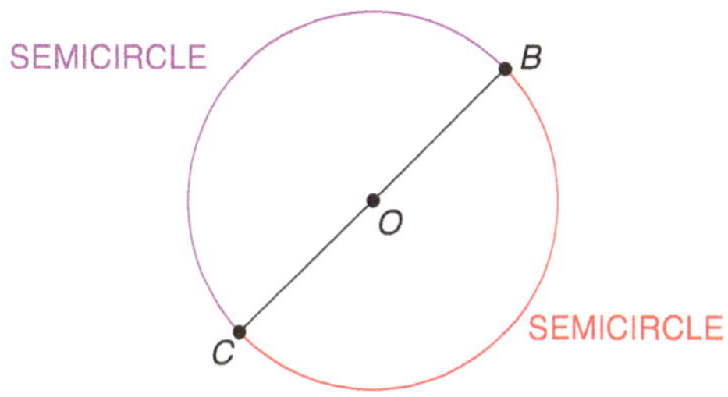

This is a proportional relationship between the arc length to the circle's circumference and its central angle to the 360° circle angle total. We can find the length of the arc is we know the central angle by using this proportion

$$\frac{\text{arc length}}{\text{circumference}} = \frac{\text{central angle}}{360°}$$

For a circle with radius r, let arc length = L, let central angle = $x°$, and since Circumference = $2\pi r$, our formula becomes

$$\frac{L}{2\pi r} = \frac{x°}{360°} \quad \text{or} \quad L = \frac{2\pi rx}{360} = \frac{\pi rx}{180}$$

Find the arc length, to the nearest hundredth of a unit, from A to B with a central angle of 110° in a circle with radius of 8 units.

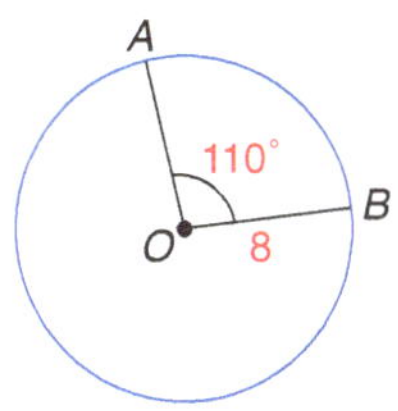

Let $\pi = 3.14$, let $r = 8$, let $x = 110$ and

$$\text{L} = \frac{\pi rx}{180} = \frac{(3.14)(8)(110)}{180} = 15.35 \text{ units}$$

Find the arc length for the major arc to this minor arc.

$$360° - 110° = 250°, \text{ so let } x = 250$$

$$\text{L} = \frac{\pi rx}{180} = \frac{(3.14)(8)(250)}{180} = 34.89 \text{ units}$$

(You could also find this by taking the circumference −15.35 units or $2(\pi)$ (8) − 15.35 = 34.89.)

SECTORS

A SECTOR is a portion of the area of a circle that is between the two angle segments (rays) of a central angle. It is kind of like a "slice of pie" from the circle.

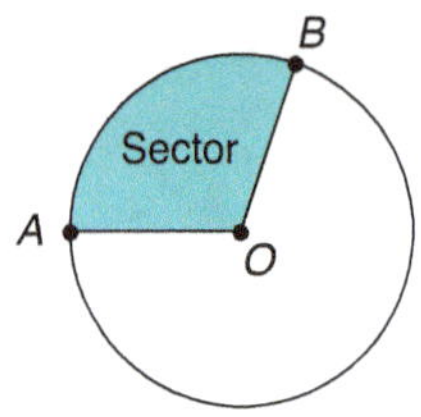

The same kind of proportional relationship happens: the area of a sector to the total area of the circle is equal to the central angle measure to the 360° total angle for the circle.

$$\text{Or } \frac{\text{Area}_{\text{sector}}}{\text{Area}_{\text{circle}}} = \frac{\text{Central angle}}{360°}$$

With a circle with radius r, A_s = the area of the sector, $x°$ = the central angle measure, and since Area of a circle is πr^2 our proportion becomes

$$\frac{A_s}{\pi r^2} = \frac{x°}{360°} \text{ or } A_s = \frac{\pi r^2 x}{360}$$

Given sector with central angle 109° in circle with radius of 15 ft.

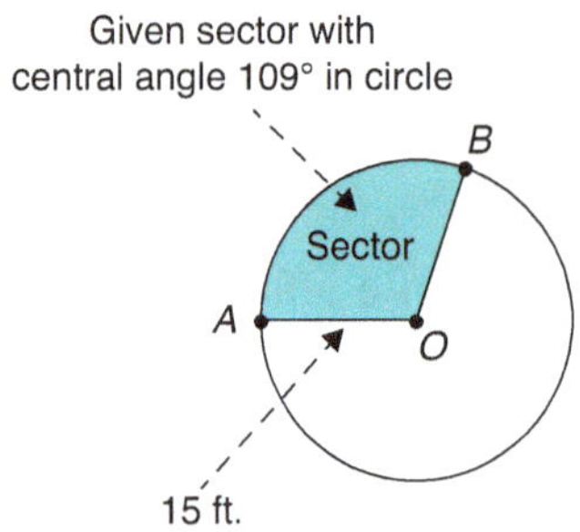

find the area to the nearest hundredth of a square foot.

Let $r = 15$, $x = 109$, and $\pi = 3.14$ $\qquad A_s = \frac{\pi r^2 x}{360} = \frac{(3.14)(15^2)(109)}{360} = 213.91$ sq. ft.

Practice

For this sector, find the arc length and the area

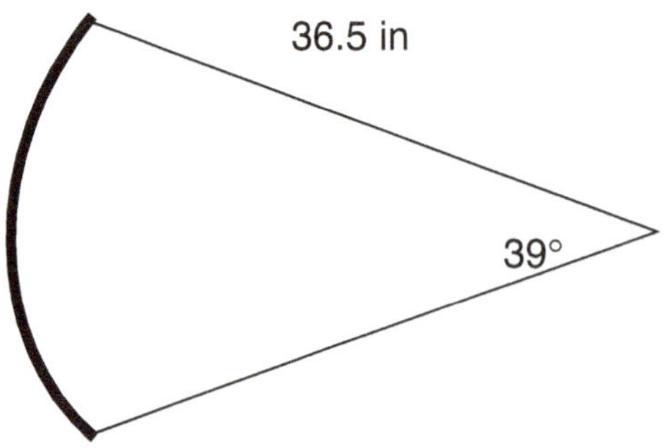

4.9 Exercises

Find measure to the nearest hundredth of a unit when appropriate.

1. For this circle with radius 10", find
   a) the arc length for this minor arc from point A to B.
   b) the area of the sector made by the arc.

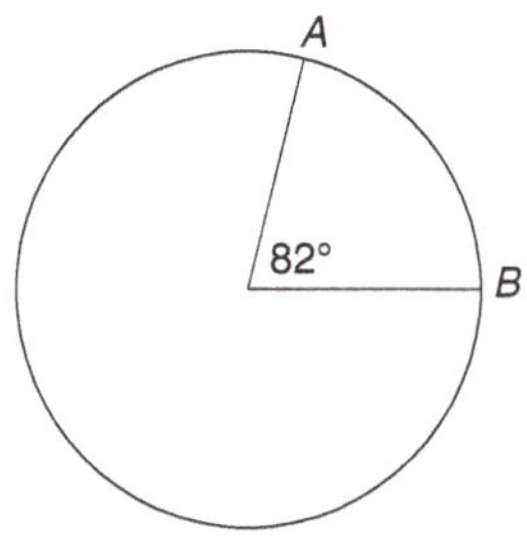

2. For this circle with diameter 21.7 mm, find
   a) the arc length for this minor arc from point C to D.
   b) the area of the sector made by the arc.

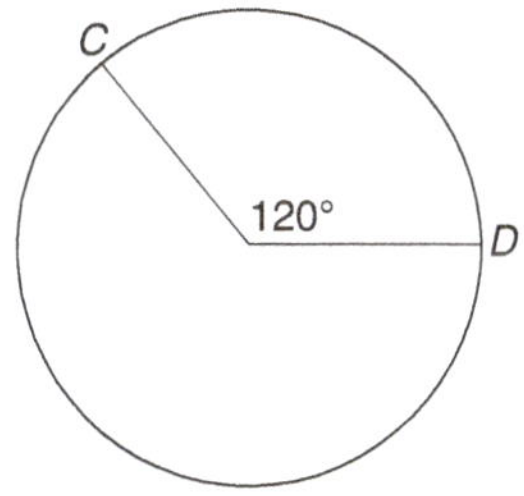

3. For this circle
   a) the arc length for this minor arc from point A to B.
   b) the area of the sector made by the arc.

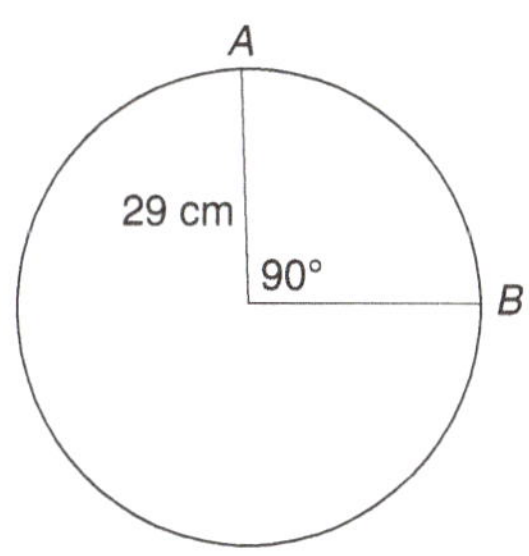

4. In a circle with a radius of 45 m, an Arc has a central angle of 30°.
   a) What is the arc length?
   b) what is the area of the sector for this arc?

5. An arc with an arc length of 42" which has a central angle of 66°.
   a) What is the radius of the circle?
   b) what is the area of the sector for this arc?

6. An arc with an arc length of 4.8 cm is on a circle with radius of 19.2 cm.
   a) What is the central angle of the arc?
   b) what is the area of the sector for this arc?

7. A patio was made that was shaped as a portion of a circle with a radius of 10 feet, according to this floor plan.

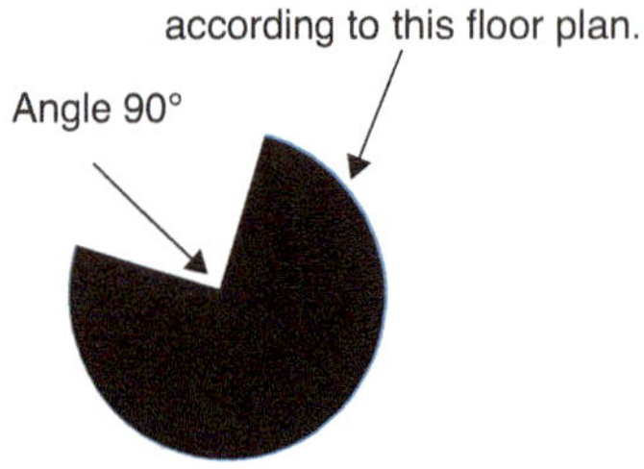

   a) What is the perimeter of the patio?
   b) What is the area of the patio?

8. A 20" pizza was cut into 8 equal slices. (the 20" is the diameter)

   a) what is the "arc length" of a slice of the pizza?
   b) what is the area of the pizza slice?

Answers

1. a) 14.31″ b) 71.56 in^2
2. a) 22.72 mm b) 123.28 mm^2
3. a) 45.55 cm b) 660.52 cm^2
4. a) 23.56 m b) 530.14 m^2
5. a) 36.46″ b) 765.56 in^2
6. a) 14.32° b) 46.07 cm^2
7. a) 67.12′ b) 235.62 ft^2
8. a) 7.85″ b) 39.27 in^2

4-10 ELLIPSES

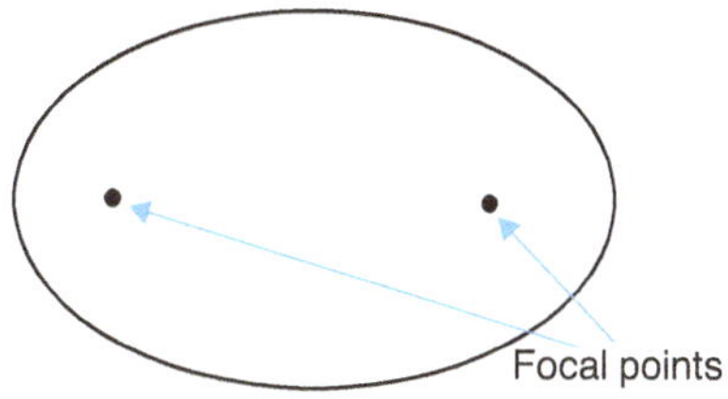

An **ELLIPSE** is an enclosed curved figure on a plane that surrounds two focal points such that the sum of the distances to each of the two focal points is constant for every point on the curve. As such, it is a generalization of a circle, which is a special type of an **ellipse** that has both focal points at the same location.

For this ellipse, $a_1 + b_1 = a_2 + b_2 = a_3 + b_3$, and so on

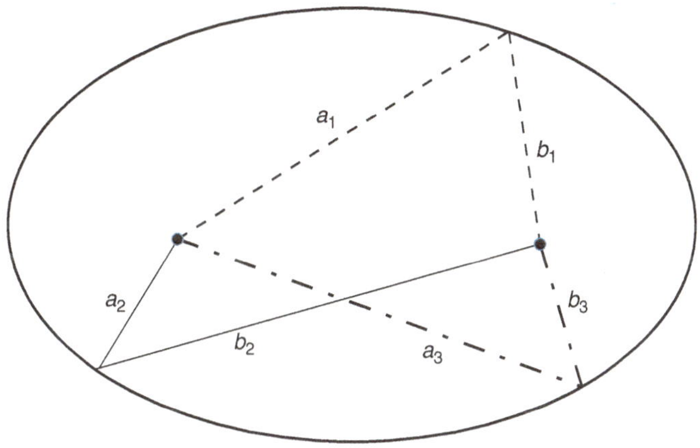

An ellipse has two lines of symmetry. They form axes of the ellipse.

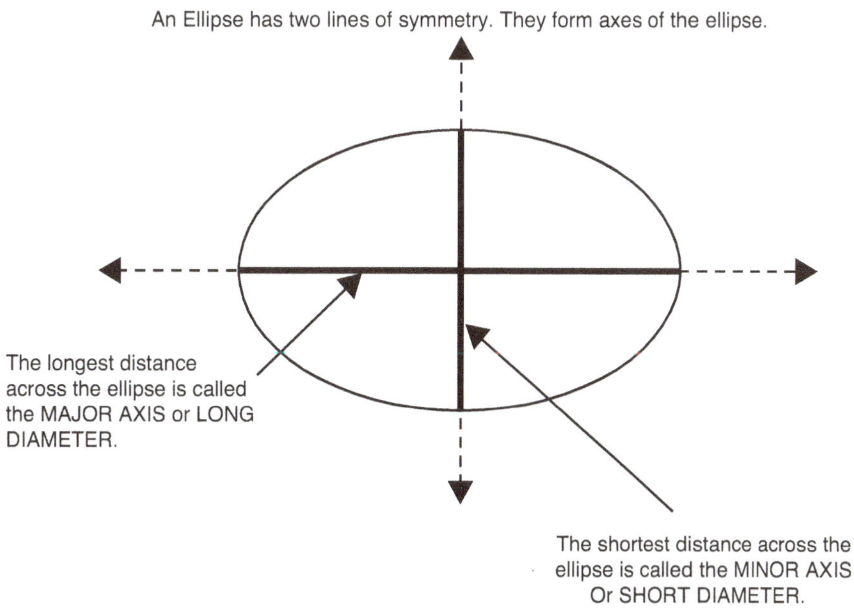

AREA AND PERIMETER

Let half of the long diameter = a and half the short diameter = b

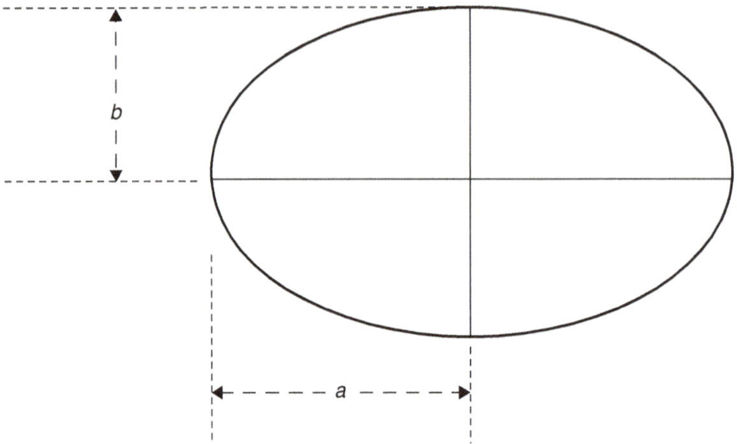

The area of the ellipse is $A = \pi ab$

The perimeter of the ellipse is $p = \pi\sqrt{2(a^2 + b^2)}$

Example

Find the area and perimeter of this ellipse

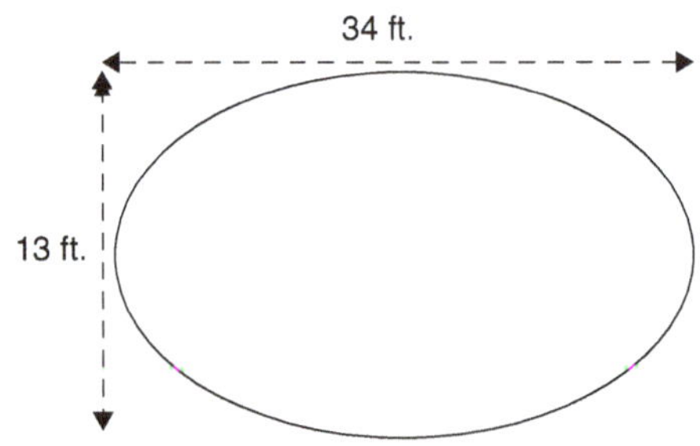

$$a = \frac{34 \text{ ft.}}{2} = 17 \text{ ft. } b = \frac{13 \text{ ft.}}{2} = 6.5 \text{ ft.}$$

Area = πab = 3.14 (17 ft.)(6.5 ft.) = 346.97 sq. ft.

Perimeter = $\pi\sqrt{2(\boldsymbol{a}^2 + \boldsymbol{b}^2)} = 3.14\sqrt{2(17^2 + 6.5^2)} = 80.82$ ft.

Find the area and perimeter of this ellipse

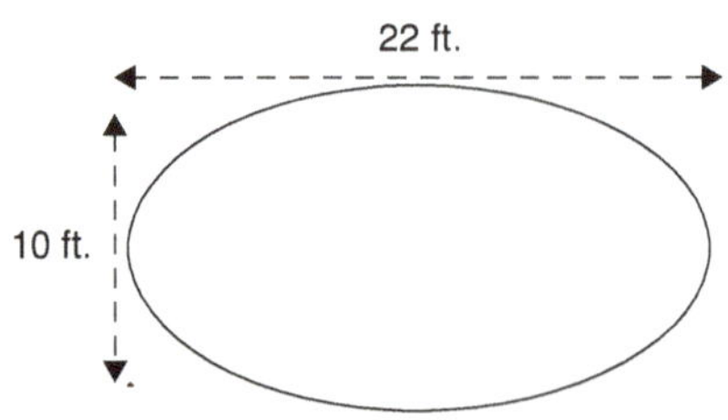

Find the area and perimeter of this ellipse

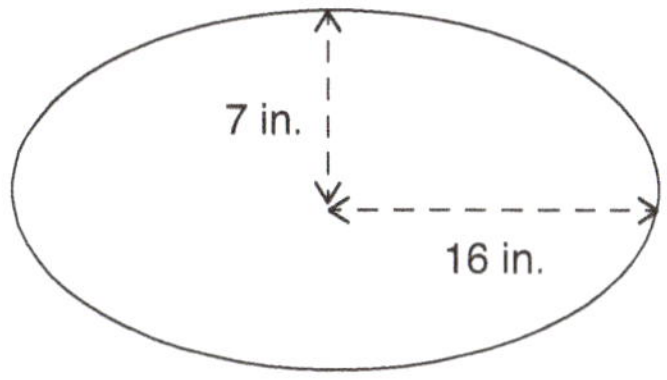

4-10 Exercises

Answer to the nearest hundredth of a unit when appropriate.

1. Find a) the area and b) perimeter of this ellipse

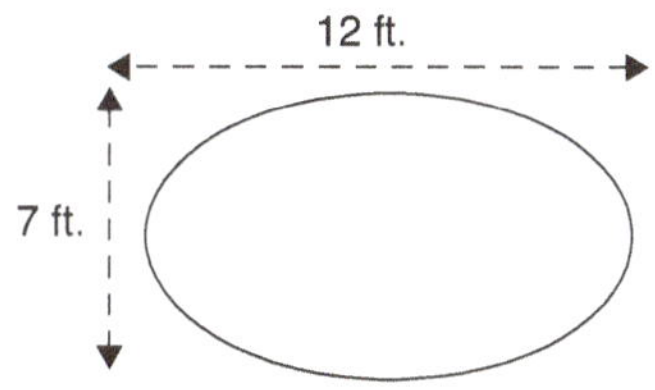

2. Find a) the area and b) perimeter of this ellipse

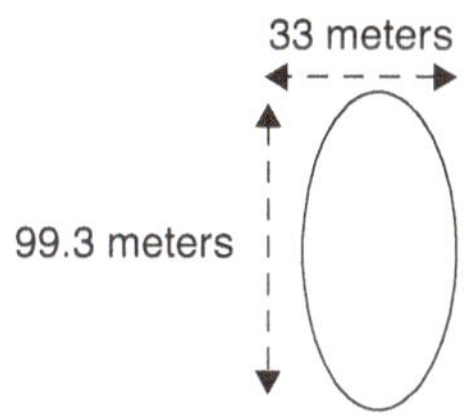

3. Find a) the area and b) perimeter of this ellipse

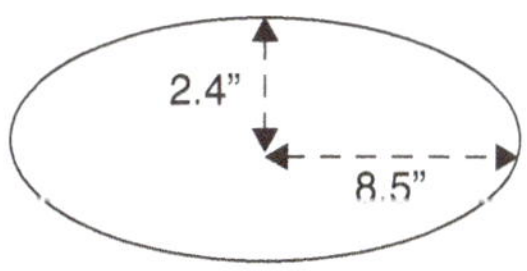

4. Find a) the area and b) perimeter of this ellipse

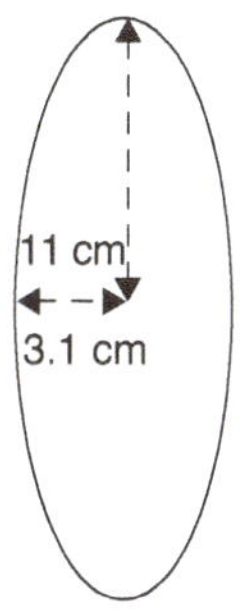

Answers

1. a) 65.97 ft^2 b) $30.86'$
2. a) 2573.67 m^2 b) 232.45 m
3. a) 64.09 in^2 b) $39.24''$
4. a) 107.13 cm^2 b) 50.78 cm

4-11 ANNULUS

Concentric circles are two (or more) circles with different radii that share the same center point on a plane.

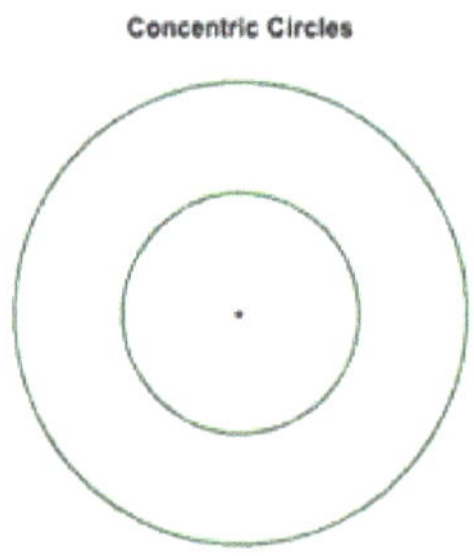

An ANNULUS is the area between two concentric circles. It appears as a "flat ring-shaped" object.

Notice, the annulus had "two circumferences" making up its perimeter, the outer circle circumference and the inner circle circumference.

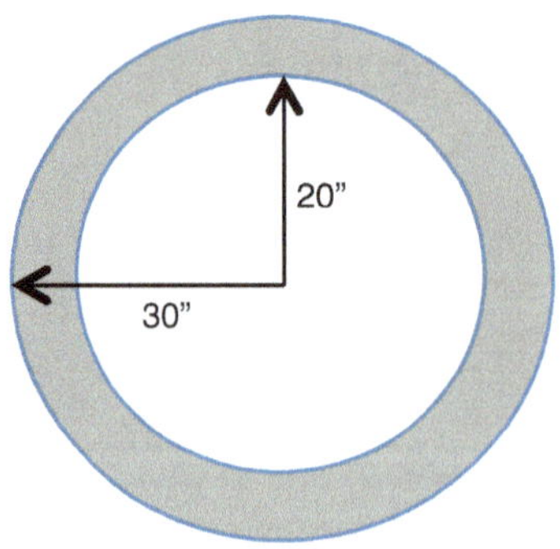

So if you have an inner circle radius of 20" and an outer circle radius of 30", The inner perimeter is $2\pi r = 2(20")\pi = 40\pi" \approx 125.7"$ and the outer perimeter is $2(30")\pi = 60\pi" \approx 188.5"$

AREA

To find the area of the annulus, we need to subtract the area of the inner circle from the area of the outer circle. For our example of an inner circle with a radius of 20" and the outer circle with a radius of 30"

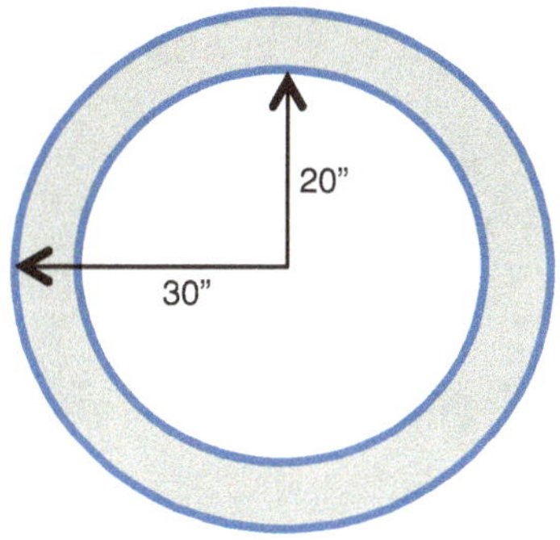

$$\text{Area of outer circle} = \pi r^2 = \pi(30in)^2 \approx 2827.4\,in^2$$

$$\text{Area of inner circle} = \pi r^2 = \pi(20in)^2 \approx 1256.6\,in^2 \text{ so}$$

$$area_{outer\ circle} - area_{inner\ circle} \approx (2827.4 - 1256.6)in^2 \approx 1570\,in^2$$

We can simplify this formula, given outer circle radius = r_0 and inner circle radius = r_i as

$$area_{annulus} = area_{outer\ circle} - area_{inner\ circle} = \pi\, r_0^2 - \pi\, r_i^2$$

or

$$area_{annulus} = \pi(r_0^2 - r_i^2)$$

Example: Find the area of the annulus for these concentric circles, with the outer circle's radius of 24 feet and the inner circle's radius of 10 feet.

$Area_{annulus} = \pi\,(r_0^2 - r_i^2)$

$= \pi\,[(24\,ft)^2 - (10\,ft)^2]$

$= \pi\,(476\,ft^2) \approx 1495.4\,ft^2$

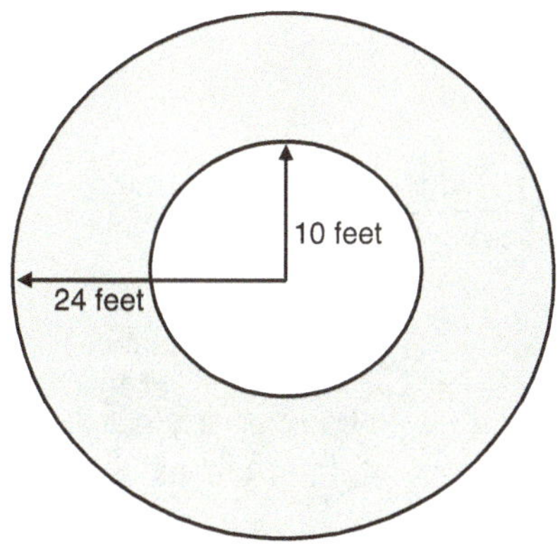

Example: A cross section of a lead pipe where the outer circle of the pipe has a diameter 12 cm and the inner circle of the pipe has a diameter of 10 cm (so it is a 1 cm thick pipe.) What is the area of the cross section of the pipe material (which is the annulus).

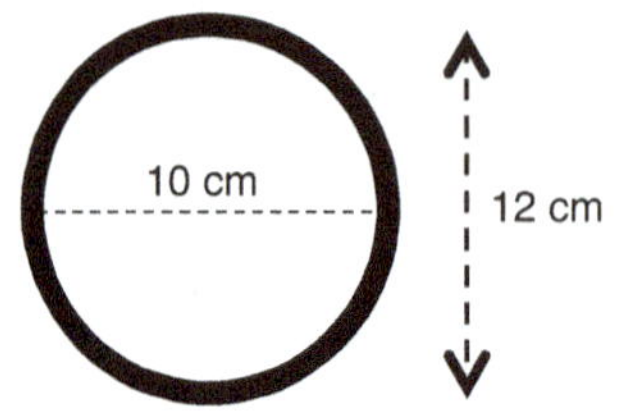

As radius of inner circle is $\frac{10\text{ cm}}{2} = 5\text{ cm}$ and radius of outer circle is $\frac{12\text{ cm}}{2} = 6\text{ cm}$

$\text{area}_{annulus} = \pi(r_o^2 - r_i^2) = 3.14\,[(6\text{ cm})^2 - (5\text{ cm})^2] = \pi(11\text{ cm}^2) \approx 34.6\text{ cm}^2$

Practice

Looking down, a marble bowl has an 3 inch circular ring edge, with an outer diameter of 18 inches and an inner diameter of 12 inches. Find the surface area of the top bowl "ring" annulus.

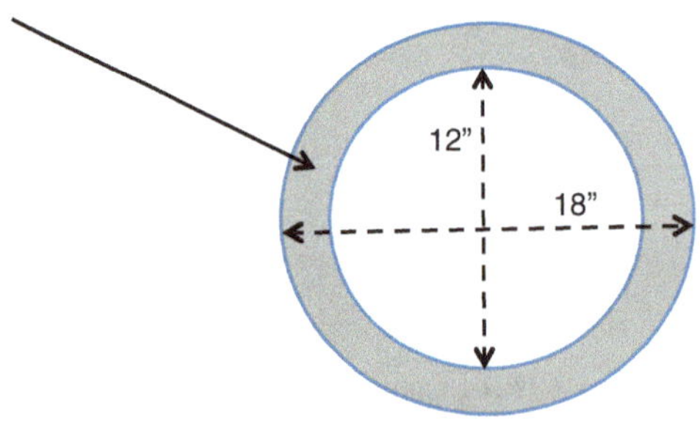

Find the area of the circle ring seal (annulus) from a canning jar, with outer radius of 12cm and inner radius of 10cm.

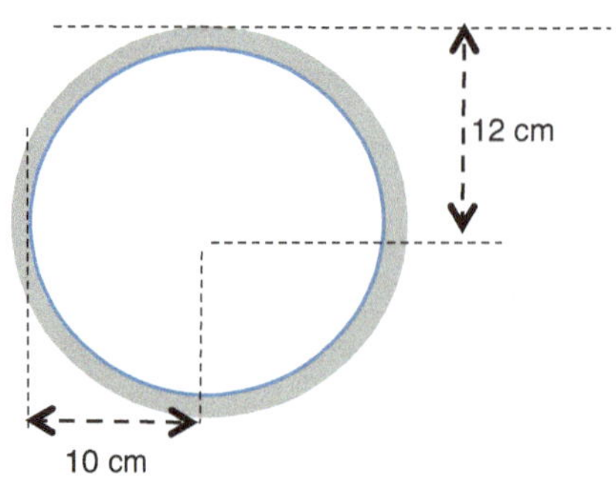

4-11 Exercises

Find the area of the following annuli

1. Outer circle diameter 5”
   Inner circle diameter 2.5”

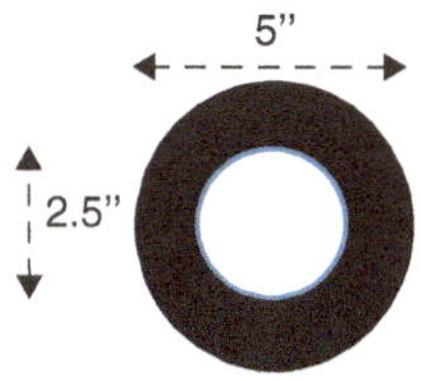

2. Outer circle radius 35 mm
   Inner circle radius 29 mm

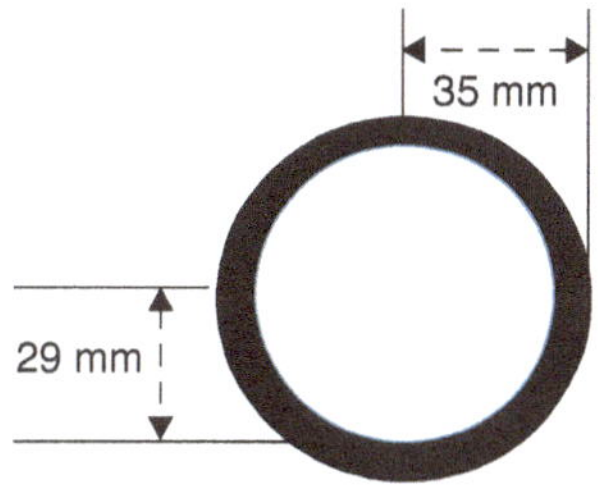

3. Outer circle radius 7.8 cm
   Thickness of the “ring” 3.4 cm

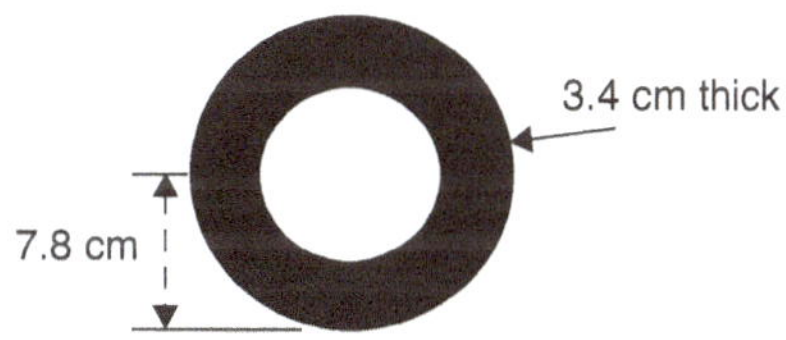

4. What is the area of an annulus for two concentric circles where one has a 47 km diameter and the other has a 11.8 km diameter?

5. If an annulus ring is 6 inches thick and the inner circle has a radius of 4 inches, what is the area of the annulus?

6. 

©Illizia/Shutterstock.com

A silver washer has an outer circle diameter of 4.6 cm and the inner circle of 2.6 cm. What is the flat area of the washer?

7. Looking down, a circular park area has a 5 feet wide circular path around it. If the outer edge of the path is a circle with a radius of 100 feet, find the surface area of the pathway.

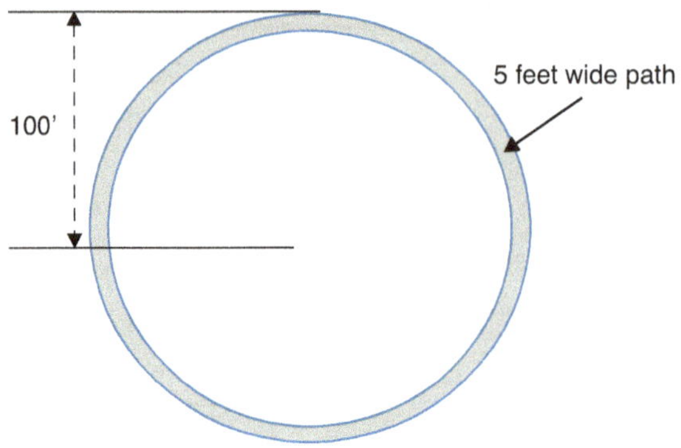

8. A circular moat surrounded a statue in the park according to the following.

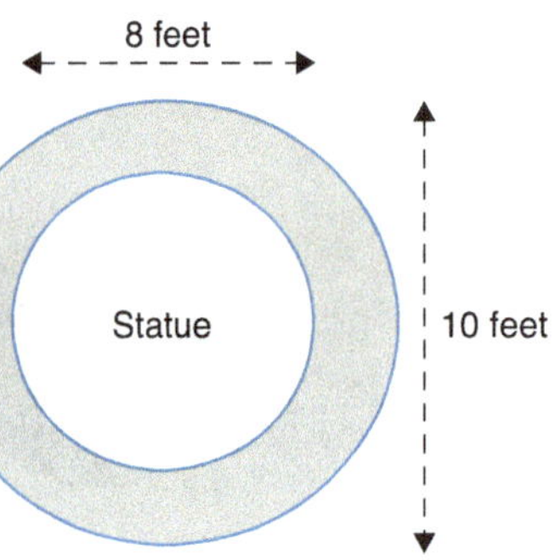

What is the area of the water in the moat?

Answers

1. 14.73 in^2
2. 1206.37 mm^2
3. 130.31 cm^2
4. 1625.59 km^2
5. 263.89 in^2
6. 11.31 cm^2
7. 3063.05 ft^2
8. 28.27 ft^2

CHAPTER 5

Three Dimensional Figures: Surface Area and Volume

5-1 SOLID GEOMETRIC DEFINITIONS AND SHAPES

Figures that are not contained in one plane are THREE-DIMENSIONAL figures. When a three-dimensional figure is enclosed it makes a SOLID figure.

The measurements we use most often with solid figures are **SURFACE AREA**, or the area making up the boundary or "wrapping" of the solid, and the **VOLUME**, the space inside the boundary of the solid figure. Surface area uses square units for measurement and volume uses cubic units for measurements.

Irregularly shaped solids are hard to measure unless we have higher math, such as calculus.

However, for a number of solid figures, because of their design and or construction, we can create formulas to measure surface area and volume, and these will prove useful as we work with solids in the industrial profession.

POLYHEDRONS

A POLYHEDRON is a solid three-dimensional enclosed figure whose surface boundary is made up of polygons. The word polyhedron comes from the Greek prefix *poly-*, which means "many," and the root word *hedron*, which refers to "surface."

Each polygon making up the surface boundary is called a FACE of the polyhedron. Each line segment where the faces meet is called an EDGE of the polyhedron. In addition, each corner point is called a VERTEX of the polyhedron.

For example, to help you identify faces, edges, and vertices, here is a polyhedron that has squares for faces. This is a special polyhedron called a CUBE.

There are an infinite number of types of polyhedrons, and doing measurements of a majority of these solid shapes either involve complicated formulas or calculating. However, we will be examining some special types of polyhedrons with some directly applicable formulas so finding volume and surface area is easy. In subsequent sections, special polyhedrons we will be looking at are PRISMS, PYRAMIDS, AND FRUSTUMS.

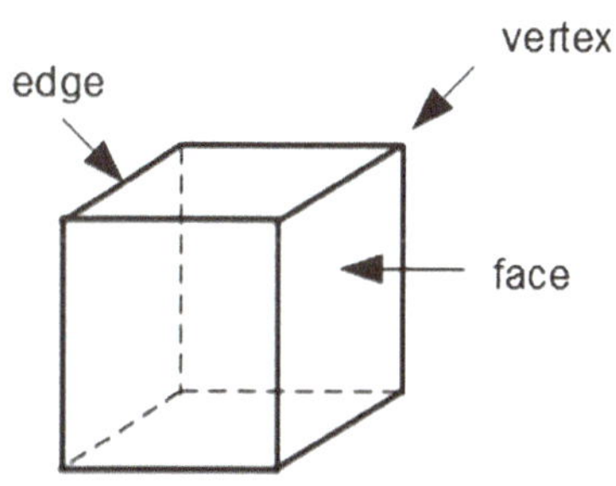

THREE-DIMENSIONAL FIGURES WITH CIRCULAR CONNECTIONS

We will also be looking at two types of enclosed figures with circular surfaces. A CYLINDER has two congruent circle bases and a CONE has one circular base. And another "circular" enclosed figure is the SPHERE, which is formed by all points equidistant from some point in space.

Their designs and constructions result in simple formulas to figure volume and surface area.

Cylinder

Cone

Sphere

5-1 Exercises

Identify which of these solid figure are polyhedrons

1. 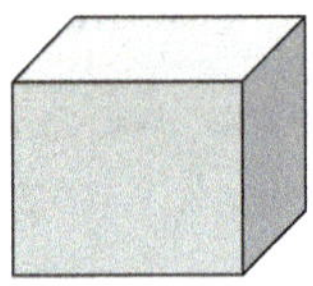

____

2. 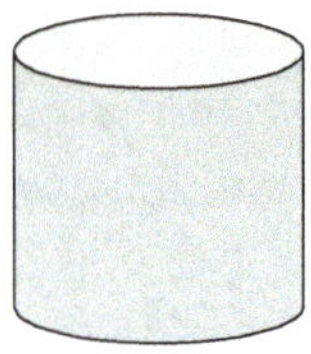

____

3. 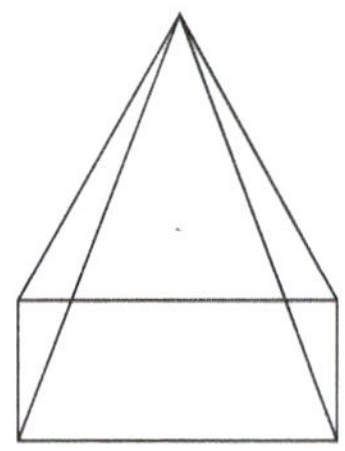

____

4. 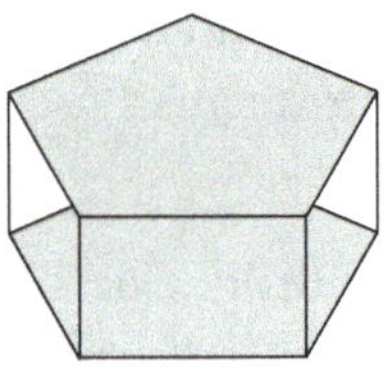

____

5. 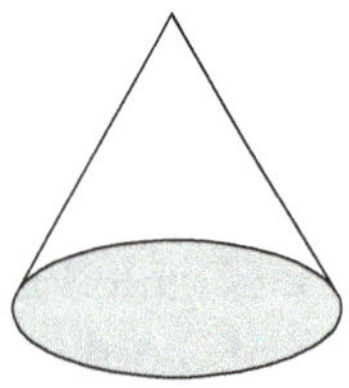

____

6. 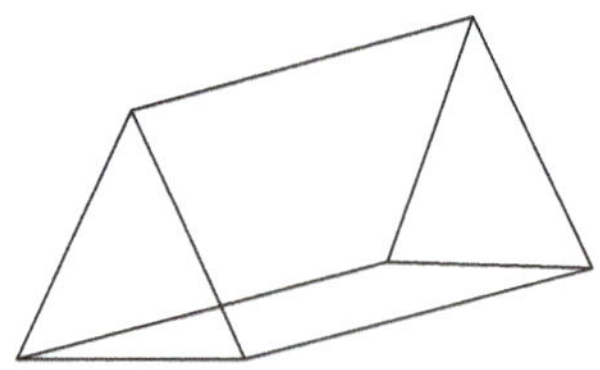

____

7. 

____

8. 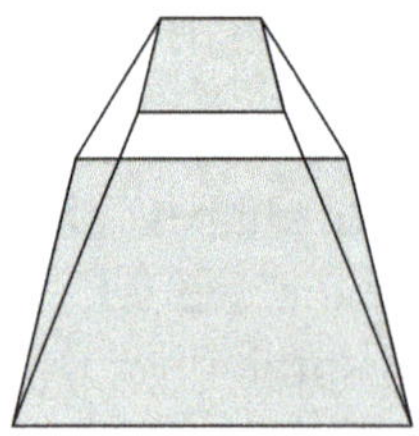

____

9. 

____

10. Identify the Face, Edge, Vertex, of this polyhedron.

a) ________

b) ________

c) ________

Answers

1. Polyhedron
2. Not a polyhedron
3. Polyhedron
4. Polyhedron
5. Not a polyhedron
6. Polyhedron
7. Not a polyhedron
8. Polyhedron
9. Polyhedron
10. a) Vertex

    b) Edge

    c) Face

5-2 PRISMS AND CYLINDER—VOLUME

A ***PRISM*** is a special polyhedron that has two parallel polygonal faces called bases. Each base is congruent. The other faces are always parallelograms. We say these faces make up the lateral part of the prism. A *prism* is named by the shape of its base. Here are two types of *prisms*.

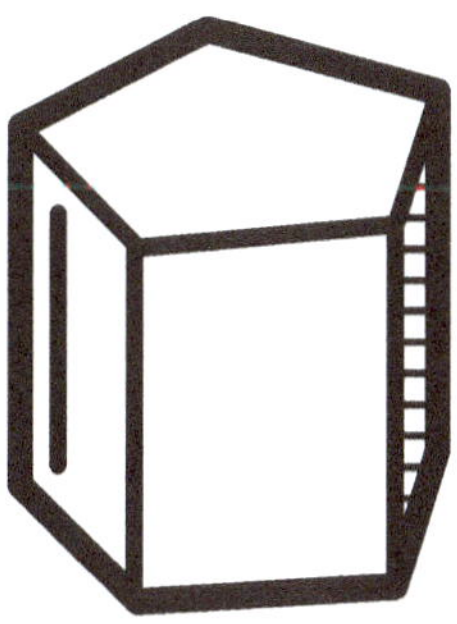

Prism

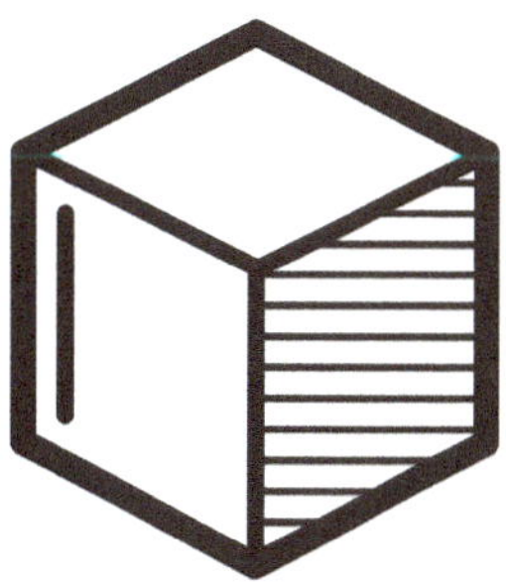

Cube

Below is a right rectangular prism. You can see the two rectangular bases. There are four other faces that are rectangular making up the lateral part of the prism.

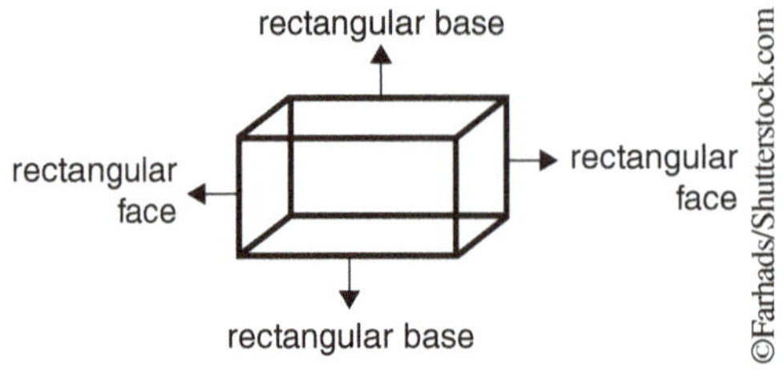

Here is a right hexagonal prism. It has two hexagonal bases and six rectangular lateral faces.

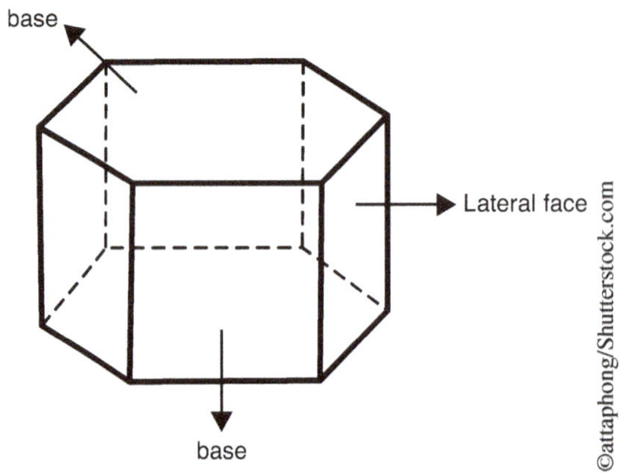

Hexagonal prism

When a prism is cut across parallel to its bases, it always has the same cross section that matches the polygonal base.

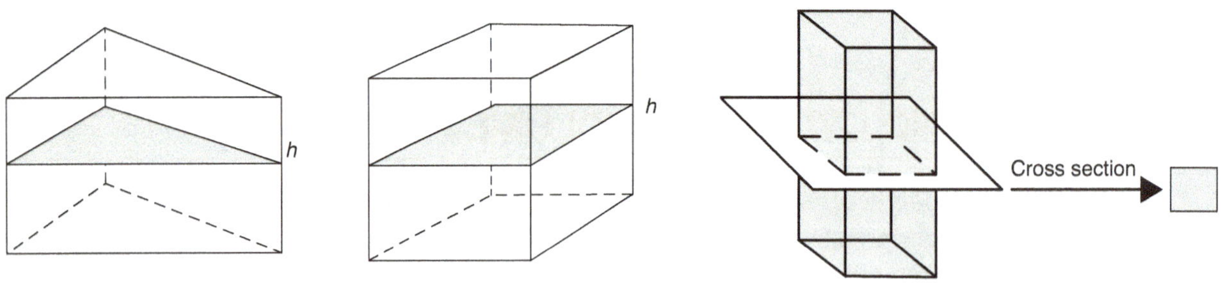

RIGHT VERSUS OBLIQUE

When the bases are right "on top of each other," we say the prism is a RIGHT PRISM. When it is a right prism, the lateral face edges are perpendicular with the base edges. And the lateral faces will be retangles.

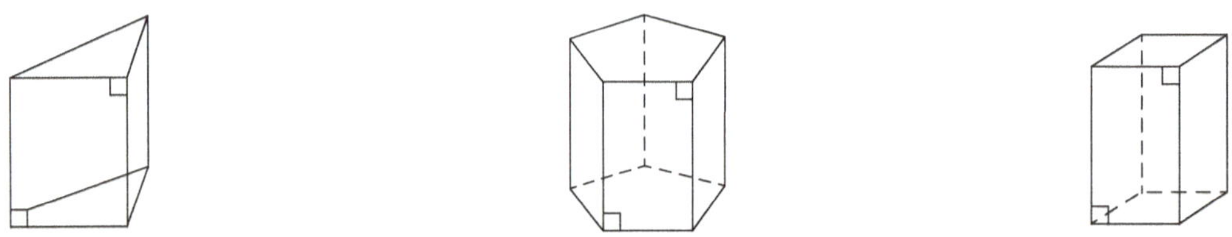

Right triangular prism   Right pentagonal prism   Right rectangular prism

When the bases are not "on top of each other," we say the prism is an OBLIQUE PRISM. The lateral edges do not meet at a right angle, and the lateral faces are parallelograms.

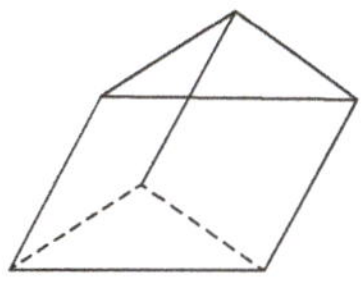

Oblique triangular prism

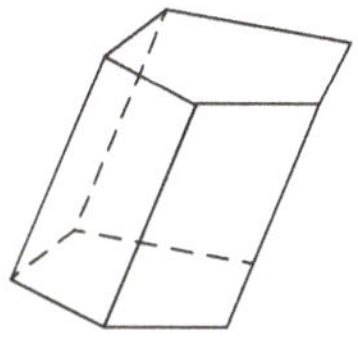

Oblique pentagonal prism

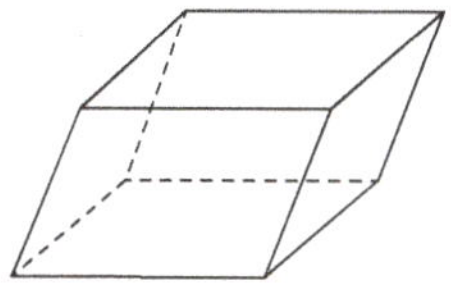

Oblique rectangular prism

CYLINDER

A CYLINDER is similar to a prism, in that, it has two parallel congruent bases, but in a cylinder, the bases are <u>circles.</u> The lateral part of the cylinder is curved (but when laid out is actually a parallelogram in shape!)

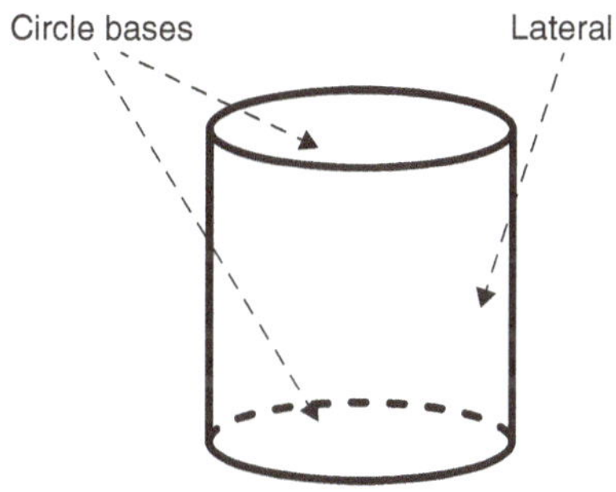

Cylinder

Like a prism, when a cylinder is cut across <u>parallel</u> to its bases, it always has the same cross section that matches the polygonal circular base.

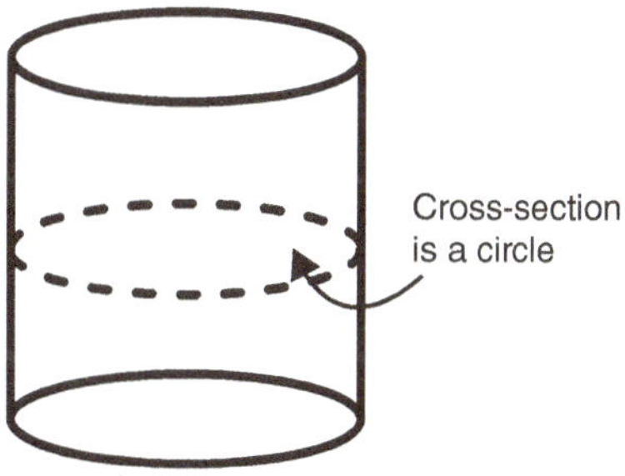

Cylinder

RIGHT VERSUS OBLIQUE

Like a right prism, when the bases are right "on top of each other," we say the cylinder is a RIGHT CYLINDER. In a right cylinder, a "segment" going from the centers of the two circle bases is perpendicular to the bases. The lateral part, when laid out, is a rectangle.

If the circle bases are not on "right on top of each other," then the cylinder is an OBLIQUE CYLINDER. The lateral part, when laid out, is a parallelogram.

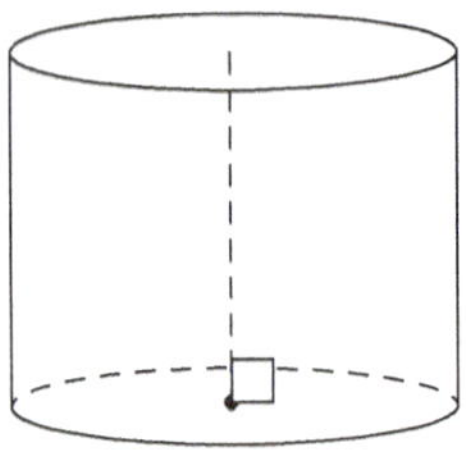

Right prism

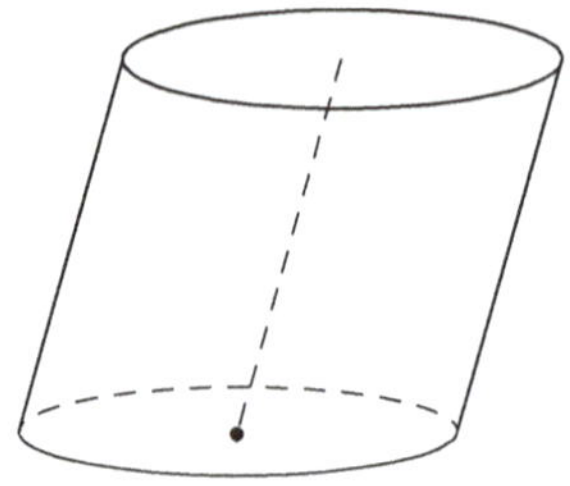

Oblique prism

VOLUME OF A PRISM

The height of the prism is how tall it is if it is sitting on one of its bases.

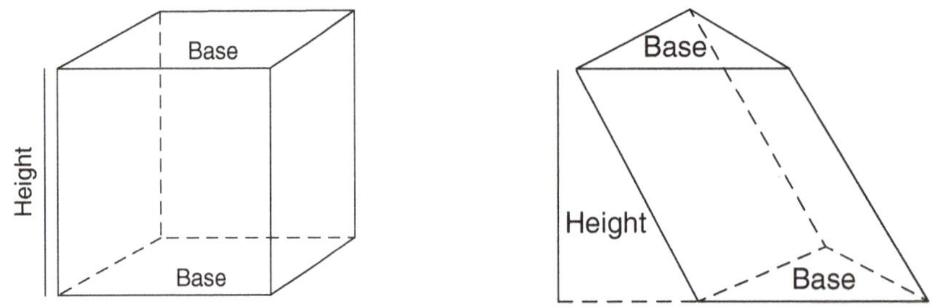

Given the base area of one of the bases, which we call B, and given the height of the prism, which we call h, the volume of any prism is

$$V = Bh$$

This volume formula works for both right and oblique prisms.

Example: Find the volume of a rectangular prism with a rectangle base that is 30 in. by 40 in. and a height of 70 in.

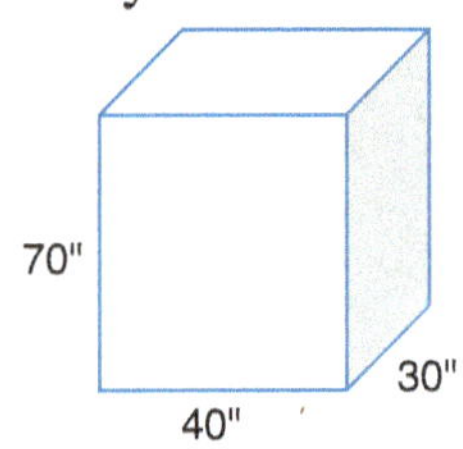

The rectangular base area is $B = 40 \text{ in.} \times 30 \text{ in.} = 1{,}200 \text{ in.}^2$ and the height $h = 70$ in.

Volume is $V = Bh = 1{,}200 \text{ in.}^2 \times 70 \text{ in.} = 84{,}000 \text{ in.}^3$

We can use this to find other volume measures.

What is this volume in cubic feet? If you recall, $1 \text{ ft.}^3 = 1728 \text{ in.}^3$

So $84{,}000 \text{ in.}^3 \cdot \dfrac{1 \text{ ft.}^3}{1{,}728 \text{ in.}^3} = 48.61 \text{ ft.}^3$

What is this volume in gallons? If you recall, $1 \text{ gal.} = 231 \text{ in.}^3$

So $84{,}000 \text{ in.}^3 \cdot \dfrac{1 \text{ gal.}}{231 \text{ in.}^3} = 363.64 \text{ gal.}$

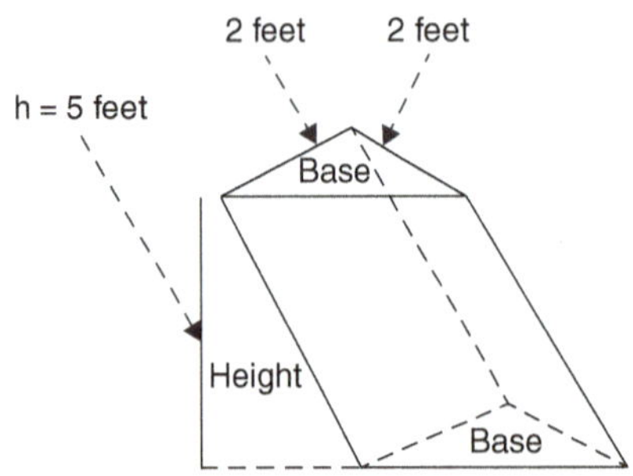

Example: Find the volume of an oblique triangular prism, with bases that are right triangles. The leg lengths of the right triangle base are each 2 ft. and a height of the prism of 5 ft.

The base area is .5 (leg length)(leg length) = .5 (2 ft.)(2 ft.) = 2 $\text{ft.}^2 = B$

$$\text{Volume} = Bh = 2\ \text{ft.}^2\ (5\ \text{ft.}) = 10\,\text{ft.}^3$$

VOLUME OF A CYLINDER

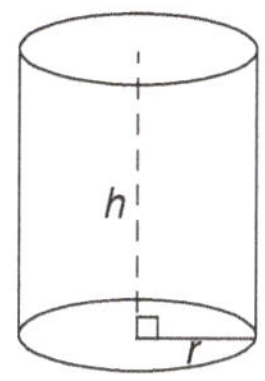

Just like a prism, the height of a cylinder is how tall the cylinder is when sitting on one of the circle bases. We also note the radius of the circle base, and this will be used in figuring the volume.

The volume of a cylinder is also the area of one of the base circle times the height. For a cylinder with base circles with radius of r, the area of one of the circle base is Area = πr^2 . So the volume is the base area time the height or

$$V = \pi r^2 h$$

This volume formula works for both right and oblique cylinders.

Example: Find the volume of a cylinder with radius of 4 in. and height of 8 in.

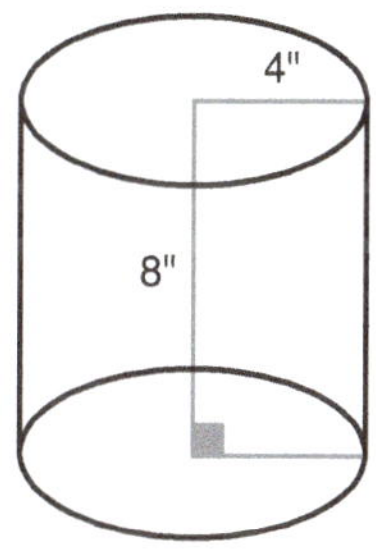

$$r = 4\ \text{in.},\ h = 8\ \text{in.}$$

$$V = \pi r^2 h = \pi \cdot (4\ \text{in.})^2 \cdot 8\ \text{in.} \approx 402.12\ \text{in.}^3$$

What is the volume of this cylinder in cubic feet?

$$1\ \text{ft.}^3 = 1728\ \text{in.}^3 \text{ so } 402.12\ \text{in.}^3 \cdot \frac{1\ \text{ft.}^3}{1{,}728\ \text{in.}^3} \approx .23\ \text{ft.}^3$$

What is the volume in gallons?

$$1\ \text{gal.} = 231\ \text{in.}^3 \text{ so } 402.12\ \text{in.}^3 \cdot \frac{1\ \text{gal.}}{231\ \text{in.}^3} \approx 1.74\ \text{gal.}$$

Example: Find the volume of the oblique cylinder with circle base radius of 5 cm and height of 15 cm.

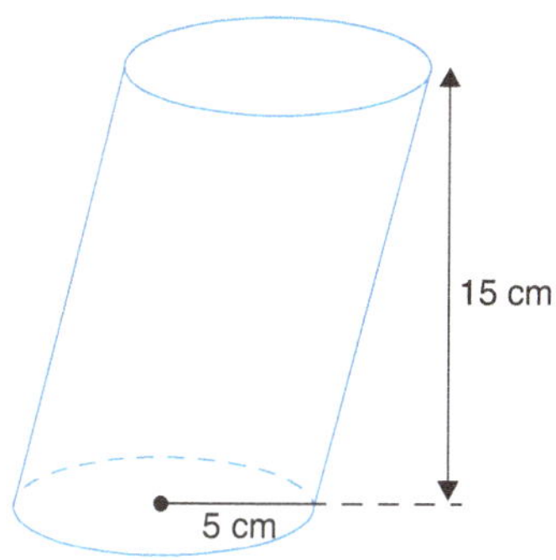

$R = 5$ cm, $h = 15$ cm

$$V = \pi r^2 h = \pi \cdot (5\ \text{cm})^2 \cdot 15\ \text{cm} \approx 1178.10\ \text{cm}^3$$

Practice

Find the volume of a right regular hexagonal prism with height of 6.8 in. and base side length of 3.6 in.

Answer: Let side length $s = 3.6$ in., and the area of the regular hexagonal base is equal to $2.598\,s^2 = 2.598\,(3.6\ \text{in.})^2 \approx 33.67\ \text{in.}^2 = B$

The height of the prism is = 6.8 in. = h

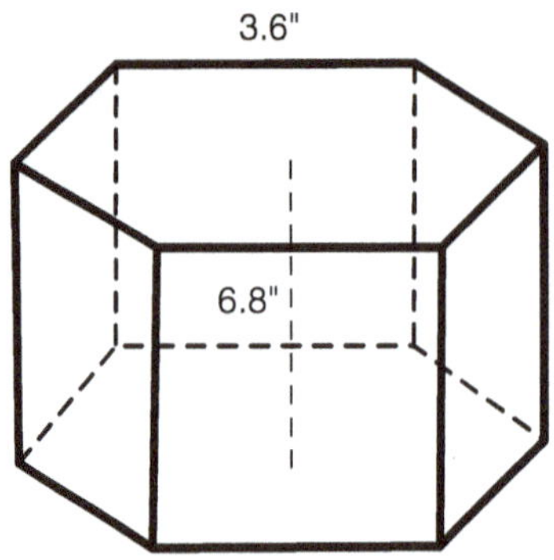

Hexagonal prism

$$V = Bh = (33.67 \text{ in.}^2)(6.8 \text{ in.}) \approx 228.96 \text{ in.}^3$$

Practice: Find the volume of this right cylinder. It has a circle base diameter of 10 ft. and a height of 21 ft.

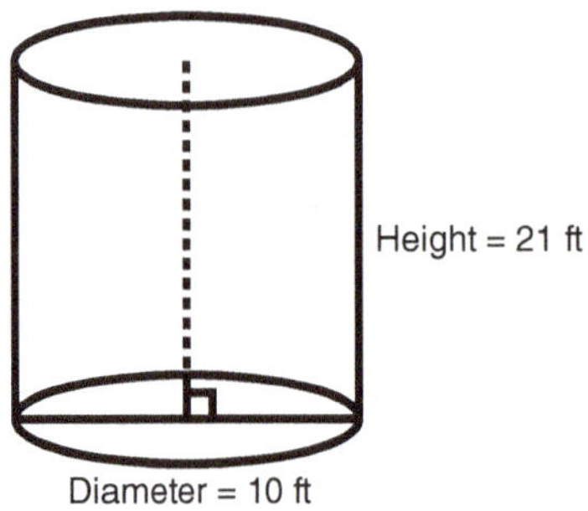

5-2 Exercises

Find answers to the nearest hundredth of a unit when appropriate.

Find the volume of the following prisms

1. 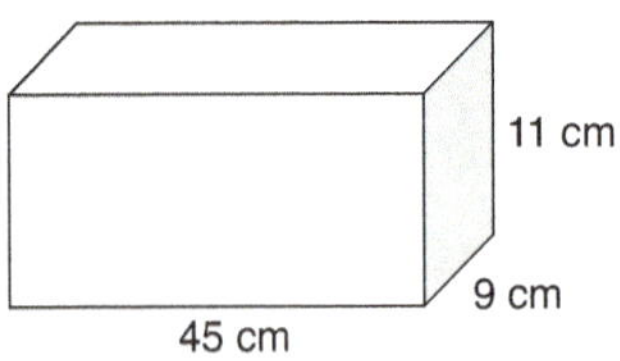

   Rectangular prism

2. 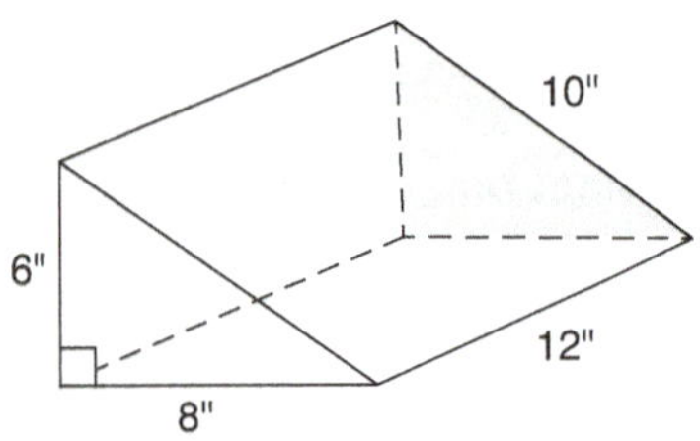

   Right triangular prism

3. 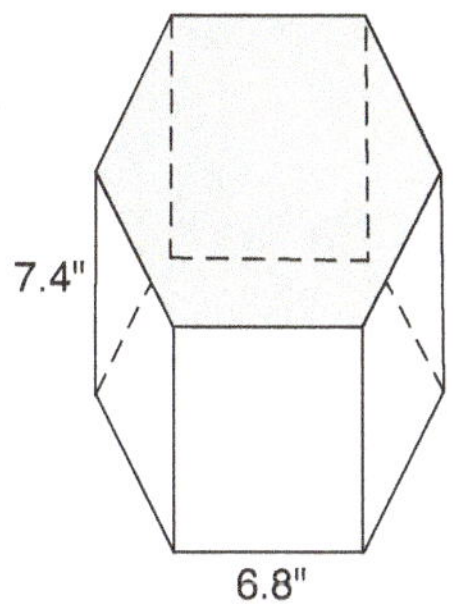

Regular hexagonal prism

Base hexagon side 6.8', height of prism is 7.4'

4. 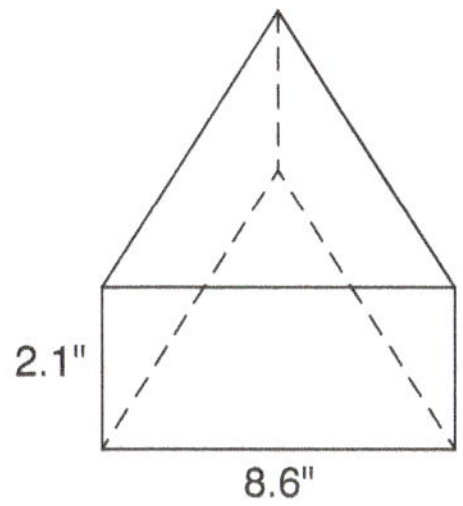

Equilateral triangular prism
Base Triangle side is 8.6", height of prism is 2.1"

5. 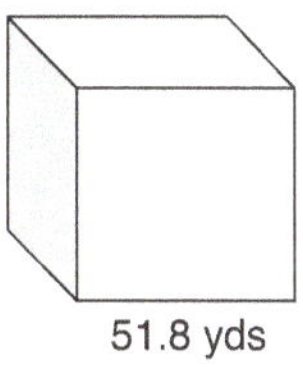

Cube

Find the volume of these cylinders 2"

6. 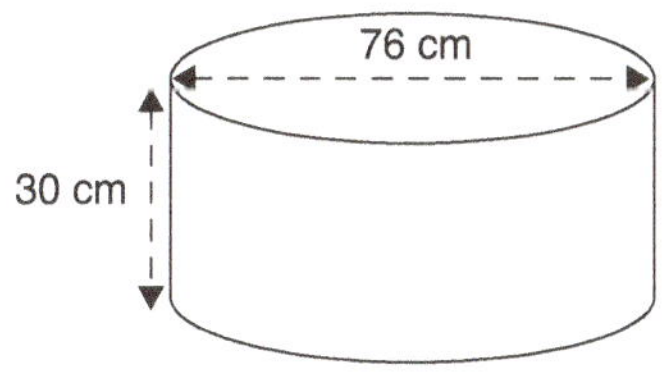

7. 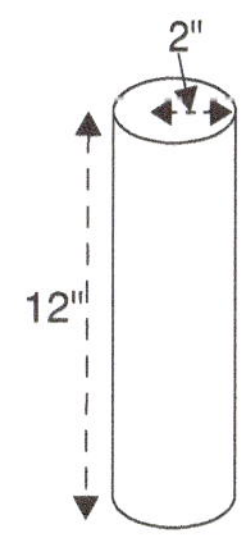

8. 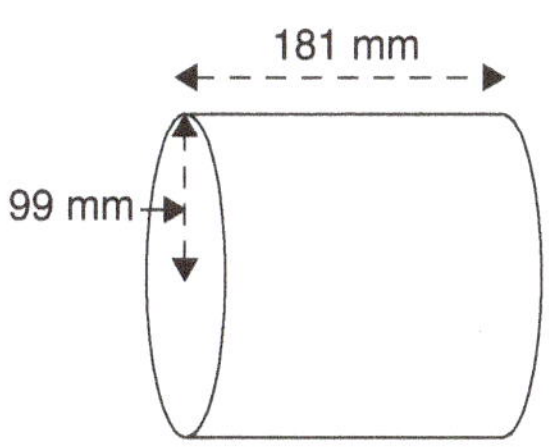

9. A shoe box has dimensions of 8" by 15" by 6". What is the volume of the box?

10. A cylindrical grain silo has a base radius of 20 feet and a height of 100 feet. What is the maximum volume of grain that can be put in the silo?

©XXXLphoto/ shutterstock.com

11. A metal nut is made as a regular octagonal prism with a base with side length of 3 cm and a height of 4 cm with a .5 cm radius cylinder hole thorough the prism. What is the volume of the nut?

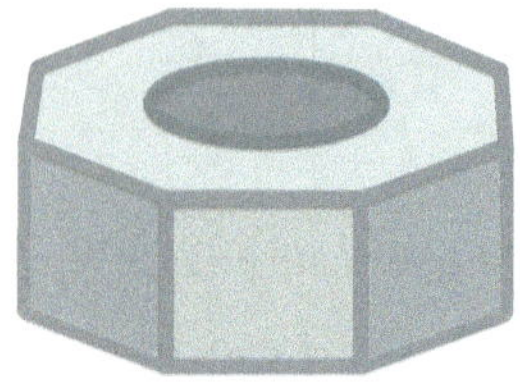

©vectorisland/ shutterstock.com

12. A storage cylinder of gasoline sits by John's business. If the cylinder's base has a diameter of 4 feet and a height of 7.5 feet, what is the maximum volume of gasoline that can fill the cylinder?

13. Cement was poured into a 120 yards by 15 yards driveway such that it was 6 inches thick. If cement cost \$25.00 per cubic yard, figure the total cost of cement that would be needed to complete the job.

©Stephenframe/ shutterstock.com

14. A special plastic machine part is designed as a cube attached to a cylinder. If the cube has an edge of 7"and the cylinder has a radius of 3 inches and a height if 4 inches, and if the plastic cost \$.05 per cubic inch, what is the total cost of plastic for the part?

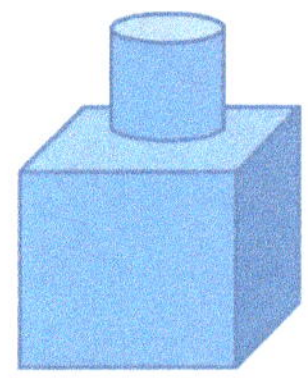

Answers

1. $45\text{ cm} \times 9\text{ cm} \times 11\text{ cm} = 4455\ \text{cm}^3$

2. $.5(6" \times 8") \times 12" = 288\ \text{in}^3$

3. $2.598\ (6.8')^2 \times 7.4' = 888.97\ \text{ft}^3$

4. $.433\ (8.6")^2 \times 2.1" = 67.25\ \text{ft}^3$

5. $(51.8\text{ yd})^3 = 138991.83\ \text{yd}^3$

6. $\pi\ (38\text{ cm})^2 \times 30\text{ cm} = 136093.79\ \text{cm}^3$

7. $\pi\ (2")^2 \times 12" = 150.80\ \text{in}^3$

8. $\pi\ (99\text{ mm})^2 \times 181\text{ mm} = 5573125.68\ \text{mm}^3$

9. $8" \times 15" \times 6" = 720\ \text{in}^3$

10. $\pi\ (20')^2 \times 100' = 125663.71\ \text{ft}^3$

11. Octagonal prism part volume = $4.828\ (3\ \text{cm})^2 \times 4\ \text{cm} = 173.81\ \text{cm}^3$; cylinder hole volume = $\pi\ (.5\ \text{cm})^2 \times 4\ \text{cm} = 3.14\ \text{cm}^3$; volume of nut = $173.81\ \text{cm}^3 - 3.14\ \text{cm}^3 = 170.67\ \text{cm}^3$

12. $\pi\ (2')^2 \times 7.5' = 94.25\ \text{ft}^3$

13. $120\ \text{yd} \times 15\ \text{yd} \times \frac{1}{6}\ \text{yd} = 300\ \text{yd}^3$; $300\ \text{yd}^3 \times \frac{\$25.00}{1\ \text{yd}^3} = \$7500.00$

14. $(7'')^3 + \pi(3'')^2 \times 4'' = 456.10\ \text{in}^3$; $456.10\ \text{in}^3 \times \frac{\$\ .05}{1\ \text{in}^3} = \$22.81$

5-3 SURFACE AREA OF PRISM AND CYLINDERS

Remember a "prism" is a polyhedron with two parallel congruent polygonal bases and lateral faces that are parallelograms. If it is a right prism, the lateral faces are rectangles. A cylinder is like a prism, except the two parallel bases are congruent circles. And the lateral part, which unfolded, is a parallelogram. If it is a right cylinder, the lateral part is a rectangle.

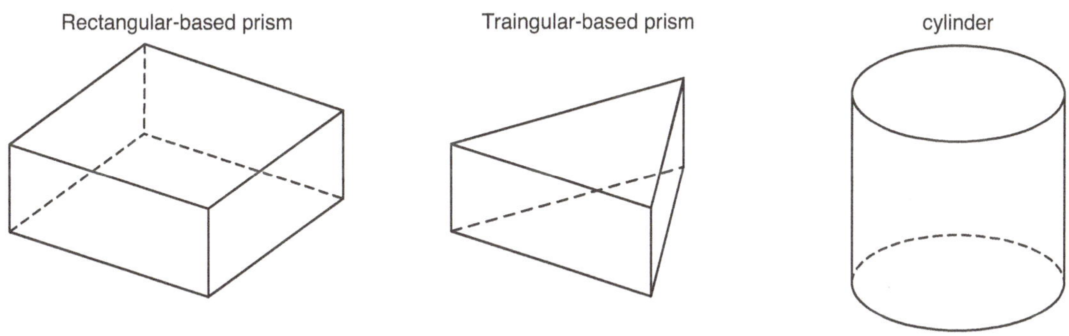

SURFACE AREA OF PRISMS

The surface area of any prism equals the sum of the areas of its faces. That means the surface area is the sum of the two base areas and the lateral faces area. As the bases in a prism are congruent so the base areas are equal, if we make B equal the base area and L.A. equal the total are of the lateral faces, then

$$\text{SURFACE AREA} = B + B + \text{L.A. or SURFACE AREA} = 2B + \text{L.A.}$$

To figure surface area, let's "unwrap" the surface boundary of a right rectangular prism. For example, given a prism with a base that is 6 cm by 4 cm and a height of 8 cm.

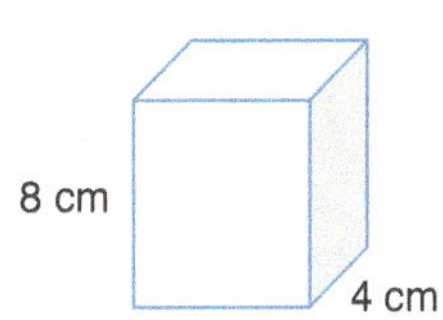

Here is what it looks like.

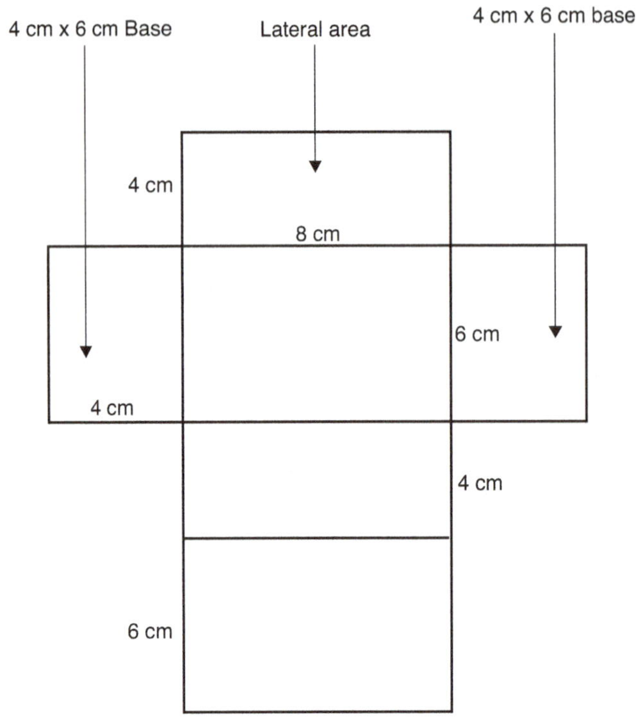

This arrangement of the surface area of a prism is called a NET.

The SURFACE AREA = $2B$ + L.A. The area of one base is 4 cm × 6 cm = $24\ \text{cm}^2$ so the area of the two bases is 2 ($24\ \text{cm}^2$) = $48\ \text{cm}^2$

The lateral area is made up of the sum of the areas of four rectangles, two with dimensions of 4 cm × 8 cm and two with dimensions of 6 cm × 8 cm. So the total lateral area = (4 cm × 8 cm) + (4 cm × 8 cm) + (6 cm × 8 cm) + (6 cm × 8 cm) = $32\ \text{cm}^2 + 32\ \text{cm}^2 + 48\ \text{cm}^2 + 48\ \text{cm}^2 = 160\ \text{cm}^2$.

So the SURFACE AREA = $2B$ + L.A. = $48\ \text{cm}^2 + 160\ \text{cm}^2 = 208\ \text{cm}^2$

Here's another way to look at the lateral area, which can be a more concrete way to figure it.

The four lateral rectangular faces make up one rectangle with a length of (4 cm + 6 cm + 4 cm + 6 cm) = 20 cm, and a width of 8 cm.

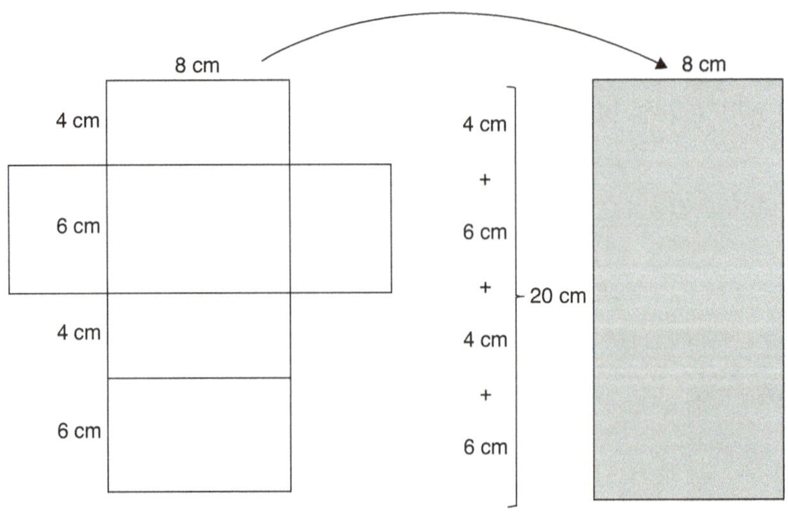

The lateral area = (4 cm + 6 cm + 4 cm + 6 cm) × 8 cm = 20 cm × 8 cm = 160 cm.

You should see that the sum (4 cm + 6 cm + 4 cm + 6 cm) = 20 cm is the **PERIMETER** of the base! And the 8 cm is the **HEIGHT** of the prism.

For any right prism, the L.A. = perimeter of the base × the height of the prism

In any right prism, the lateral area is always a rectangle with dimensions of the perimeter of the base by height, *h*, of the prism. So, if we have a right prism, we can use the formula for the surface area as

SURFACE AREA = 2(*B*) + (PERIMETER OF BASE) (*h*).

Example: Find the surface area of a right triangular prism with height of 10 cm, with a right triangular base with legs of 16 cm and 12 cm

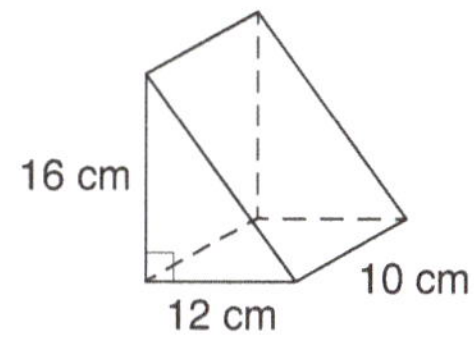

The area of one of the right triangle bases is .5 $(\text{leg length}_1)(\text{leg length}_2)$ = .5(16 cm)(12 cm) = 96 $\text{cm}^2 = B$

To find the Lateral area, we need the perimeter of the base. For the perimeter of the triangle base, we need the hypotenuse of the right triangle. Using the Pythagorean theorem, the hypotenuse of the right triangle is $\sqrt{16^2 + 12^2}$ = 20 cm

So the perimeter of the triangle base is 16 cm + 12 cm + 20 cm = 48 cm.

And height of the prism, h = 10 cm

So for the surface area of this prism

S.A. = 2(*B*) + (perimeter of base) (*h*) = $2(96\ \text{cm}^2) + (48\ \text{cm})(10\ \text{cm}) = 192\ \text{cm}^2 + 480\ \text{cm}^2 = 672\ \text{cm}^2$

SURFACE AREA OF CYLINDER

The surface area of a right cylinder works the same way as a right prism.

S.A. = 2(*B*) + (perimeter of base) (*h*).

Here is an unwrapping NET of a cylinder with radius *r* and height *h*.

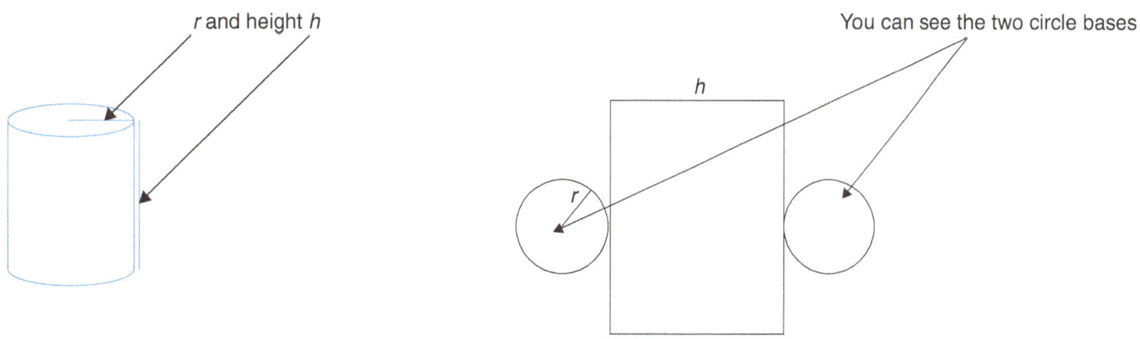

The rectangle lateral area has one side of length *h* and the other side is the "perimeter" of the circle base, or the circumference!

If the radius of the circle base is *r*, then the base area of one of the circle bases is $B = \pi r^2$, and the perimeter of the base is the circumference of the circle of = $2\pi r$

So the surface area of a cylinder with radius r and height h is

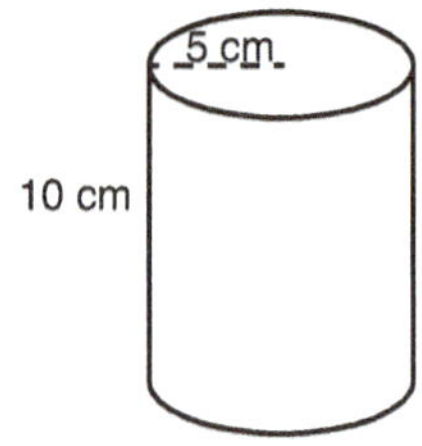

$$\text{SURFACE AREA} = 2(\pi r^2) + 2\pi rh$$

Find the surface area of this cylinder. $r = 5$ cm, $h = 10$ cm

$$\text{S.A.} = 2(\pi r^2) + 2\pi rh = 2(\pi)(5\text{ cm})^2 + 2\pi(5\text{ cm})(10\text{ cm}) =$$
$$157.08\text{ cm}^2 + 314.16\text{ cm}^2 = 471.24\text{ cm}^2$$

Practice

Find the surface area of a right regular hexagonal prism, with base edge length 2 ft. and height 5 ft.

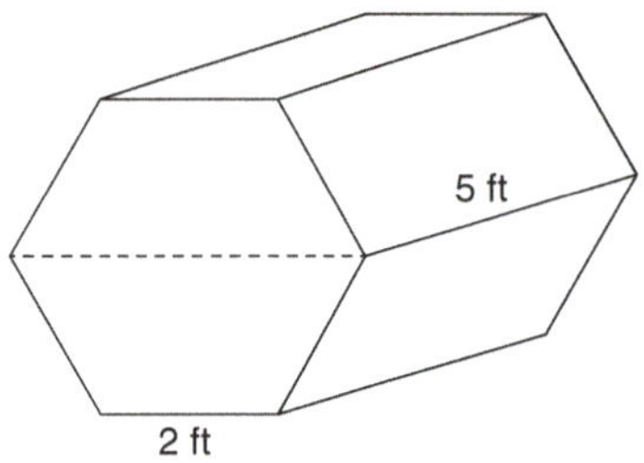

Find the surface area of a right cylinder with base diameter 7 in. and height 10 in.

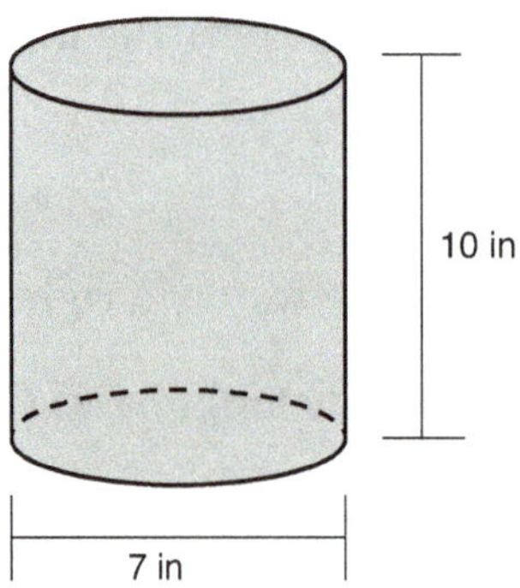

5-3 Exercises

Find answers to the nearest hundredth of a unit when appropriate.

Find the surface area of the following prisms

1. 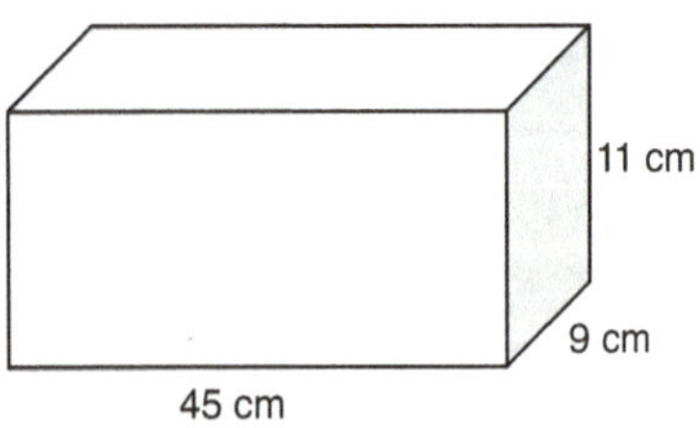

Rectangular prism

2.

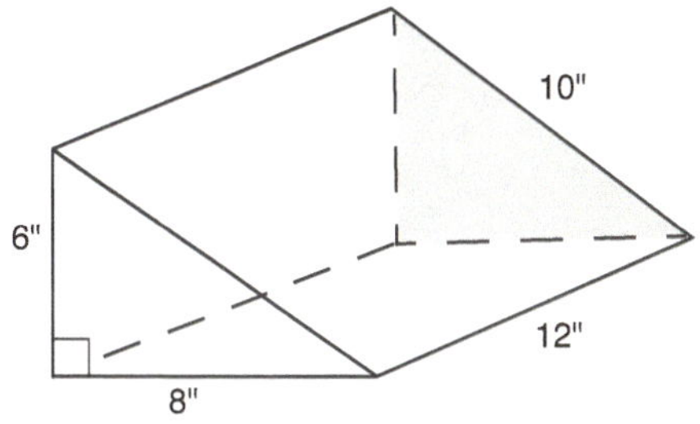

Right triangular prism

3.

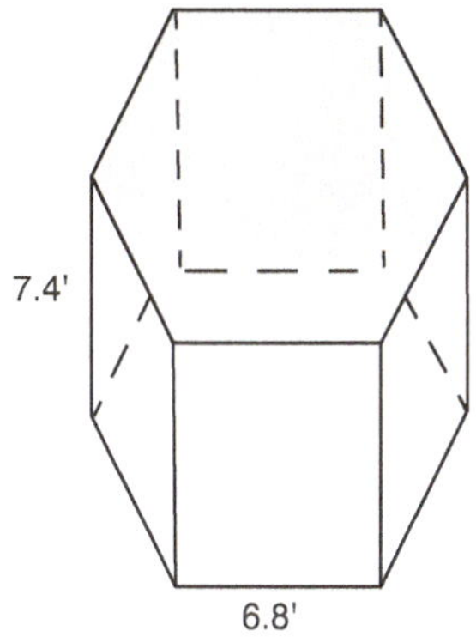

Regular Hexagonal prism
Base hexagon side 6.8', height of prism is 7.4'

4.

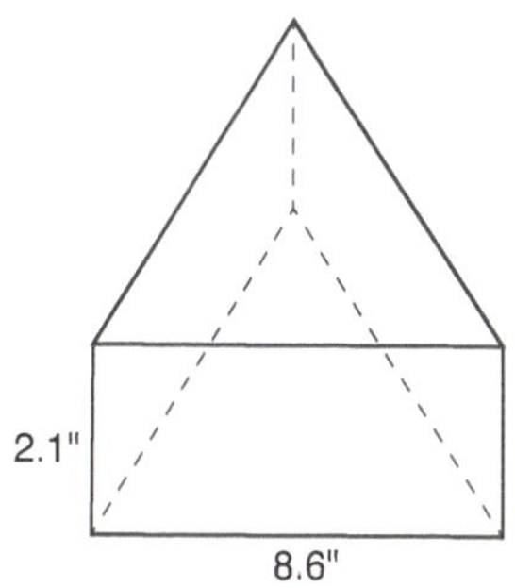

Equilateral triangular prism
Base Triangle side is 8.6", height of prism is 2.1"

5.

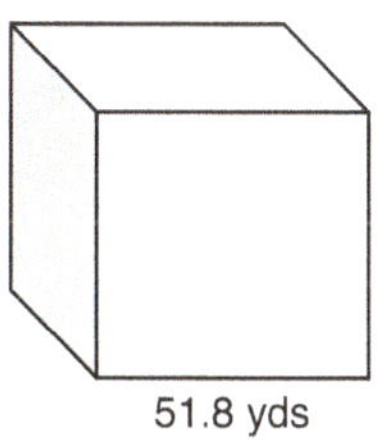

Cube

Find the surface area of these cylinders

6.

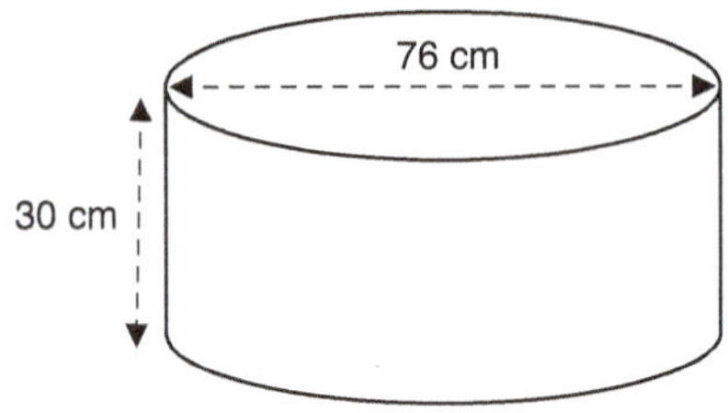

7.

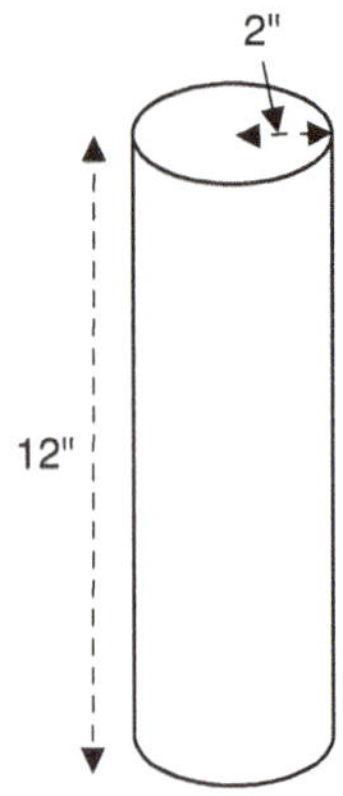

8.

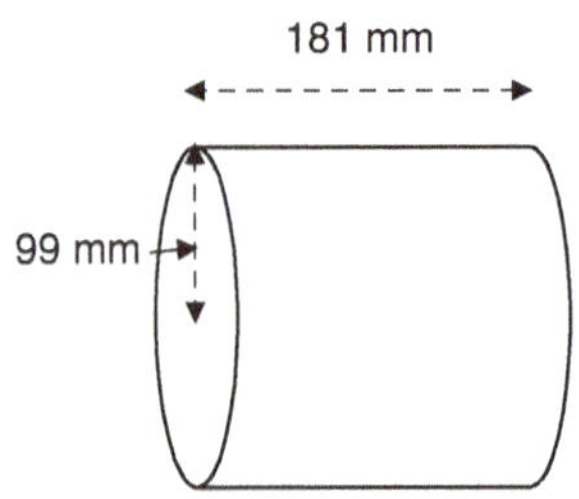

9. A rectangular suitcase has dimensions of 4' by 2.5' by 1.5'. What is the surface area of the suitcase?

10. A metal cylindrical storage can has a base radius of 2.5 feet and a height of 6 feet. What is the square footage of metal needed to make the can?

©3DMAVR/Shutterstock.com

11. This business sign is a regular hexagonal prism with a base side length of 3.5 feet and a thickness of 9 inches. What is the surface area of the sign?

©Who is Danny/Shutterstock.com

12. A Water tower that is cylindrical has a base diameter of 12.1 meters and a height of 30 meters. The tower is painted on its bases and lateral area. How many square meters of paint is needed to complete the job?

13. A popular candy manufacture designed an equaliarteral triangular prism package for a new candy bar. The triangle base side is 4 inches and the package is 12 inches long. A special cellophane covering is put on the package. How much surface area of celophane is need for a package?

14. A special plastic machine part is designed as a cube attached to a cylinder. If the cube has an edge of 7” and the cylinder has a radius of 3 inches and a height if 4 inches, What is the surface area of the part?

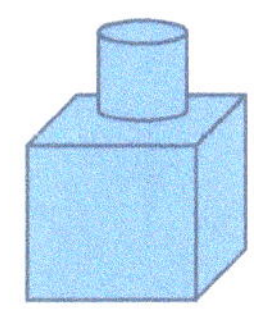

Answers

1. $2(45 \text{ cm} \times 9\text{cm}) + 2(45 \text{ cm} + 9 \text{ cm})(11 \text{ cm}) = 1998 \text{ cm}^2$
2. $2(.5 \times 6" \times 8") + (6" + 8" + 10")(12") = 336 \text{ in}^2$
3. $2(2.598)\ (6.8')^2 + 6(6.8')(7.4') = 542.18 \text{ ft}^2$
4. $2(.433)\ (8.6")^2 + 3(8.6")(2.1') = 118.23 \text{ in}^2$
5. $6(51.8 \text{ yd})^2 = 16099.44 \text{ yd}^2$
6. $2[\pi \times (38 \text{cm})^2] + \pi \times 76 \text{ cm} \times 30 \text{ cm} = 16235.75 \text{ cm}^2$
7. $2[\pi \times (2")^2] + \pi \times 4" \times 12" = 175.93 \text{ in}^2$
8. $2[\pi \times (99 \text{ mm})^2] + \pi \times 198 \text{ mm} \times 181 \text{ mm} = 174169.90 \text{ mm}^2$
9. $2(4' \times 2.5') + 2(4' + 2.5')\ (1.5') = 39.5 \text{ ft}^2$
10. $2[\pi \times (2.5')^2] + \pi\ x\ 5'\ x\ 6' = 133.52 \text{ ft}^2$
11. $2(2.598)\ (3.5')^2 + 6(3.5')(.75') = 79.40 \text{ ft}^2$
12. $2[\pi \times (6.05 \text{ m})^2] + \pi \times 12.1 \text{ m} \times 30 \text{ m} = 1370.38 \text{ m}^2$
13. $2(.433)\ (4")^2 + 3(4')(12") = 157.86 \text{ in}^2$
14. Cube surface area $= 6(7")^2 - \pi(3")^2 = 265.73 \text{ in}^2$

    Cylinder surface area $= \pi\ (3")^2 + \pi \times 6" \times 4" = 103.67 \text{ in}^2$

    Total surface area $= 265.73 \text{ in}^2 + 103.67 \text{ in}^2 = 369.40 \text{ in}^2$

5-4 SPHERES

A ***SPHERE*** is a three-dimensional figure that is made up of the set of points in space equidistant from a point called the *center*. of the sphere. The distance from the center to the points on the sphere is called the *radius* of the sphere. The center point defines the starting point of the sphere, and the radius defines the position of all the spherical points about the center.

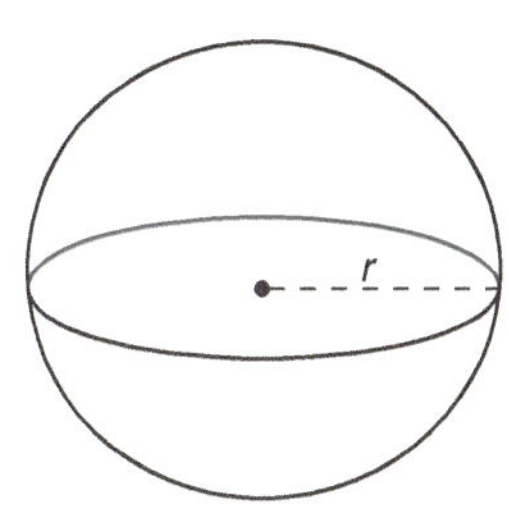

A cross section of the sphere that goes through the center of the sphere will be circular is shape. It is called a GREAT CIRCLE of the sphere. It has the same radius as the sphere.

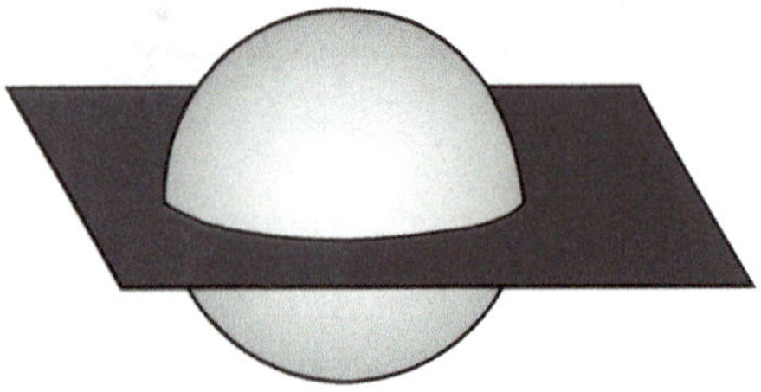

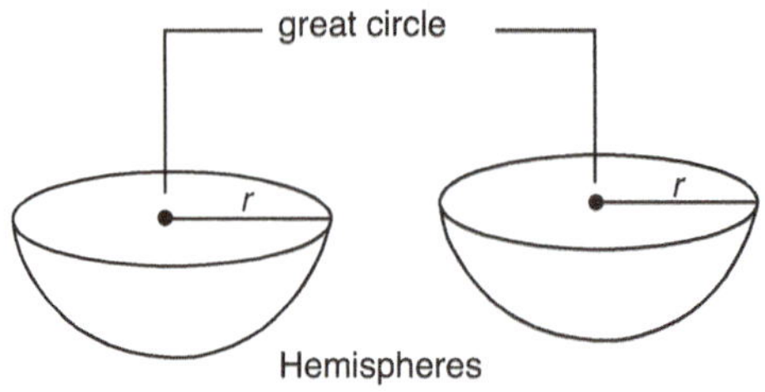

Cross sections of the sphere that are parallel to one of the great circle cross sections will also be circular but with the smaller radii.

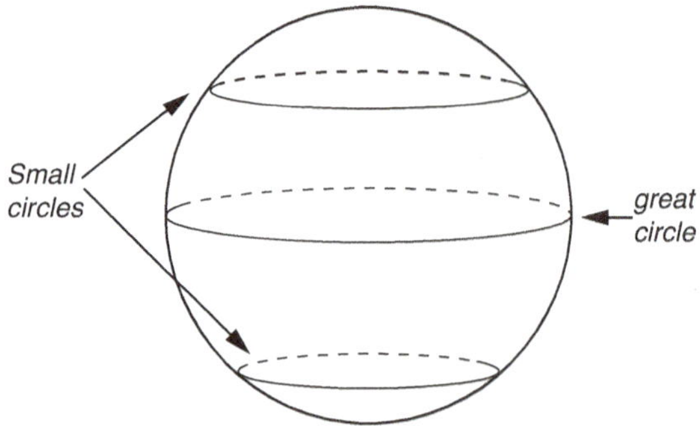

VOLUME AND SURFACE AREA OF SPHERES

The volume of a sphere with radius r can be found using the formula

$$V = \frac{4}{3}\pi r^3$$

And the surface area of a sphere with radius r is

$$\text{S.A.} = 4\pi r^2$$

Find the volume and surface area of this sphere

$$\text{Volume} = \frac{4}{3}\pi r^3 = \frac{4}{3}\pi(30\text{ m})^3 = 113097.64\text{ m}^3$$

$$\text{Surface area} = 4\pi r^2 = 4\pi(30\text{ m})^2 = 11309.73\text{ m}^2$$

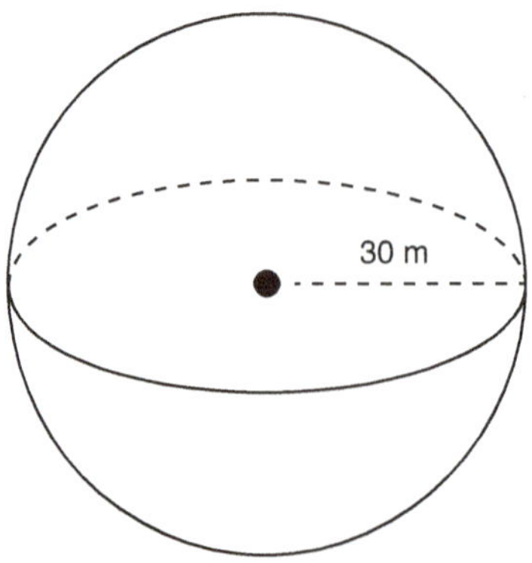

Practice: Find the volume and surface area of a sphere with diameter of 38 ft.

5-4 Exercises

Find answers to the nearest hundredth of a unit, when appropriate.

Find a) the surface area b) the volume of these spheres.

1.

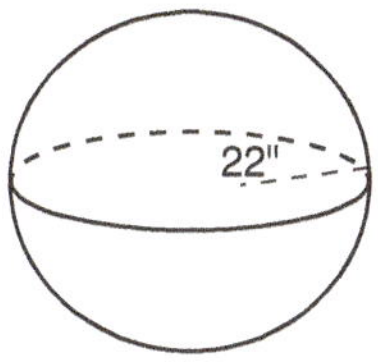

2.

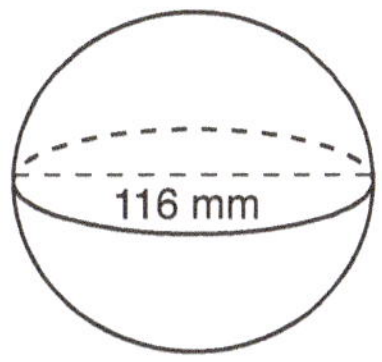

3.

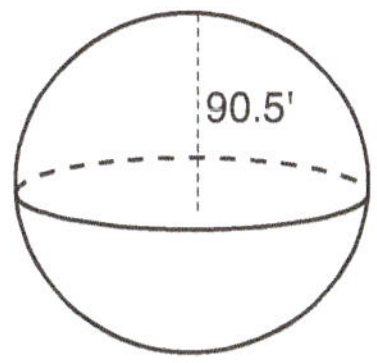

4. A sphere has a radius of 7.9 inches. Find the a) surface area and b) volume.

5. A sphere has a diameter of 3 meters. Find the a) surface area and b) volume.

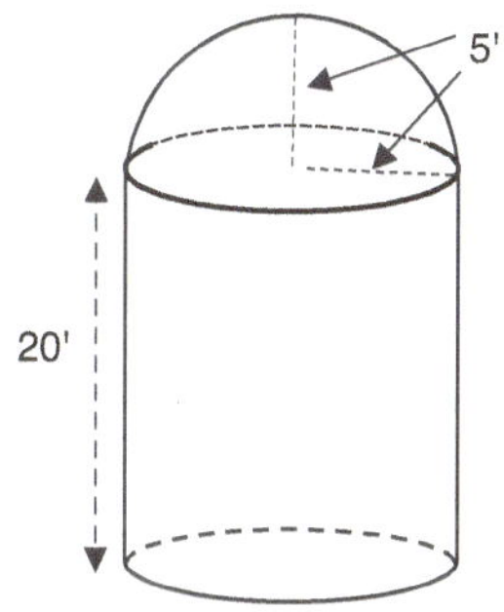

6. A steel water container is constructed with as a hemisphere on a right cylinder.

Given these measurements, find a) the volume and b) the square footage of metal needed to construct it (note, the upper circle cylinder base/hemisphere circle base is not part of the surface area.)

7. An artist created a sculpture that has three spheres on top of a cube. If the radii of the three spheres are 25 cm, 15 cm, and 8 cm and the cube has an edge of 140 cm,

©wacoma/Shutterstock.com

a) find the overall volume of the statute. ______________

b) if the sculpture was completely painted, how many cm^2 of paint would be needed (that is, find the surface area of the spheres and the cube.) _________________

8. A plastic plug is designed as a hemisphere with 2 congruent cylinders at the base.

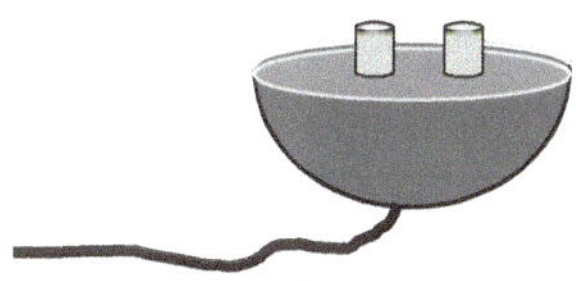

©NPeter/Shutterstock.com

If the radius of the hemisphere is 5 cm and the cylinders has a radii of .5 cm and height of 1.5 cm. Find the a) volume of the plug and b) the surface area of the plug.

9. The earth is primarily spherical in shape. It has a diameter of about 7972 miles.

Find the approximate a) surface area in square feet and b) volume in cubic feet of the earth.

Answers

1. a) $4 \times \pi \times (22\text{ in})^2 = 6082.12\text{ in}^2$ b) $\frac{4}{3} \times \pi \times 22^3 = 44602.24\text{ in}^3$

2. a) $4 \times \pi \times (58\text{ mm})^2 = 42273.27\text{ mm}^2$ b) $\frac{4}{3} \times \pi \times (58\text{ mm})^3 = 817283.23\text{ mm}^3$

3. a) $4 \times \pi \times (90.5')^2 = 102921.72\text{ ft}^2$ b) $\frac{4}{3} \times \pi \times (90.5')^3 = 3104805.13\text{ ft}^3$

4. a) $4 \times \pi \times (7.9'')^2 = 784.27\text{ in}^2$ b) $\frac{4}{3} \times \pi \times (7.9'')^3 = 2065.24\text{ in}^3$

5. a) $4 \times \pi \times (1.5\text{ m})^2 = 28.27\text{ m}^2$ b) $\frac{4}{3} \times \pi \times (1.5\text{ m})^3 = 14.14\text{ m}^3$

6. a) Cylinder vol = $\pi \times (5')^2 \times 20' = 1570.80\text{ ft}^3$
   Hemisphere vol = $.5[\frac{4}{3} \times \pi \times (5')^3] = 261.80\text{ ft}^3$
   Total volume = $1570.80\text{ ft}^3 + 261.80\text{ ft}^3 = 1832.60\text{ ft}^3$
   b) Surface area of cylinder = $\pi \times (5')^2 + (\pi \times 2 \times 5' \times 20') = 706.86\text{ ft}^2$
   Surface area of hemisphere = $.5[4 \times \pi \times (5')^2] = 157.08\text{ ft}^2$
   total surface area = $706.86\text{ ft}^2 + 157.08\text{ ft}^2 = 863.94\text{ ft}^2$

7. a) Volume of the spheres = $\frac{4}{3} \times \pi \times (25\text{ cm})^3 + \frac{4}{3} \times \pi \times (15\text{ cm})^3 + \frac{4}{3} \times \pi \times (8\text{ cm})^3$ $= 81731\text{ cm}^3$; volume of cube = $(140\text{ cm})^3 = 2744000\text{ cm}^3$
   total volume = $81731\text{ cm}^3 + 2744000\text{ cm}^3 = 2825731.67\text{ cm}^3$
   b) Surface area of spheres = $4 \times \pi \times (25\text{ cm})^2 + 4 \times \pi \times (15\text{ mm})^2 + 4 \times \pi \times (8\text{ cm})^2$ $= 11485.66\text{ cm}^2$; surface area of cube = $6\,(140\text{cm})^2 = 11760\text{ cm}^2$
   Total surface area = $11485.66\text{ cm}^2 + 11760\text{ cm}^2 = 129085.66\text{ cm}^2$

8. a) Volume of hemisphere = $.5\left[\frac{4}{3} \times \pi \times (5\text{ cm})^3\right] = 261.80\text{ cm}^3$
   volume of two cylinders = $2\,[(\pi \times (.5\text{ cm})^2 \times 1.5\text{ cm}] = 2.36\text{ cm}^3$
   total volume = $261.80\text{ cm}^3 + 2.36\text{ cm}^3 = 264.16\text{ cm}^3$
   b) Surface area of hemisphere = $.5\,[4 \times \pi \times (5\text{ cm})^2] - 2[\pi \times (.5\text{ cm})^2] = 155.51\text{ cm}^2$
   Surface area of the two cylinders = $2[(\pi \times (.5\text{ cm})^2) + (\pi \times 2 \times .5\text{ cm} \times 1.5\text{ cm})]$ $= 11.00\text{ cm}^2$; total surface area = $155.51\text{ cm}^2 + 11.00\text{ cm}^2 = 166.51\text{ cm}^2$

9. Radius of earth is 3986 miles
   a) Surface area = $4 \times \pi \times (3986\text{ mi})^2 = 199656959.3$ sq. miles or = $1.997 \times 10^8\text{ miles}^2$
   b) Volume = $\frac{4}{3} \times \pi \times (3986\text{ mi})^3 = 265277546000$ cu. miles or = $2.653 \times 10^{11}\text{ miles}^3$

5-5 PYRAMIDS AND CONES

PYRAMIDS

A PYRAMID is a three-dimensional polyhedron that has one polygonal base and lateral faces that are triangles. The triangle faces start at the base then meets at a single point, called the APEX of the pyramid.

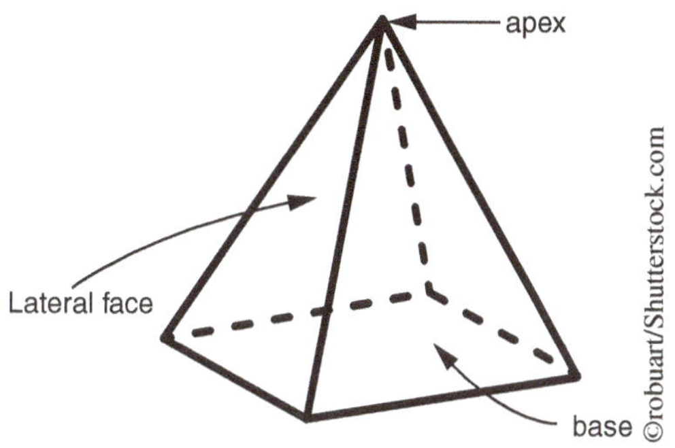

Example

The pyramid takes its name from the polygon bases.

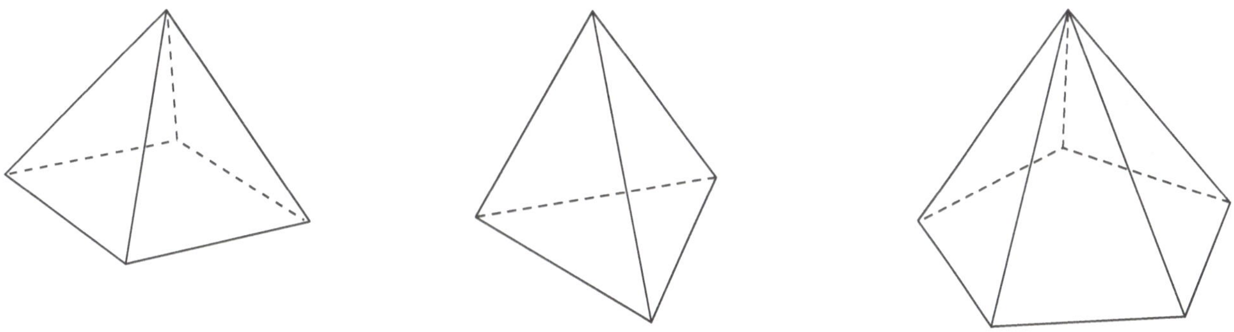

Rectangular pyramid Triangular pyramid Pentagonal pyramid

A right pyramid has the apex directly over the true center of the base. In an oblique pyramid, the apex is not directly over the center of the base. The height of the pyramid is the distance from the apex to the base plane.

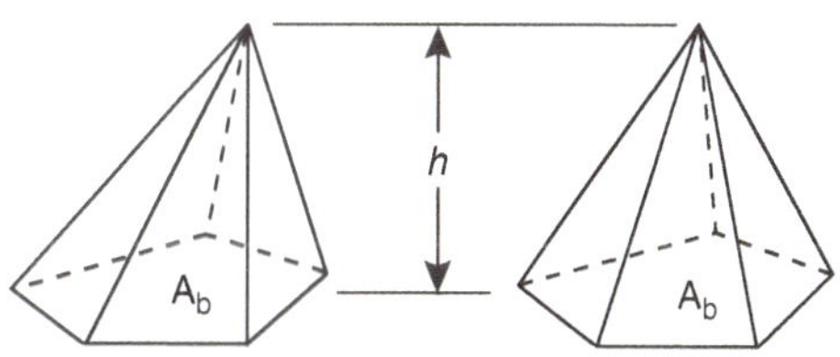

Oblique pyramid Right pyramid

A REGULAR PYRAMID has a regular polygon for a base.

VOLUME OF A PYRAMID

Given a prism with base that has area B, height h, and a pyramid with a congruent base area of B and the same height h, it is true that the prism is always **three times** the volume as the pyramid.

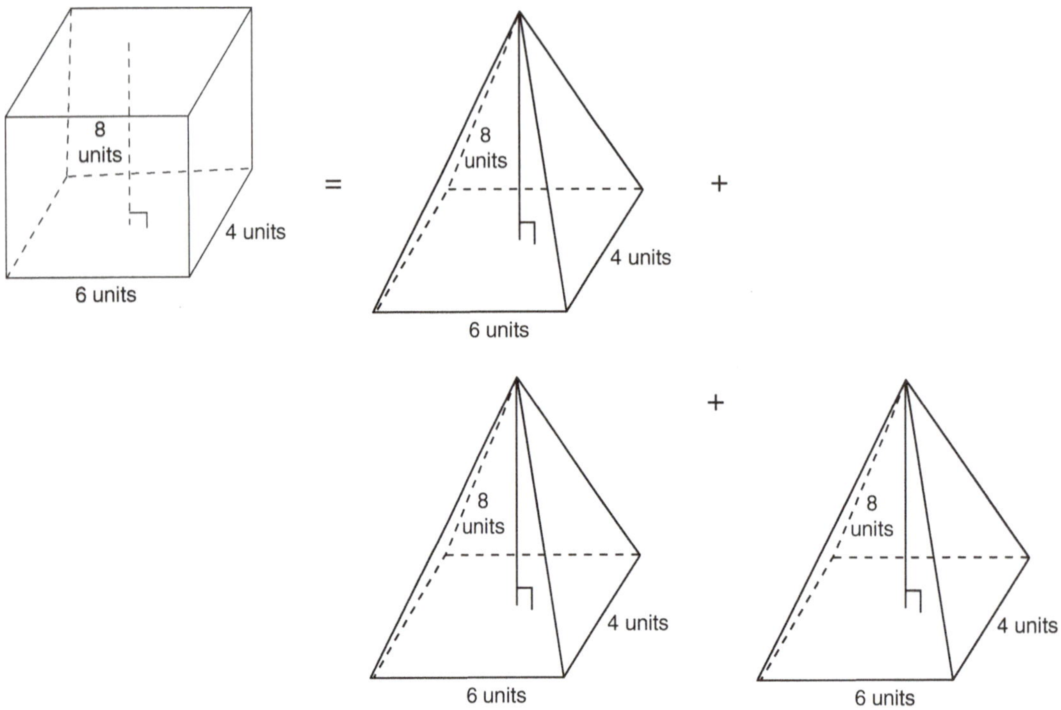

As the formula of a prism with base area B and height h is $V = Bh$

Then it follows that the volume of a pyramid with base area B and height h is

$$V_{\text{pyramid}} = \frac{1}{3} Bh$$

Find the volume of a rectangular pyramid with a base having a length of 8in. and a width of 7in. and the pyramid height is 6 in.

$$V_{\text{pyramid}} = \frac{1}{3} Bh = \frac{1}{3} (8'' \cdot 7'')(6'') = 112 \text{ in.}^3$$

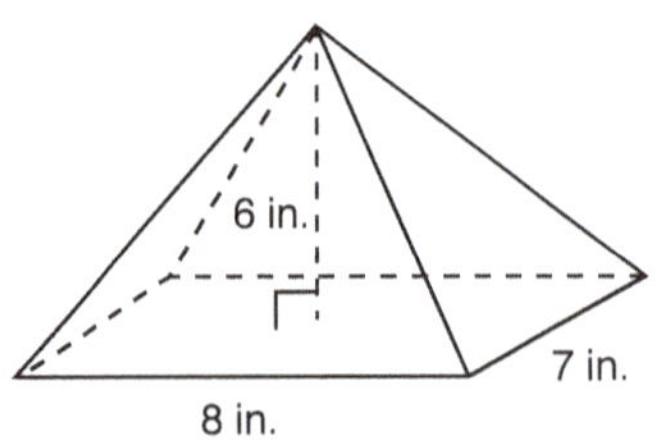

CONE

A CONE is a three-dimensional figure that is similar to a pyramid, but instead of one polygonal base, the cone has a circle base. The lateral part is curved up from the base to a point, called the apex of the cone.

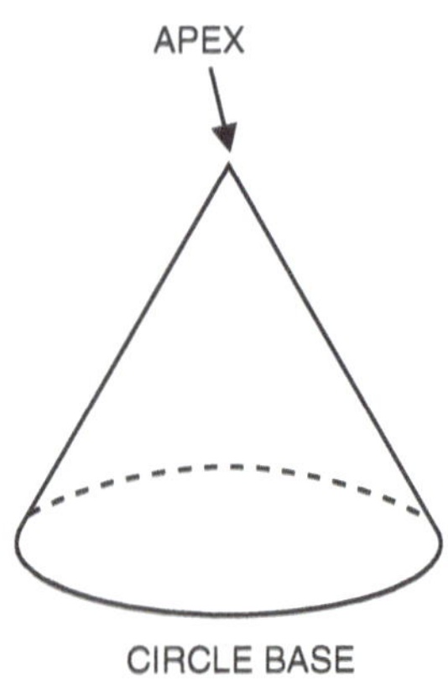

A right cone has the apex directly over the true center of the circle base. In an oblique cone, the apex is not directly over the true center of the circle base. The height of the cone is the distance from the apex to the base plane.

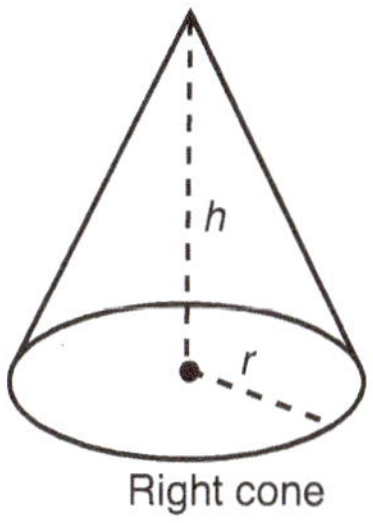

Right cone

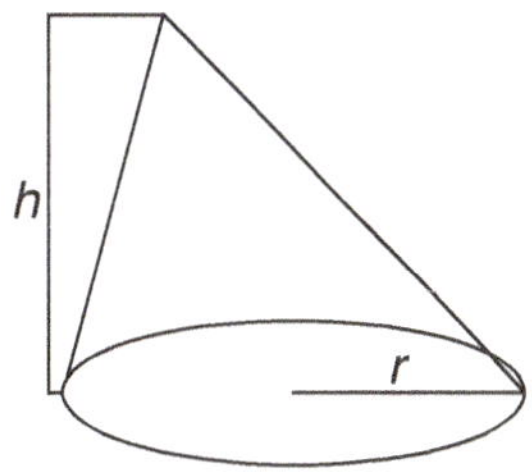

Oblique cone

VOLUME OF A CONE

Similar to a pyramid, it is also true that given a cylinder with radius r and height h, it is three times the volume of a cone with the same radius r and height h.

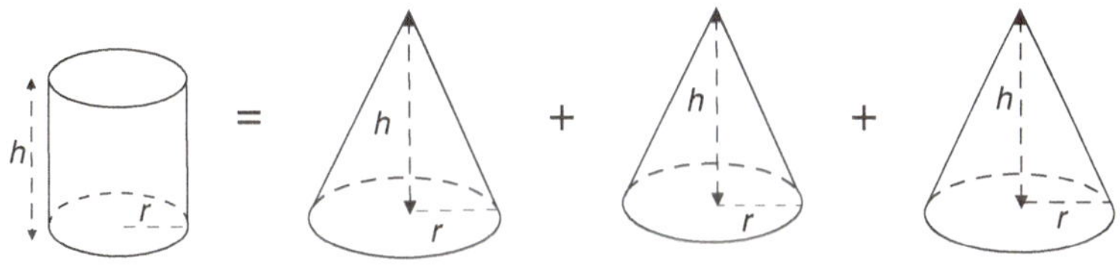

As the volume of a cylinder with radius r and height h is $V = \pi r^2 h$

Then the volume of a cone with radius r and height h is

$$V_{\text{cone}} = \frac{1}{3}\pi r^2 h$$

Find the volume of a cone with radius of 5.7 cm and height of 12 cm.

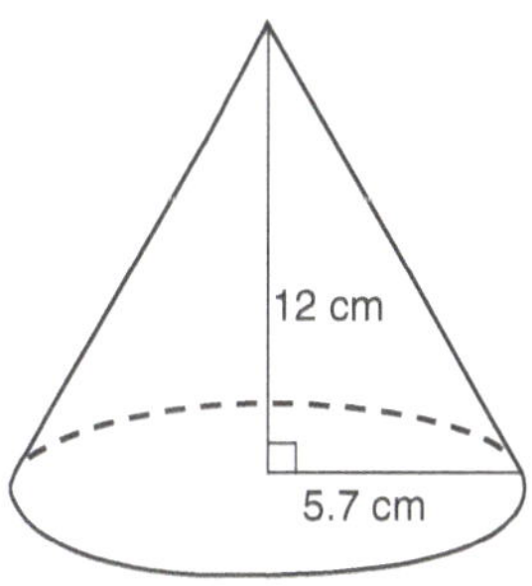

$$V_{\text{cone}} = \frac{1}{3}\pi r^2 h = \frac{1}{3}\pi\,(5.7\,\text{cm})^2\,(12\text{ cm}) = 408.28\ \text{cm}^3$$

Practice

Find the volume of a right hexagonal pyramid with a regular hexagon base with side length 50 cm and a height of 30 cm.

Find the volume of a right come with a diameter of 62 in. and a height of 23 in.

5-5 Exercises

Find answers to the nearest hundredth of a unit when appropriate.

Find the volume of these right pyramids.

1. Rectangular pyramid

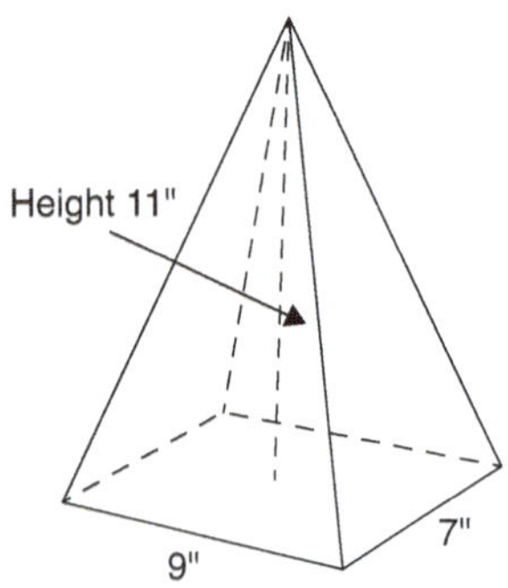

2. Regular hexagonal pyramid

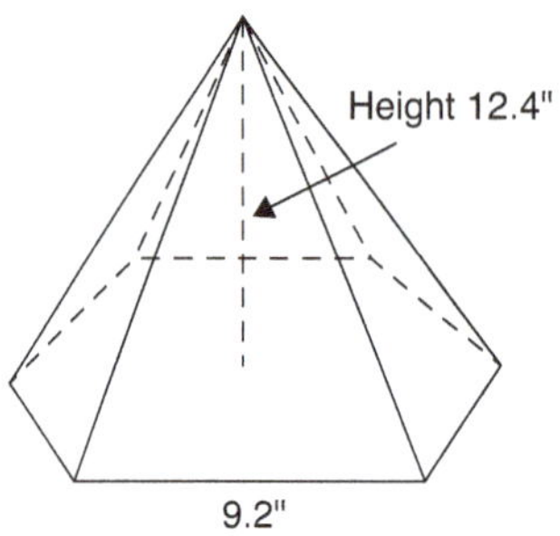

3. Right triangular pyramid

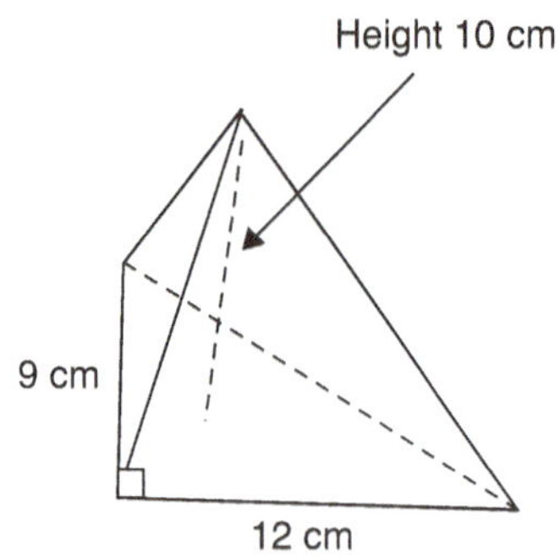

4. Find the Volume of a right equilateral triangular pyramid with base side length of 78 mm and a height of 54 mm.

5. A figure is made up of two right square pyramids that share the same 4.5" sided base. The upper pyramid has a height of 6" and the lower pyramid has a height of 9". Find the figure's volume.

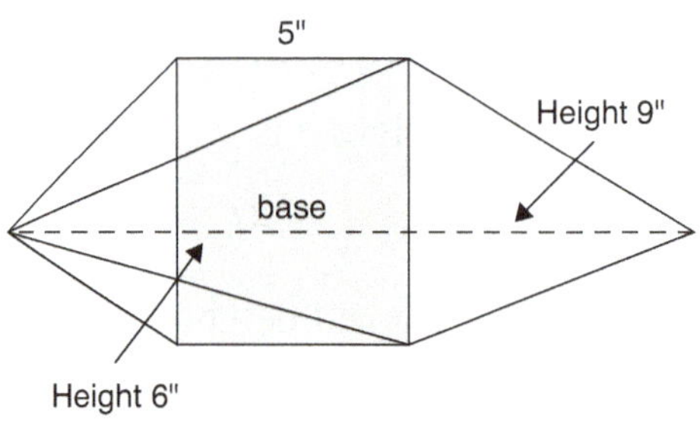

6. Rosie designed a candy dish made up of a regular hexagonal prism and regular hexagonal pyramid. If the side of the regular hexagonal base is 11 cm and the height of the prism is 6 cm and the height of the pyramid is 7 cm, what is the volume?

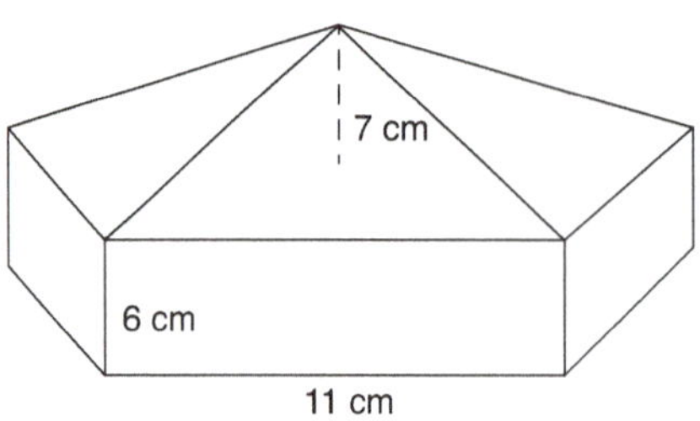

Find the volume for these right cones.

7.

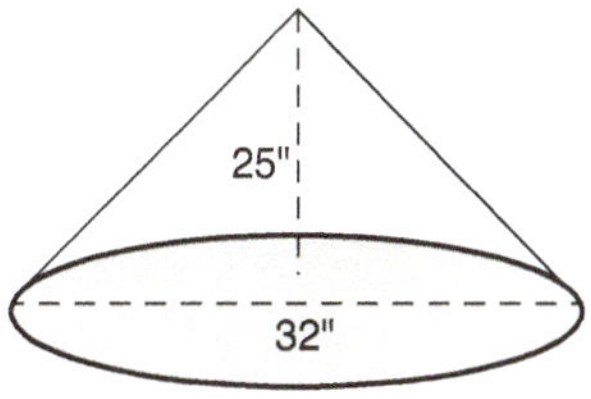

8.

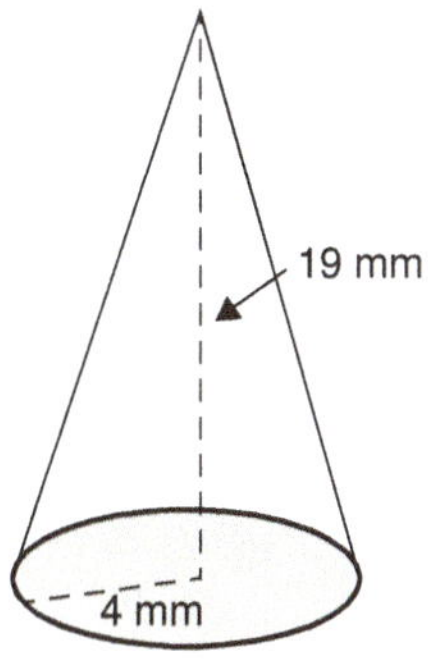

9.

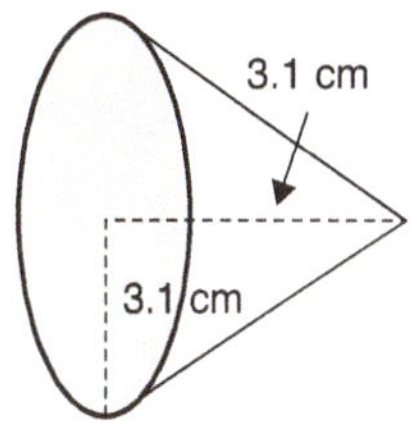

10. A right cone is 220 feet high and has a diameter of 100 feet. Find its volume

11. A mechanical tubing was fashioned out of two right cones and a cylinder. Find its volume.

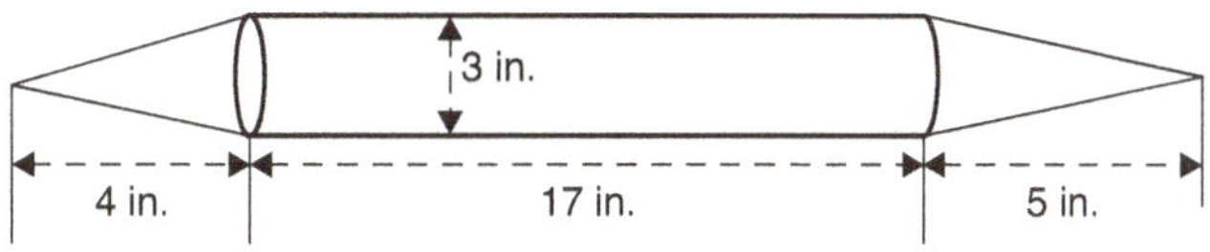

12. Jen and Berry's Ice Cream Shoppe has a huge ice cream cone art piece in front of the store. It is made up of a right cone and a hemisphere. Find its volume

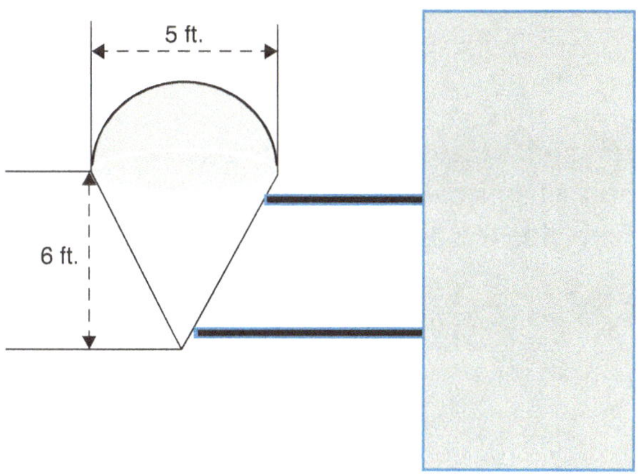

Answers

1. $\frac{1}{3} \times 9" \times 7" \times 11" = 231 \text{ in}^3$

2. $\frac{1}{3} \times 2.598\ (9.2')^2 \times 12.4' = 908.90 \text{ ft}^3$

3. $\frac{1}{3} \times .5 \times 9 \text{ cm} \times 12 \text{ cm} \times 10 \text{ cm} = 180 \text{ cm}^3$

4. $\frac{1}{3} \times .433\ (78 \text{ mm})^2 \times 54 \text{ mm} = 47418.70 \text{ mm}^3$

5. $\frac{1}{3} \times (5")^2 \times 6" + \frac{1}{3} \times (5")^2 \times 9" = 125 \text{ in}^3$

6. $2.598\ (11 \text{ cm})^2 \times 6 \text{ cm} + \frac{1}{3} \times 2.598\ (11 \text{ cm})^2 \times 7 \text{ cm} = 2619.09 \text{ cm}^3$

7. Radius is 16", $\frac{1}{3} \times \pi \times (16")^2 \times 25" = 6702.06 \text{ in}^3$
8. $\frac{1}{3} \times \pi \times (4 \text{ mm})^2 \times 19 \text{ mm} = 318.35 \text{ mm}^3$
9. $\frac{1}{3} \times \pi \times (3.1 \text{ cm})^2 \times 3.1 \text{ cm} = 31.20 \text{ cm}^3$
10. Radius is 50', $\frac{1}{3} \times \pi \times (50')^2 \times 220' = 575958.65 \text{ ft}^3$
11. Radius is 1.5", $\pi \times (1.5")^2 \times 17" + \frac{1}{3} \times \pi \times (1.5")^2 \times 4" + \frac{1}{3} \times \pi \times (1.5")^2 \times 5" = 141.37 \text{ in}^3$
12. Radius is 2.5', $\frac{1}{2} \times \frac{4}{3} \times \pi \times (2.5')^3 + \frac{1}{3} \times \pi \times (2.5')^2 \times 6' = 71.99 \text{ ft}^3$

5-6 SURFACE AREA OF PYRAMIDS AND CONES

The surface area of any pyramid can be found by the sum of the polygonal base area and the triangular lateral face areas. And the surface area of a cone is the sum of the circle base area and the lateral area. If the pyramid is oblique and/or not regular (i.e., it does not have a regular polygon for a base), one must calculate each face separately. And if the cone is oblique, it requires more complicated figuring to find the surface area. However, for right regular pyramids and for right cones, we can find surface area in formulaic manner.

RIGHT REGULAR PYRAMID

In right regular pyramids, all triangular faces are congruent isosceles triangles. The height of one of the isosceles triangular face form what is called the SLANT HEIGHT of the pyramid. For example, for this right square pyramid with side length of 4 in. and height of 6 in.

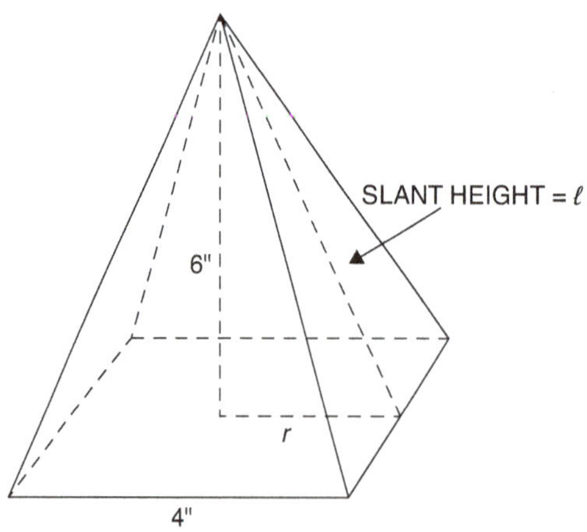

The **SLANT HEIGHT** is the height of one of the triangle faces. To figure the slant height, ℓ, you should see that this is a hypotenuse of a right triangle with legs that are the height of the pyramid and the length from the center of the polygonal base to the side of the polygon base, which we will call r. In this case, the height of the pyramid is 6 in. and the distance r is half the 4 in. base length, so $r = 4\text{in}/2 = 2\text{in}$.

Using the Pythagorean theorem, with leg lengths of 6" and 2", we get slant height $\ell = \sqrt{6^2 + 2^2} = \sqrt{40} = 6.32$ in.

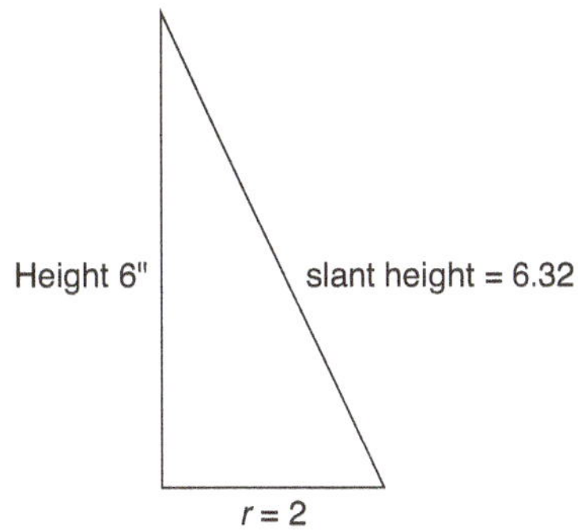

This means one of the isosceles triangle faces has a base length of 4in. and height of 6.32 in. For this square pyramid, we have a square base and four isosceles triangles faces, so the surface area = (Base area) + 4 times (isosceles triangle Base area)

or surface area = $(4 \text{ in.})^2 + 4\,(.5 \cdot 4 \text{ in.} \cdot 6.32 \text{ in.}) = 66.56 \text{ in.}^2$

For a right regular polygon, if we have the slant height, ℓ, and the side length, s, of the regular polygonal base with n number of sides, then we can find the surface area by

$$\text{SURFACE AREA} = \text{Base Area} + n\,(.5 \cdot s \cdot \ell)$$

Example, given a right hexagonal pyramid with base side length of 5 cm and a slant height of 7 cm, find the surface area

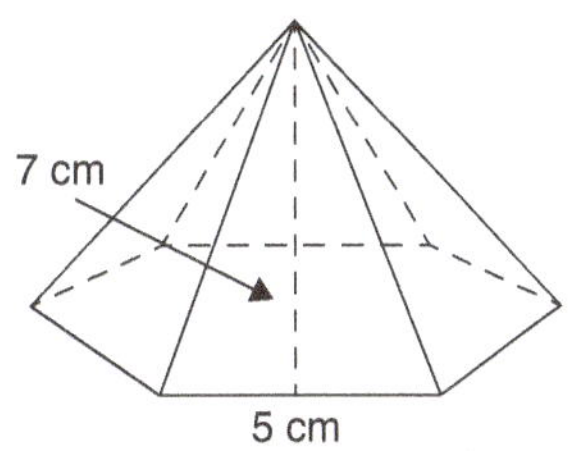

As $n = 6$, the

$$\text{SURFACE AREA} = \text{Base Area} + n\,(.5\ s \cdot l\,) = 2.598\,(5 \text{ cm})^2 + 6\,(.5 \cdot 5 \text{ cm} \cdot 7 \text{ cm}) = 169.95 \text{ cm}^2$$

Practice

For a square pyramid with height of 8 in. and side length of 3 in.

Find the surface area of an equilateral triangular pyramid with a side length of 8.4 cm and a slant height of 19 cm.

RIGHT CONE

The length from the apex to a point on the circle of the base is the slant height.

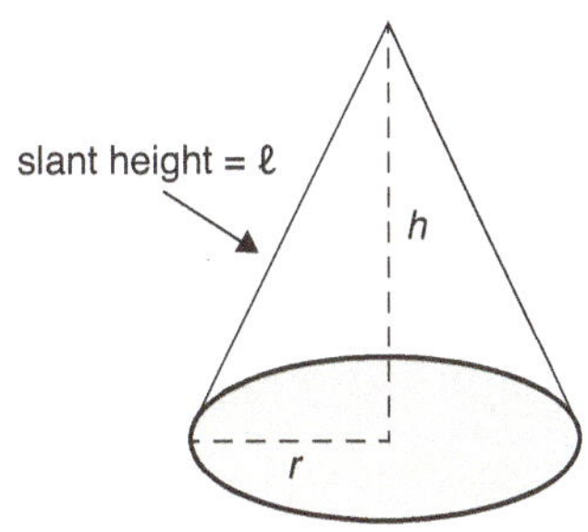

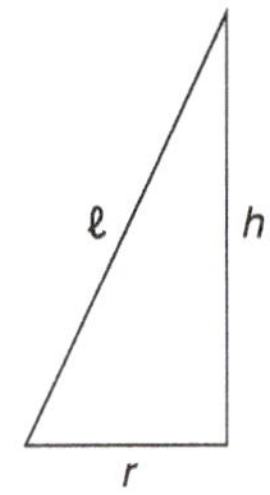

You can see that the slant height is the hypotenuse of a right triangle with legs the height of the cone and the radius of the cone.

If we know the height of the right cone, h, and the radius of the circle base, r, using the Pythagorean theorem. The slant height $\ell = \sqrt{h^2 + r^2}$

Example: Find the slant height of this right cone with radius 10" and height 14"

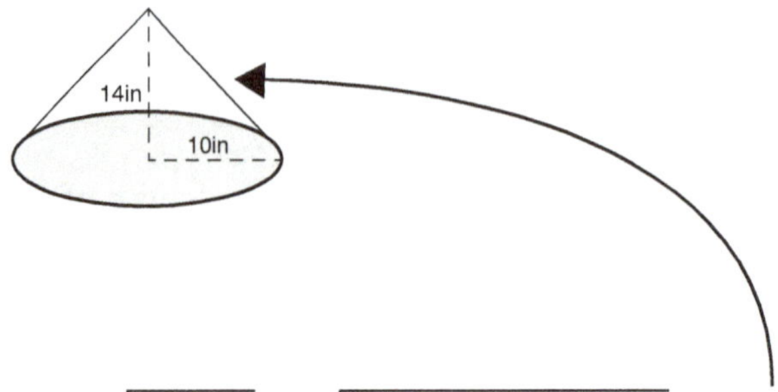

$$\text{Slant height} = \ell = \sqrt{h^2 + r^2} = \sqrt{(14\text{ in.})^2 + (10\text{ in.})^2} = 17.20\text{ in.}$$

SURFACE AREA OF RIGHT CONE

The surface area of a right cone is the sum of the circle base area and the lateral area.

The lateral area for a right cone is figured by π times the radius times the slant height.

Or the lateral area $= \pi r\sqrt{h^2 + r^2} = \pi r\ell$

And the circle base area $= \pi r^2$

Then if we know the radius r and the height h, then

$$\text{SURFACE AREA} = \pi r^2 + \pi r\sqrt{h^2 + r^2}$$

And if we know the slant height ℓ, and the radius of the circle base, r,

$$\text{SURFACE AREA} = \pi r^2 + \pi r\ell$$

Examples

Find the surface area of a right cone with a diameter of 55 in. and a height of 45 in.

$$r = 55/2\text{ in.} = 27.5\text{ in. } h = 45\text{ in.}$$

Surface area $= \pi r^2 + \pi r\sqrt{h^2 + r^2} = \pi(27.5\text{ in.})^2 + \pi(27.5\text{ in.})\sqrt{(45\text{ in.})^2 + (27.5\text{ in.})^2}$ $= 6932.03\text{ in.}^2$

Find the surface area of right cone with a radius of 6 cm and a slant height of 11 cm.

$$r = 6\text{ cm } \ell = 11\text{ cm}$$

$$\text{Surface area} = \pi r^2 + \pi r\ell = \pi(6\text{ cm})^2 + \pi(6\text{ cm})(11\text{ cm}) = 320.44\text{ cm}^2$$

Practice

A right cone has a radius of 50 m and a height of 70 m.

Find the surface area.

A right cone has a diameter of 4.56 in. and a slant height of 6.9 in.

Find the surface area.

5-6 Exercises

Find answers to the nearest hundredth of a unit when appropriate.

Find the Surface area of these right pyramids.

1. Square pyramid

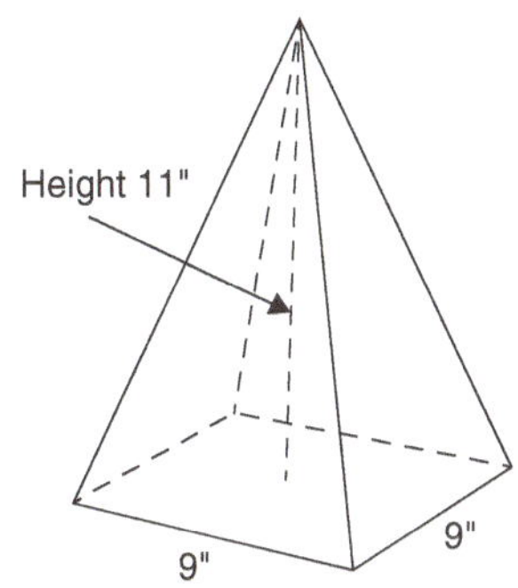

2. Regular hexagonal pyramid

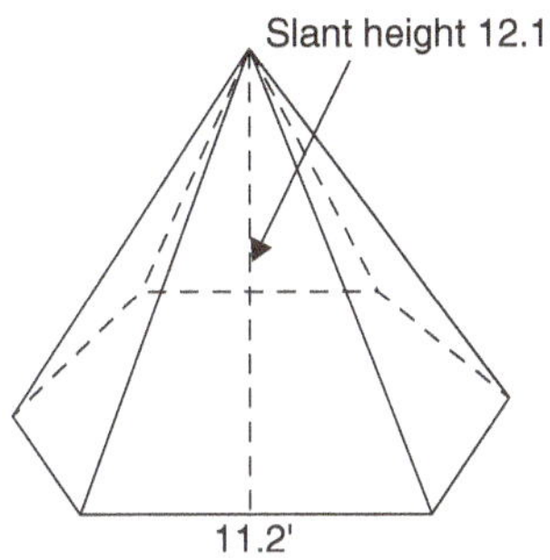

3. Equilateral triangular pyramid

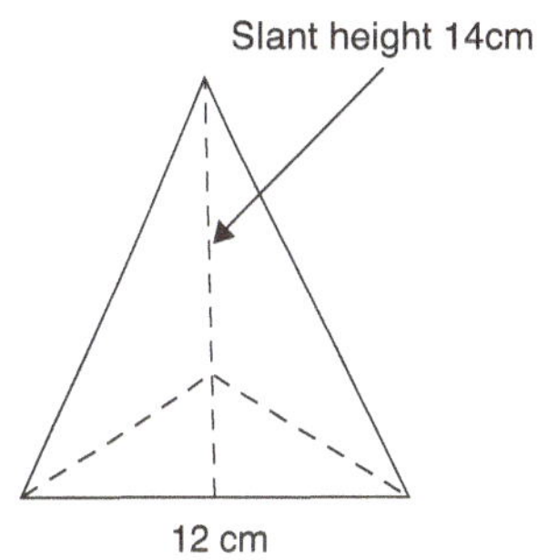

4. Find the surface area of a right square pyramid with base side length of 78 mm and a height of 54 mm.
5. Find a regular octagonal pyramid with base side length of 7 meters and a slant height of 9 meters.
6. A figure is made up of two right square pyramids that share the same 5" sided base. The upper pyramid has a height of 6" and the lower pyramid has a height of 9". Find the figure's surface area.

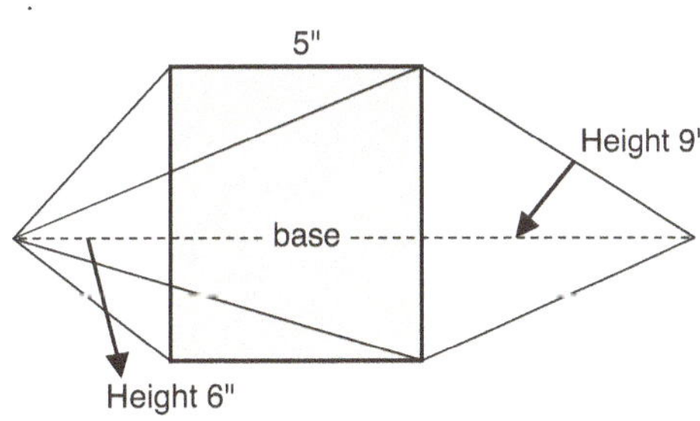

7. Rosie designed a candy dish made up of a regular hexagonal prism and regular hexagonal pyramid. If the side of the regular hexagonal base is 11 cm and the height of the prism is 6 cm and the slant height of the pyramid is 8 cm, what is the surface area?

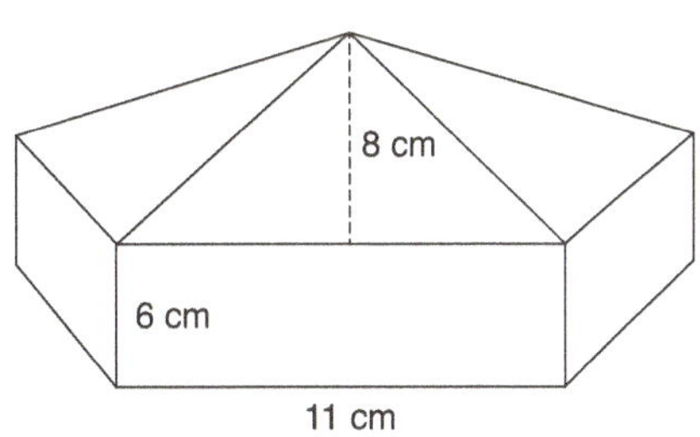

Find the surface area for these right cones.

8.

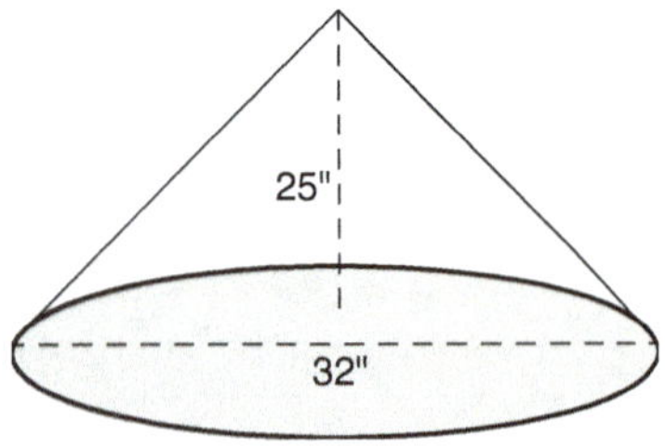

9.

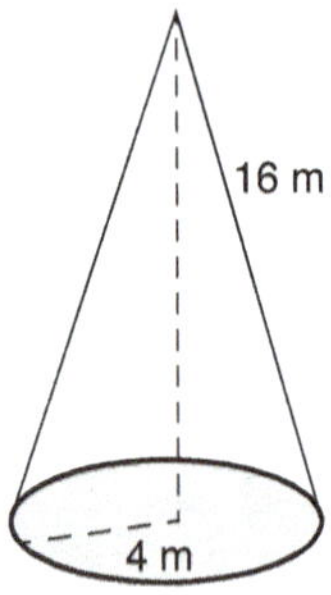

10.

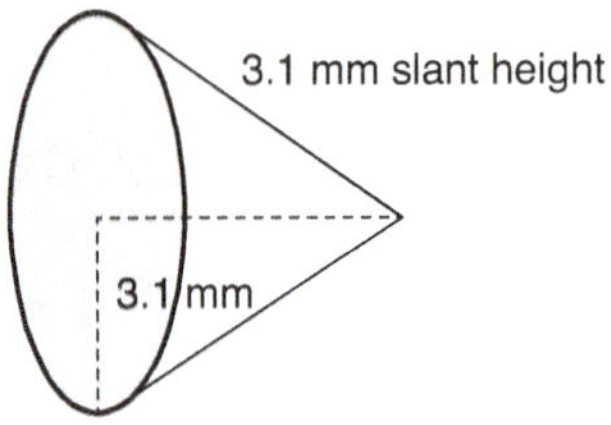

11. A right cone is 220 feet high and has a diameter of 100 feet. Find its surface area.

12. A mechanical instrument was fashioned out of two right cones and a cylinder. Find its surface area.

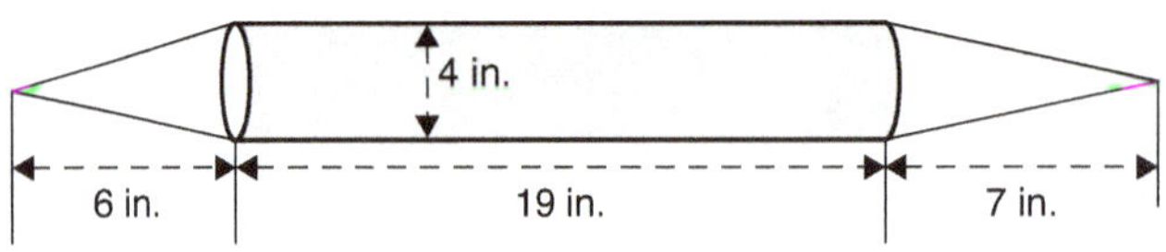

13. Jen and Berry's Ice Cream Shoppe has a huge ice cream cone art piece in front of the store. It is made up of a right cone and a hemisphere. Find its surface area.

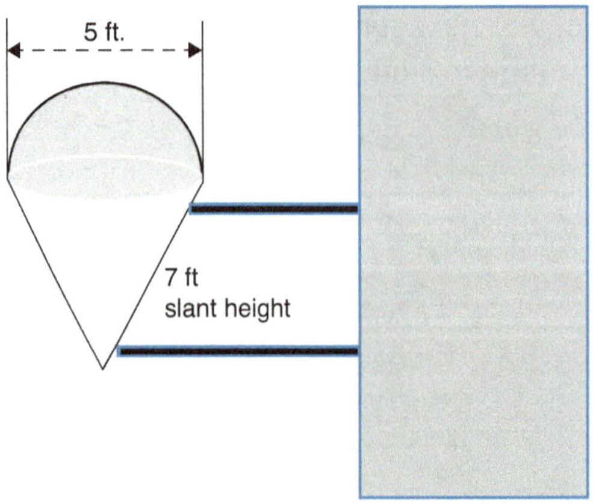

Answers

1. $r = \frac{9"}{2} = 4.5"$, $h = 11"$, $\ell = \sqrt{11^2 + 4.5^2}" = 11.88'$
   $(9")^2 + 4\,(.5 \times 9 \times 11.88) = 294.84 \text{ in}^2$

2. $2.598 \times (11.2')^2 + 6(.5 \times 11.2' \times 12.1') = 732.45 \text{ ft}^2$

3. $.433 \times (12 \text{ cm})^2 + 3(.5 \times 12 \text{ cm} \times 14 \text{ cm}) = 314.35 \text{ cm}^2$

4. $r = \frac{78 \text{ mm}}{2} = 39 \text{ cm}$, $h = 54 \text{ mm}$ $\ell = \sqrt{54^2 + 39^2} \text{ mm} = 66.61 \text{ mm}$
   $(78 \text{ mm})^2 + 4(.5 \times 78 \text{ mm} \times 66.61 \text{ mm}) = 16475.16 \text{ mm}^2$

5. $4.828 \times (7 \text{ m})^2 + 8(.5 \times 7 \text{ m} \times 9 \text{ m}) = 488.57 \text{ m}$

6. 6" height pyramid, $r = \frac{5"}{2} = 2.5"$, $h = 6"$, $\ell = \sqrt{6^2 + 2.5^2} = 6.5"$
   9" height pyramid, $r = 2.5"$ $h = 9"$, $\ell = \sqrt{9^2 + 2.5^2} = 9.34"$
   $4(.5 \times 5" \times 6.5") + 4\,(.5 \times 5" \times 9.34") = 158.4 \text{ in}^2$

7. $2.598 \times (11\text{cm})^2 + 66 \text{ cm} \times 6 \text{ cm} + 6\,(.5 \times 11 \text{ cm} \times 8 \text{ cm}) = 974.36 \text{ cm}^2$

8. $r = \frac{32"}{2} =$ is $16"$ $h = 25"$
   $\pi \times (16")^2 + \pi \times 16" \times \sqrt{25^2 + 16^2}" = 2296.21 \text{ in}^2$

9. $r = 4 \text{ m}$, $\ell = 16\text{m}$
   $\pi \times (4 \text{ m})^2 + \pi \times 4 \text{ m} \times 16 \text{ m} = 251.33 \text{ m}^2$

10. $r = 3.1 \text{ mm}$, $\ell = 3.1 \text{ mm}$; $\pi \times (3.1 \text{ mm})^2 + \pi \times 3.1 \text{ mm} \times 3.1 \text{ mm} = 60.38 \text{ mm}^2$

11. $r = \frac{100'}{2} = 50$ $h = 220'$
    $\pi \times (50')^2 + \pi \times 50' \times \sqrt{220^2 + 50^2}' = 43292.76 \text{ ft}^2$

12. $r = \frac{4"}{2} = 2"$
    $(\pi \times 2" \times \sqrt{6^2 + 2^2}") + (\pi \times 4" \times 19") + (\pi \times 4" \times \sqrt{7^2 + 2^2}") = 324.24 \text{ in}^2$

13. $r = \frac{5'}{2} = 2.5'$
    $[.5 \times 4 \times \pi \times (2.5')^2] + (\pi \times 2.5' \times 7') = 94.25 \text{ ft}^2$

5-7 FRUSTUM

A FRUSTUM is a "bottom" portion of a pyramid and/or cone that happens when it is divided by a parallel cross section of the base. It results in two similar bases (i.e., same shape, different size).

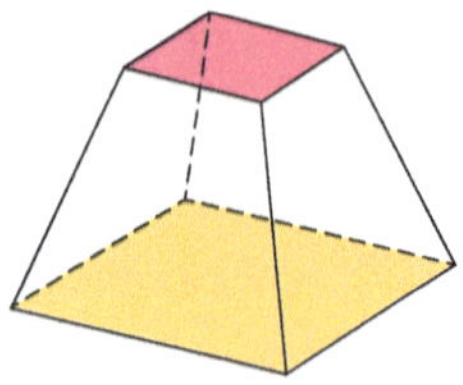

Pyramidal frustum

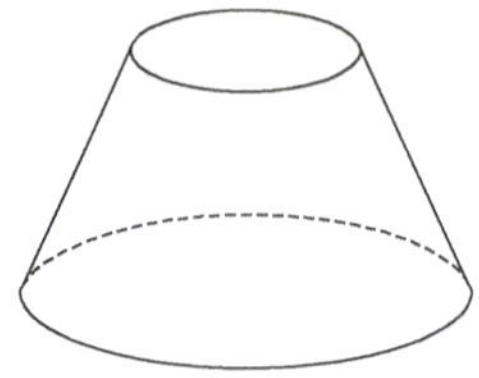

Conical frustum

Frustums are also called "truncated" pyramids or cones.

VOLUME OF A PYRAMIDAL FRUSTUM

For a pyramidal frustum with larger base area B and smaller base area b, and distance between the two bases is the height, h, the volume of the frustum is

$$V_{\text{frustum}} = \frac{1}{3}h\left(\mathrm{B} + \mathrm{b} + \sqrt{\mathrm{B} \cdot \mathrm{b}}\right)$$

Example

This rectangular frustum has one base with dimensions 5 in. by 4 in. and another base with dimensions 7.5 in. by 6 in. The height of the frustum is 3 in.

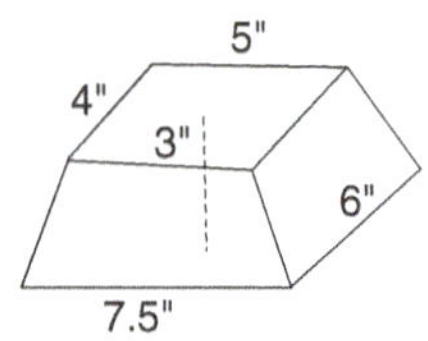

The larger base area is $B = (7.5\text{ in.})\ (6\text{ in.}) = 45\text{ in.}^2$. The smaller base area is $b = (5\text{ in.})\ (4\text{ in.}) = 20\text{ in.}^2$. Height of frustum is $h = 3$ in.

$$V_{\text{frustum}} = \frac{1}{3}h\left(B + b + \sqrt{B \cdot b}\right) =$$

$$\frac{1}{3} \cdot 3\text{ in.}\ \left(45\text{ in.}^2 + 20\text{ in.}^2 + \sqrt{45\text{ in.}^2 \cdot 20\text{ in.}^2}\right)$$

$$= \frac{1}{3} \cdot 3\text{ in.}\ \left(45\text{ in.}^2 + 20\text{ in.}^2 + \sqrt{900\text{ in.}^4}\right) = \frac{1}{3} \cdot 3\text{ in.}\ \left(45\text{ in.}^2 + 20\text{ in.}^2 + 30\text{ in.}^2\right)$$

$$= \frac{1}{3} \cdot 3\text{ in.}\ \left(95\text{ in.}^2\right) = 95\text{ in.}^3$$

Example:

A hole is dug in a form of a frustum with bigger square base with side lengths 6.4 m and the smaller base with side lengths 3.6 m, and the depth of the hole (or the height) of 4 m. What is the volume of the hole?

Area of bigger base $B = (6.4\text{ m})^2 = 40.96\text{ m}^2$ Area of smaller base $b = (3.6\text{ m})^2 = 12.96\text{ m}^2$

Height $h = 4$ m

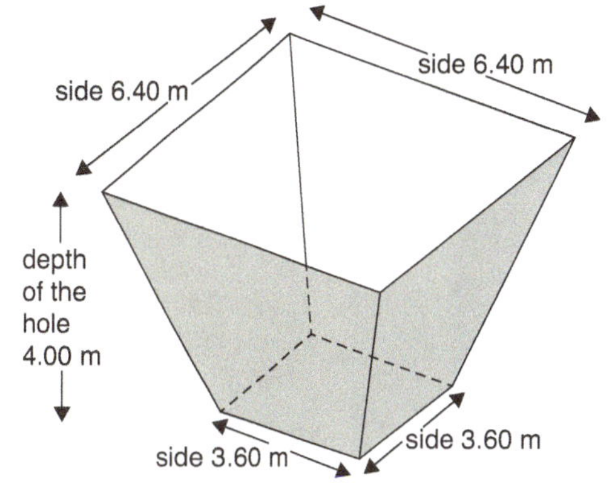

$$V_{\text{frustum}} = \frac{1}{3}h\left(B+b+\sqrt{B\cdot b}\right) =$$

$$\frac{1}{3}\cdot 4\text{ m}\cdot\left(40.96\text{ m}^2+12.96\text{ m}^2+\sqrt{40.96\text{ m}^2\cdot 12.96\text{ m}^2}\right)$$

$$= 102.61\text{ m}^3$$

VOLUME OF A CONICAL FRUSTUM

For a conical frustum, the same formula applies as with a pyramidal frustum. As the two bases for a pyramidal frustum are circles, we can use the respective radii for the two bases in figuring the base areas.

If the larger base radius is R and the smaller base radius is r, then the area of the larger circle base is $B = \pi R^2$ and the area of the smaller circle base is $b = \pi r^2$. The distance between the two bases is the height, h. Our frustum formula $V_{\text{frustum}} = \frac{1}{3}h\left(B+b+\sqrt{B\cdot b}\right)$

Is now $V_{\text{frustum}} = \frac{1}{3}h\left(\pi R^2+\pi r^2+\sqrt{\pi R^2\cdot\pi r^2}\right)$

We can further simplify our formula by manipulation

$$V_{\text{frustum}} = \frac{1}{3}h\left(\pi R^2+\pi r^2+\sqrt{\pi R^2\cdot\pi r^2}\right) \text{ multiply under the radical}$$

$$= \frac{1}{3}h\left(\pi R^2+\pi r^2+\sqrt{\pi^2R^2r^2}\right) \text{ simplify the square root}$$

$$= \frac{1}{3}h\left(\pi R^2+\pi r^2+\pi Rr\right) \text{ factor out } \pi$$

$$= \frac{1}{3}h\,\pi\left(R^2+r^2+Rr\right) \text{ or}$$

$$= \frac{\pi}{3}h\left(R^2+r^2+Rr\right) \text{ and as } \frac{\pi}{3} \text{ is } \approx 1.047$$

$$= 1.047\,h\left(R^2+r^2+Rr\right)$$

Our final formula for a cylindrical frustum is

$$V_{\text{frustum}} = 1.047\,h\left(R^2+r^2+Rr\right)$$

Example: Find the volume of conical frustum that is 4.5 in. high with the bigger base with 4 in. radius and the smaller base with 1.8in. radius.

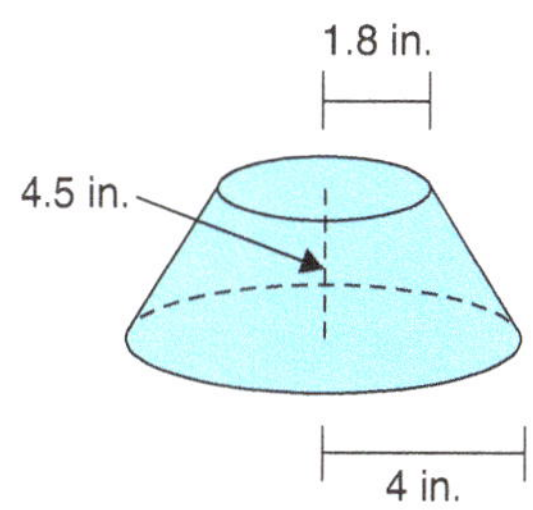

$$V_{\text{frustum}} = 1.047\,h\left(R^2+r^2+Rr\right)$$

$$= 1.047\,(4.5\text{ in.})\left((4\text{ in.})^2+(1.8\text{ in.})^2+(4\text{ in.}\cdot 1.8\text{ in.})\right) = 124.57\text{ in.}^3$$

Practice

Find the volume of a pyramidal frustum with right triangular bases that is 43 cm high and with the larger base triangle with leg lengths of 20 cm and 30 cm and the smaller base triangle with leg length of 10 cm and 15 cm.

Find the volume of a conical frustum with circle base diameters of 60 ft. and 10 ft. and a height of 40 ft.

5-7 Exercises

1. Find the volume of this rectangular pyramidal frustum.

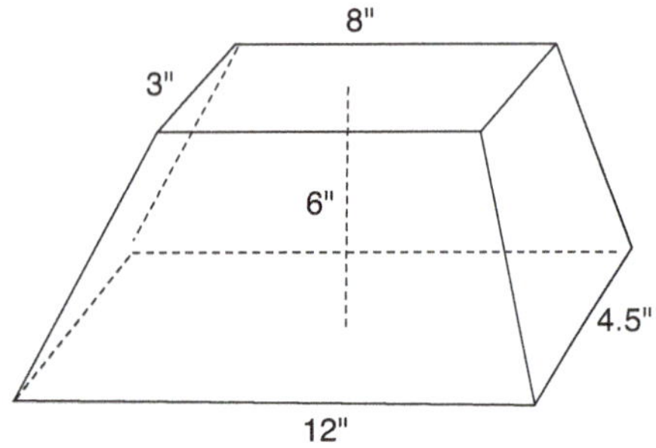

2. Find the volume of this conical frustum.

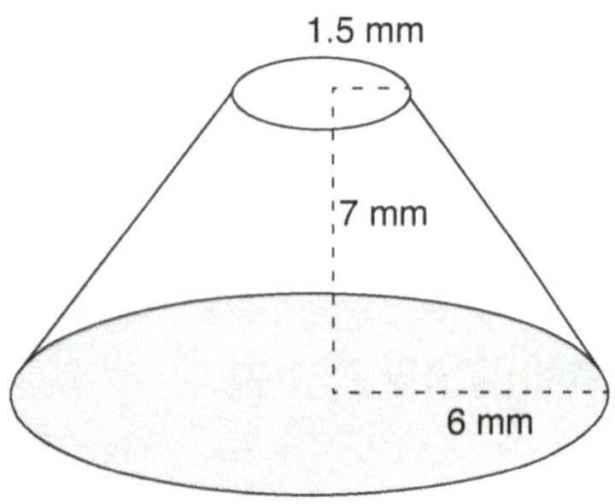

3. A regular hexagonal pyramidal frustum has a bigger base side length of 40.5 inches, a smaller base side length of 22.2 inches, and a height of 5 inches. Find its volume.

4. A conical frustum has a smaller circle base diameter of 2 feet, a bigger base diameter of 10 feet and a height of 12 feet. Find its volume.

5. A right triangular pyramidal frustum has a smaller base with legs 15 cm and 20 cm, a larger base with legs 30 cm and 40 cm and a height of 10 cm. Find its volume.

Answers

1. $B = 12" \times 4.5" = 54 \text{ in}^2,\ b = 3" \times 8" = 24 \text{ in}^2$

$$\frac{1}{3} \times 6" \times (54 \text{ in}^2 + 18 \text{ in}^2 + \sqrt{54 \times 24} \text{ in}^2) = 228 \text{ in}^3$$

2. $\frac{1}{3} \times 7 \text{ mm} \times (\pi(6 \text{ mm})^2 + \pi\,(1.5 \text{ mm})^2 + \sqrt{\pi 6^2 \times \pi 1.5^2} \text{ mm}^2) = 346.66 \text{ mm}^3$

3. $B = 2.598\,(40.5")^2 = 4261.37 \text{ in}^2\ b = 2.598\,(22.2'')^2 = 1280.40 \text{ in}^2$

$$\frac{1}{3} \times 5" \times (4261.37 \text{ in}^2 + 1280.40 \text{ in}^2 + \sqrt{4261.37\ x\ 1280.40} \text{ in}^2) = 13129.39 \text{ in}^3$$

4. $\frac{1}{3} \times 12' \times (\pi(1')^2 + \pi\,(5')^2 + \sqrt{\pi 1^2 \times \pi 5^2}\ \text{ft}^2) = 389.56\ \text{ft}^3$

5. $B = .5(30\ \text{cm} \times 40\ \text{cm}) = 600\ \text{cm}^2 \quad b = .5(15\ \text{cm} \times 20\ \text{cm}) = 150\ \text{cm}^2$

$\frac{1}{3} \times 10\ \text{cm} \times (600\ \text{cm}^2 + 150\ \text{cm}^2 + \sqrt{600 \times 150}\ \text{cm}^2) = 3500\ \text{cm}^3$

5-8 WEIGHT, MASS, DENSITY

WEIGHT/MASS

WEIGHT is the measure for the force of gravity. We have seen our customary system use POUNDS, OUNCES, and TONS for weight measures. NEWTONS are units of force in the metric system and are used a lot in science and engineering.

(c) Dima Zel/Shutterstock.com

MASS and WEIGHT/FORCE are different dimensions. Where WEIGHT/FORCE is a measure of "attraction" (or "repulsion"), MASS is a measure of the amount of a substance. MILLIGRAMS, GRAMS, and KILOGRAMS are the common units used for mass in the metric system.

For example:

On earth, a moon-bound astronaut ready to blast off in his spaceship has a weight of 180 lb. and a mass of 80.63 kg. Halfway to the moon, the astronaut has left the gravitational pull of the earth, so is virtually weightless, and therefore has a weight of 0 lb. However, his or her mass is still 80.63 kg. When on the moon, he or she is "captured" by the gravitational pull of the moon and now weighs about 30 lb. But . . . his or her mass is still 80.63 kg.

Because most of us will be only concerned with the gravity here on earth, mass in metric is therefore used for weight measures, and we can make conversion between customary and metric, as in 1 kg = 2.205 lb., 1 oz. = 28.35 g, and so on. However, we need to be aware that given different gravitational positions in the universe, this doesn't hold . . .

DENSITY

Density is a comparison of a substances mass to volume, and is figured in formula as

$$\text{DENSITY} = \frac{\text{Mass}}{\text{Volume}} \text{ or } D = \frac{\text{m}}{\text{V}}$$

If we "manipulate" our formula, we find that

$$\text{Volume} = \frac{\text{Mass}}{\text{Density}} \text{ or } V = \frac{\text{M}}{\text{D}} \text{ and}$$

$$\text{Mass} = \text{Volume times Density or } M = V \cdot D$$

Density is kind of like how "heavy" or "filling" a substance is given its volume.

As a rule (but not always!!), solids have more density (or are more dense) than liquids, and liquids are more dense than gases.

For example, the density of steel is measured at $\frac{7.84\text{ g}}{1\text{ cm}^3}$, and because 1 cubic centimeter equals the volume of 1 milliliter then $\text{Density}_{\text{steel}} = \frac{7.84\text{ g}}{1\text{ mL}}$.

NOTE: You will see a lot of density figure as $\frac{\text{g}}{\text{mL}}$ or $\frac{\text{g}}{\text{cm}^3}$, although density can be expressed in any weight/mass to volume as long as we stay in measurement family, such as pounds/gallon or kilogram/Liter.

The higher the number of mass/weight units to a volume unit, the denser the substance. And the lower the number of mass/weight units to a volume unit, the less dense the subject. We can use this to compare density of different substances.

The density of gasoline is $\frac{.68\text{ g}}{\text{cm}^3}$ so gasoline is less dense than steel.

The density of air is $\frac{.00129\text{ g}}{\text{cm}^3}$ so air is less dense than steel or gasoline.

The density of silver is $\frac{10.5\text{ g}}{\text{cm}^3}$ so silver is denser that steel, gasoline or air.

The "great divider" for density is WATER. That is because water has the special density of $\frac{1\text{ g}}{1\text{ cm}^3}$ or $\frac{1\text{ g}}{1\text{ mL}}$. That is 1 cubic centimeter (or 1 milliliter) of water weighs (or has mass of) 1 gram.

The following is true:

A substance with density greater than $\frac{1\text{ g}}{\text{cm}^3}$ will be more dense than water, and will therefore "sink" in water.

A substance with density less than $\frac{1\text{ g}}{\text{cm}^3}$ will be less dense than water, and will therefore "float" on water.

So as $\text{Density}_{\text{steel}} = \frac{7.84\text{ g}}{\text{cm}^3} > \frac{1\text{ g}}{\text{cm}^3}$, then steel will sink.

As $\text{Density}_{\text{gasoline}} = \frac{.68\text{ g}}{\text{cm}^3} < \frac{1\text{ g}}{\text{cm}^3}$, the gasoline will float on water.

There are solid, liquid, and gas substances that have unusual density characteristics. For example, mercury is in liquid form and has a

$$\text{Density of}_{\text{mercury}} = \frac{13.6\text{ g}}{\text{cm}^3}$$

As $\frac{13.6\text{ g}}{\text{cm}^3} > \frac{1\text{ g}}{\text{cm}^3}$ mercury will therefore sink in water.

The density of copper, a solid, is

$$\text{Density of}_{\text{copper}} = \frac{8.96\text{ g}}{\text{cm}^3}\text{, and } \frac{8.96\text{ g}}{\text{cm}^3} > \frac{1\text{ g}}{\text{cm}^3} \text{ and therefore copper will sink in water.}$$

However, as $\frac{8.96\text{ g}}{\text{cm}^3} < \frac{13.6\text{ g}}{\text{cm}^3}$ Then Density of$_{\text{copper}}$ < Density of$_{\text{mercury}}$

So copper, a solid, will FLOAT on mercury, a liquid!!

Another "weird" characteristic is water to ice. The density of water turned into ice is Density$_{\text{ice}} = \frac{.91\text{ g}}{\text{cm}^3}$ and since $\frac{.91\text{ g}}{\text{cm}^3} < \frac{1\text{ g}}{\text{cm}^3}$ then the Density$_{\text{ice}}$ < Density$_{\text{water}}$

So that is why ice, which is water in solid form, floats on water!!

USING DENSITY TO SOLVE PROBLEMS

Use the density chart in back, and use the formulas

$$\text{DENSITY} = \frac{\text{Mass}}{\text{Volume}} \text{ or } D = \frac{m}{V};$$

$$\text{Volume} = \frac{\text{Mass}}{\text{Density}} \text{ or } V = \frac{M}{D}$$

$$\text{Mass} = \text{Volume times Density or } M = V \cdot D$$

Figure this Example

What is the weight in grams of a 15 cm by 4 cm by 5 cm steel plate

As Mass = Volume times Density or $M = V \cdot D$, we figure the volume of the plate as (15 cm) (4 cm) (5 cm) = 300 cm^3

The Denisity $_{\text{steel}} = \frac{7.8\text{ g}}{\text{cm}^3}$.

$$M = V \cdot D = 300\text{ cm}^3 \left(\frac{7.8\text{ g}}{\text{cm}^3}\right) = 2340\text{ g}$$

Figure this example

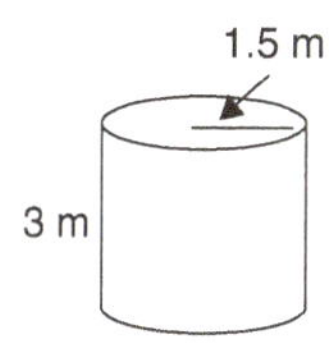

What is the weight in <u>tons</u> of a brass cylinder with height of 3 m and a base radius of 1.5 m

Volume is $\pi r^2 h = \pi (1.5\text{ m})^2 (3\text{ m}) = 21.21\text{ m}^3$

Density of brass is $\frac{8.55\text{ g}}{\text{cm}^3}$. To find the weight, we need the volume in cubic centimeters.

And there are 1,000,000 cm^3 in 1 m^3 so $21.21\text{ m}^3 \left(\frac{1{,}000{,}000\text{ cm}^3}{\text{m}^3}\right) = 21{,}210{,}000\text{ cm}^3$

$$M = D \cdot V = \frac{8.55\text{ g}}{\text{cm}^3}(21{,}210{,}000\text{ cm}^3) = 181{,}345{,}500\text{ g.}$$

To convert to tons, we need to

- Convert 181,345,500 g to kilograms, using 1,000 g = 1 kg
- Then use 1 kg = 2.205 lb.
- Then convert to tons using 1 tn. = 2,000 lb.

$$181{,}345{,}500\text{ g}\left(\frac{1\text{ kg}}{1{,}000\text{ g}}\right)\left(\frac{2.205\text{ lb.}}{1\text{ kg}}\right)\left(\frac{1\text{ tn.}}{2{,}000\text{ lb.}}\right) = 199.93\text{ tn.}$$

PRACTICE: find the weight in kilograms of a brick wall made of three hundred 30 cm by 10 cm by 5 cm bricks. Also find the weight in pounds.

Here are some densities of common substances in g/cm^3

Substance	Density (g/cm^3)
Air	0.0013
Wood (cherry)	0.43
Wood (walnut)	0.59
Wood (southern pine)	0.65
Wood (sugar maple)	0.69
Wood (birch)	0.71
Wood (mahogany)	0.71
Wood (oak)	0.85
Ice	0.91
Water	1
Rubber	1.5
Anthracite coal	1.55
Brick	2.1
Granite	2.75
Aluminum	2.7
Zinc	4.05
Titanium	4.5
Iron	6.98
Steel	7.8
Brass	8.55
Bronze, phosphor	8.8
Copper	8.9
Nickel	8.9
Silver	10.5
Lead	13.5
Mercury	13.57
Gold	19.3
Platinum	21.45

5-8 Exercises

Using the density table provided in chapter 5-8, find the following, to nearest hundredth, when appropriate.

1. 

©Rashevskyi Viacheslav/ Shutterstock.com

   a) Find the weight in grams of a 4 cm by 12 cm by 3 cm SILVER bar.

   b) if the value of silver is \$0.54 per gram, how valuable is the silver plate?

2. 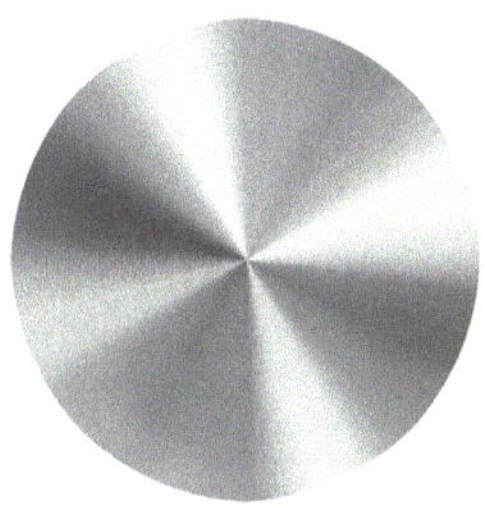

©Tonsci/Shutterstock.com

   a) Find the weight in kilograms of a circular STEEL disk with a diameter of 2 meters and a thickness of 10 cm.

   b) find the weight in pounds.

3. a) Find the weight in kilograms of a 16 cm square plate that is 3 cm in thickness if it is made out of BRONZE.

   b) if bronze is values at \$1.28 per pound, what is this plate worth?

4. Calculate the weight in kilograms of a cylindrical IRON stove that has an outer height of 1 meter and diameter .5 meter in diameter, and is 8 cm in thickness.

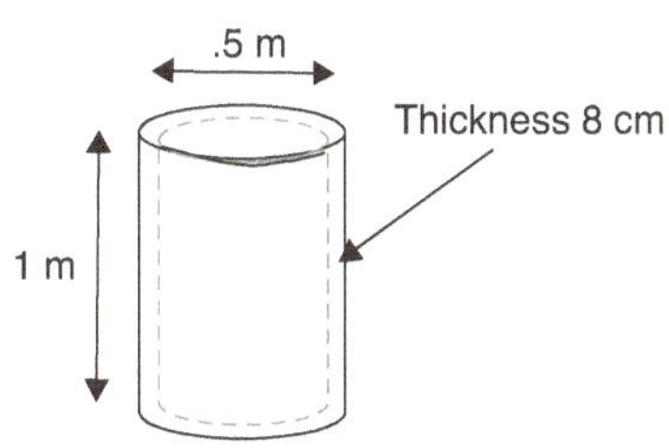

5. A machine part that is a right regular hexagonal pyramid is made out of BRASS. It has a 6 cm base side and has a height of 3.8 cm. If three hundred of the parts are produced, find the total weight in kilograms of the parts.

6. An ice bowl is fashioned out of one hemisphere out of another hemisphere. The bowl is carved out of ice, with the "outer sphere" with a radius of 15 cm and the "inner sphere" a radius of 11 cm. What does the bowl weigh in grams?

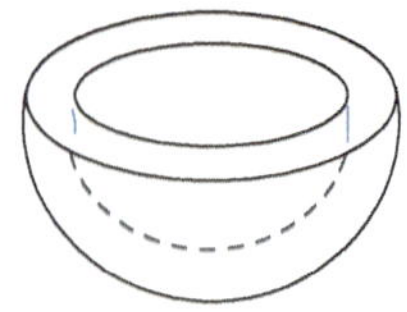

7. A mobile is made by an artist has a series of identical spherical ornaments made out of mahogany wood attached at the end of strings. If there are 20 spheres and each sphere has a diameter of 4 cm, how much did the spheres weigh in total grams?

©123Dartist/
Shutterstock.com

8. A right cone shape that has a base diameter of 3 inches and a height of 3 inches and is made of solid GOLD. If gold is \$41.76 per gram, how much is the cone worth?

9. Workers made a brick wall out of four hundred 10 cm by 13 cm by 25 cm bricks. Find the total weight in kilograms of the wall.

©AlexKoloskov/
Shutterstock.com

Answers

1. a) Volume is $4\text{cm} \times 12\text{cm} \times 3\text{ cm} = 144\text{ cm}^3$, density of silver is 10.5 g/cm^3, $144\text{ cm}^3 \times 10.5\text{ g/cm}^3 = 1512\text{ g}$.

   b) $\$0.54/ 1\text{ gram} \times 1512\text{ g} = \$816.48$

2. a) Volume is $\left(\frac{200}{2}\text{cm}\right)^2 \times \pi \times 10\text{ cm} = 314159.27\text{ cm}^3$, density of steel is $7.8\text{g}/1\text{ cm}^3$,

   $314159.27\text{ cm}^3 \times 7.8\text{ g}/1\text{ cm}^3 = 2450442.31\text{ g}$, $\frac{2450442.31\text{ g}}{1000} = 2450.44\text{ kg}$

   b) $2450.44\text{ kg} \times \frac{2.205\text{ lb}}{1\text{kg}} = 5403.22$ pounds

3. a) volume is $(16\text{cm})^2 \times 3\text{ cm} = 768\text{ cm}^3$, density of bronze is 8.8g/cm^3,

   $768\text{ cm}^3 \times 8.8\text{g/cm}^3 = 6758.4\text{ g}$ $\frac{6758.4\text{ g}}{1000} = 6.76\text{ kg}$

   b) $6.76\text{ kg} \times \frac{2.205\text{ lb}}{1\text{kg}} = 14.91\text{ lb}$. $14.91\text{ lb} \times \frac{\$1.28}{\text{lb}} = \$19.08$

4. dimension outer cylinder is 100 cm height, 25 cm radius, inner cylinder is 100 – 2(8) = 84 cm height and (25–8) = 17cm radius. Volume outer cylinder – volume inner cylinder = stove volume

   $[100\text{ cm} \times (25\text{ cm})^2 \times \pi] - [84\text{ cm} \times 17\text{ cm})^2 \times \pi] =$

   $196349.54\text{ cm}^3 - 76265.32\text{ cm}^3 = 120084.24\text{ cm}^3$, density of iron is 6.98g/cm^3

   $120084.24\text{ cm}^3 \times 6.98\text{ g/cm}^3 = 838188.00\text{ g}$. $\frac{838188.00\text{ g}}{1000} = 838.19\text{ kg}$

5. One pyramid part Volume = $\frac{1}{3} \times 2.598 \times (6\text{ cm})^2 \times 3.8\text{ cm} = 118.47\text{ cm}^3$

$300 \times 118.47\text{ cm}^3 = 35541.00\text{ cm}^3$; density of brass is 8.55g/cm^3; $35541.00\text{ cm}^3 \times 8.55\text{g/cm}^3$

$= 303875.55\text{g}$

$\frac{303875.55\mathbf{g}}{1000} = 303.88\text{ kg}$

6. volume Outer hemisphere – volume inner hemisphere = volume of bowl

$\left[\frac{1}{2}\right] \times \frac{4}{3} \times (15\text{ cm})^3 \times \pi] - \left[\frac{1}{2} \times \frac{4}{3} \times (11\text{ cm})^3 \times \pi\right]$

$= 7068.58\text{ cm}^3 - 2787.64\text{ cm}^3 = 4280.94\text{ cm}^3$, density of ice = $.91\text{g/cm}^3$

$4280.94\text{cm}^3 \times \cdot \frac{91\text{ g}}{\text{cm}^3} = 3895.66\text{ g}$

7. volume of the 20 spheres = $20\left[\frac{4}{3} \times \left(\frac{4\text{ cm}}{2}\right)^3 \times \pi\right] = 670.21\text{ cm}^3$,

density of mahogany is $.71\text{g/cm}^3$, $670.21\text{ cm}^3 \times .71\text{ g/cm}^3 = 475.85\text{ g}$

8. volume of gold cone = $\frac{1}{3} \times (\pi \times (1.5")^2 \times 3") = 7.07\text{ in}^3$, $7.07\text{ in}^3 \times \frac{16.387\text{ cm}^3}{1\text{ in}^3} = 115.86\text{ cm}^3$

Density of gold is 19.3 g/cm^3, $115.86\text{ cm}^3 \times 19.3\text{ g/cm}^3 = 2236.10\text{ g}$, $2236.10\text{ g} \times \frac{\$41.76}{1\text{g}}$

$= \$93379.54$

9. Volume of wall = 400 (10cm x 13cm x 25cm) = 1300000 cm^3,
density of brick = 2.1 g/cm^3, $1300000\text{ cm}^3 \times 2.1\text{ g/cm}^3 = 2730000\text{ g}$ $\frac{2730000\text{ g}}{1000} = 2730\text{ kg}$

CHAPTER 6

Trigonometry

6-1 SINE, COSINE, AND TANGENT RATIOS

TRIGONOMETRY (from Greek trigōnon, "triangle" and metron, "measure") is a branch of mathematics that studies relationships involving lengths and angles of triangles. Trigonometry is an extremely useful tool in the industrial profession, as it can be applied to find measures such as heights, distances, and angle measures, and can be used to model periodic phenomena, such as the harmonic motion of waves.

TRIGONOMETRIC FUNCTIONS

Trigonometric functions (also called the **circular functions**) are functions of an angle. They relate the angles of a right triangle to the lengths of its sides. The most familiar trigonometric functions are the SINE, COSINE, and TANGENT functions.

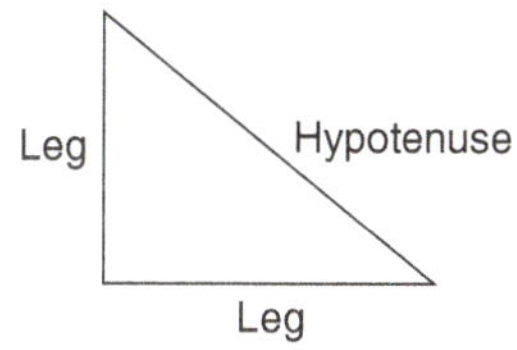

As you recall, here is a right triangle showing legs and hypotenuse.

Given one of the nonright triangle angles, we can identify one of the legs as either "opposite leg" or "adjacent leg." For example,

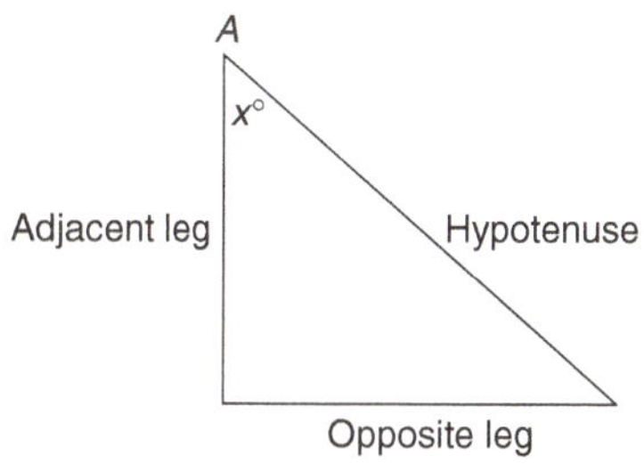

if the measure of angle A of this right triangle is $x°$,

The leg that forms the angle A with the hypotenuse is called the **ADJACENT LEG,** and the leg opposite across the triangle from angle A is called the **OPPOSITE LEG**. Use the respective lengths for the sides of the right triangle, we form ratios that correspond to the angle measure, like this

$$\text{The SINE relationship is } \sin x° = \frac{\text{opposite leg}}{\text{hypotenuse}}$$

$$\text{The COSINE relationship is } \cos x° = \frac{\text{adjacent leg}}{\text{hypotenuse}}$$

And the TANGENT relationship is $\tan x^\circ = \dfrac{\text{opposite leg}}{\text{adjacent leg}}$

Example: For this right triangle, find sine, cosine, and tangent of angle *A*

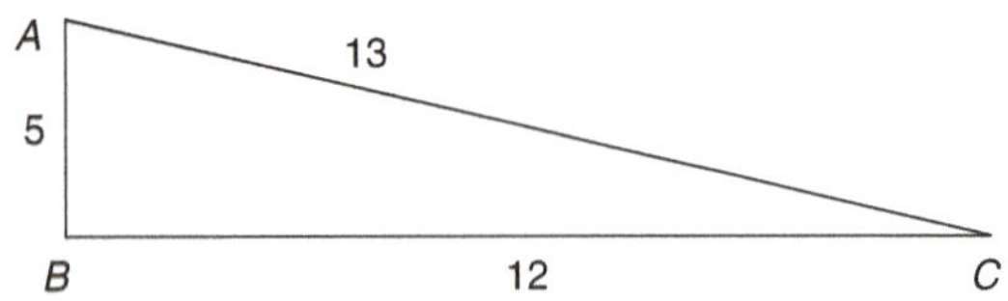

For angle *A*, opposite leg = 12

adjacent leg = 5

hypotenuse = 13

$$\sin A = \frac{\text{opposite leg}}{\text{hypotenuse}} = \frac{12}{13} = .9231$$

$$\cos A = \frac{\text{adjacent leg}}{\text{hypotenuse}} = \frac{5}{13} = .3846$$

$$\tan A = \frac{\text{opposite leg}}{\text{adjacent leg}} = \frac{12}{5} = 2.40$$

What about angle *C*? For angle *C*, opposite leg = 5, adjacent leg = 12, hypotenuse = 13

And

$$\sin C = \frac{\text{opposite leg}}{\text{hypotenuse}} = \frac{5}{13} = .3846$$

$$\cos C = \frac{\text{adjacent leg}}{\text{hypotenuse}} = \frac{12}{13} = .9231$$

$$\tan C = \frac{\text{opposite leg}}{\text{adjacent leg}} = \frac{5}{12} = .4167$$

Practice: For these right triangles, find the sin, cos, and tan of angles *A* and *C*

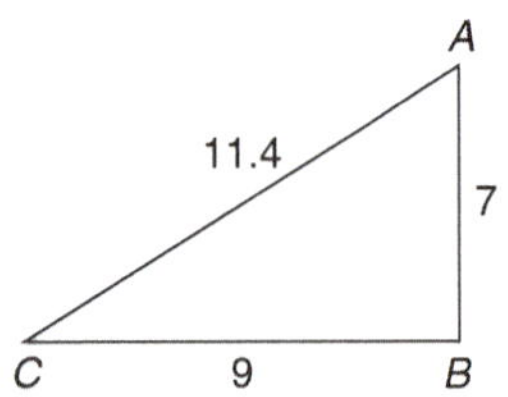

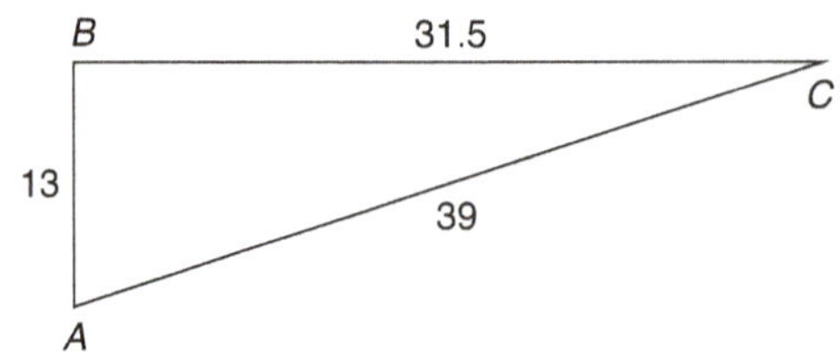

sin *A* _____ cos *A* _____ tan *A* ______ sin *A* _____ cos *A* _____ tan *A* ______

sin *C* _____ cos *C* _____ tan *C* ______ sin *C* _____ cos *C* _____ tan *C* ______

Given any right triangles with the same corresponding nonright angle measures will have the same sine, cosine, and tangent ratio value. SIMILAR right triangles, which means they will retain the same shape (angles) but different sizes have corresponding leg and/or hypotenuse lengths that are in proportion. This is why the sine, cosine, and tangent ratios will always give you the same value as long as the angle measures for your nonright angles are the same.

For example, here are two right triangles with the same corresponding angle measures. These triangles are similar.

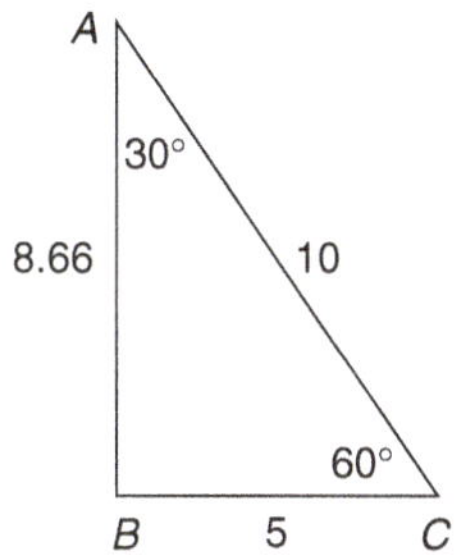

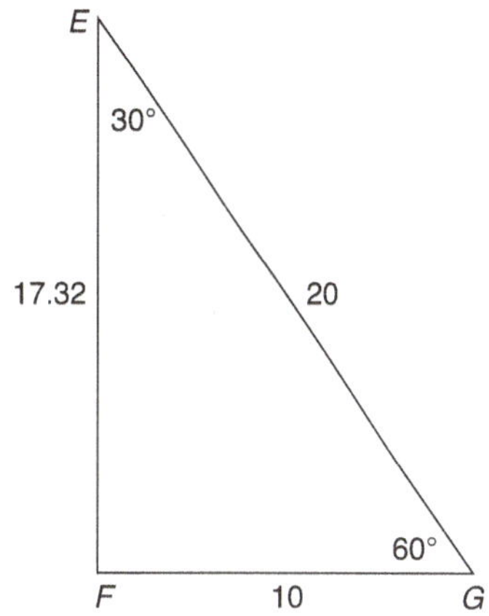

They have the same corresponding angles (30°, 60°, 90°) so have the same shape. The corresponding side lengths form equal ratios. The leg length 8.66 corresponds to 17.32, leg length 5 corresponds to 10, and hypotenuse leg length 10 corresponds to 20.

The ratios of small triangle lengths to big triangle lengths are

$\frac{8.66}{17.32} = \frac{5}{10} = \frac{10}{20}$ which all equal $\frac{1}{2}$ so are in proportion.

As angle A corresponds to angle E and are both have angle measure 30°

You will find the trigonometric function ratios hold for both triangles. Here is sin, cos, tan

$$\sin A = \frac{\text{opposite leg}}{\text{hypotenuse}} = \frac{5}{10} = .500 = \sin 30^\circ \qquad \sin E = \frac{\text{opposite leg}}{\text{hypotenuse}} = \frac{10}{20} = .500 = \sin 30^\circ$$

$$\cos A = \frac{\text{adjacent leg}}{\text{hypotenuse}} = \frac{8.66}{10} = .866 = \cos 30^\circ \qquad \cos E = \frac{\text{adjacent leg}}{\text{hypotenuse}} = \frac{17.32}{20} = .866 = \cos 30^\circ$$

$$\tan A = \frac{\text{opposite leg}}{\text{adjacent leg}} = \frac{5}{8.66} = .577 = \tan 30^\circ \qquad \tan E = \frac{\text{opposite leg}}{\text{adjacent leg}} = \frac{10}{17.32} = .577 = \tan 30^\circ$$

And given this, all angles, from 0° to 180°, can be calculated for sin, cos, and tan and recorded in tables, and better yet, in <u>your scientific calculator</u>!!!

Why is that important? Remember when we did the Pythagorean theorem? We needed at least two sides to figure the third side length. Now with Trig functions, we can use an angle of a right triangle and one of the sides to find the other two sides. We will get to that later.

Also, we can find the angles of the right triangle, given the side lengths. We will explore that now . . .

Let's use the calculator to figure sin, cos, and tan of a given angle.

Find sin, cos, and tan of 66°.

First, hit the DRG button and make sure DEG is showing. We can calculate trig functions for other types of measure, such as radians and gradians, but we will stick with degrees for now.

For sin calculation, hit the SIN button then 66. You should get .9135. So sin 66° = .9135

For cos calculation, hit the COS button followed by 66 to get .4067 so cos 66° = .4067

And do the same thing with the TAN button the 66. You should get tan 66° = 2.246

Practice, use your calculator to find sin, cos, and tangent for angles

12° ______________________

131° ______________________

2.45° ______________________

90° ______________________

179° ______________________

What if we are given a ratio value for sin, cos, or tan . . . Can we find the corresponding angle measure? Yes, we will use the inverse operation of sin, cos, and tan. We refer these operations as the "arcsin," the "arcos," and the "arctan" of the ratio. And we are just undoing the sin, cos, and tan ratio back to the angle measure.

On the calculator, to find the "arc functions," you hit 2nd button then the trig function button followed by the ratio.

For example, let's say the ratio for the triangle angle is $\frac{\text{opposite leg}}{\text{hypotenuse}} = .5200$. As this is the sin function ratio, to find the angle that produces this ratio, we need arcsin .5200.

On the calculator hit 2nd SIN .5200 ENTER.

This gives you 31.33. This means that 31.33° angle on a right triangle results in the ratio $\frac{\text{opposite leg}}{\text{hypotenuse}}$ of .5200. And this means that sin 31.33° = .5200.

If the ratio $\frac{\text{adjacent leg}}{\text{hypotenuse}}$ = .321, this is the cos ratio, so to find the angle we need arccos .321, we hit 2nd COS .321 ENTER and it produces 71.28. So, cos 71.28° = .321.

And it follows to find arctan, use 2nd TAN $\frac{\text{opposite leg}}{\text{adjacent leg}}$ ratio ENTER

Practice: Find arcsin .4590 arccos .9921 arctan 3.521

Now let's put it all together to find the angle measures of A and C in the right triangles below:

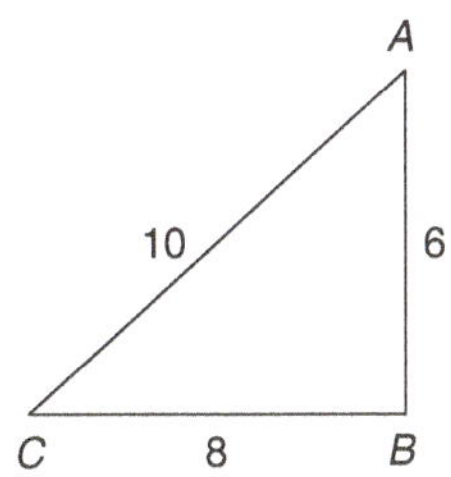

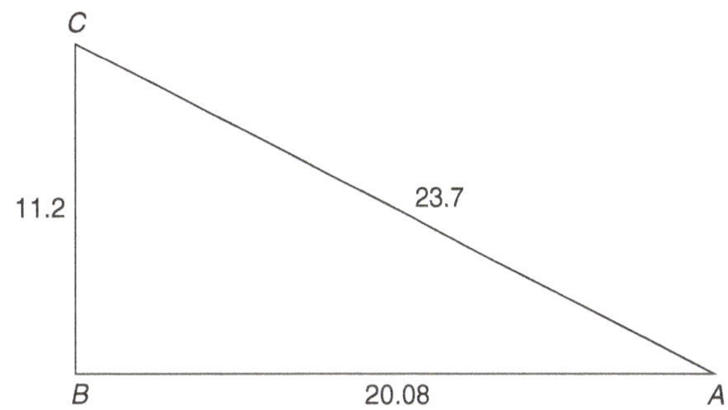

NOTE: Here is a technique for remembering the trigonometric ratios.

The SINE relationship is $\sin x° = \frac{\text{opposite leg}}{\text{hypotenuse}}$

The COSINE relationship is $\cos x° = \frac{\text{adjacent leg}}{\text{hypotenuse}}$

And the TANGENT relationship is $\tan x° = \frac{\text{opposite leg}}{\text{adjacent leg}}$

USE the anagram *"SOH CAH TOA"*

S O H refers to Sin = $\frac{\text{Opposite leg}}{\text{Hypotenuse}}$

C A H refers to Cos = $\frac{\text{Adjacent leg}}{\text{Hypotenuse}}$

T O A refers to Tan = $\frac{\text{Opposite leg}}{\text{Adjacent leg}}$

6-1 Assignments

Identify the a) adjacent side length b) opposite side length c) hypotenuse length of the following angles for these right triangles.

1. angle *A*

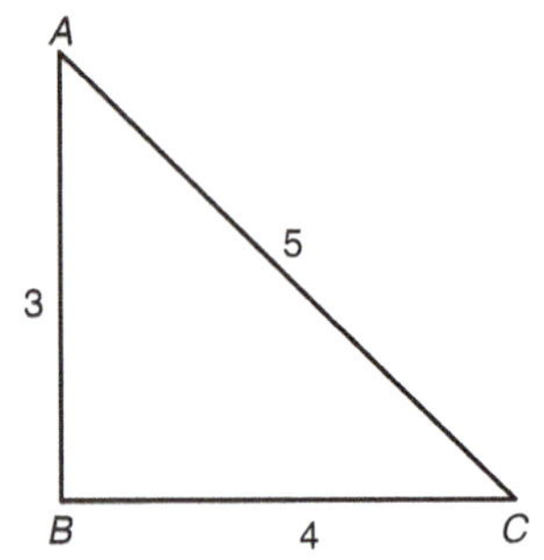

2. angle *F*

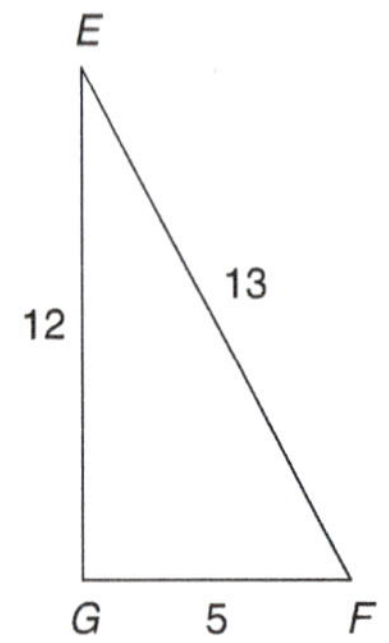

3. angle *J*

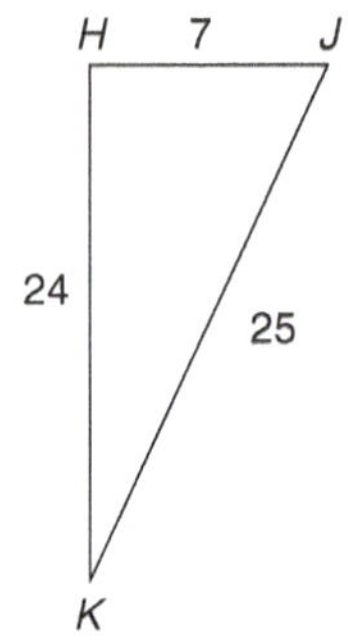

Give the trigonometric ratio for these right triangles. Keep in fractions.

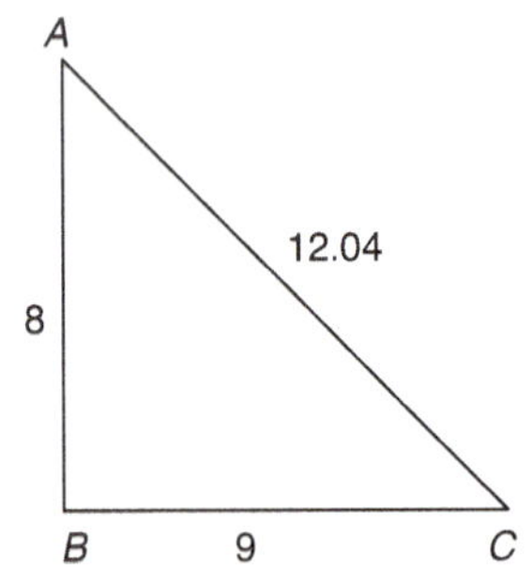

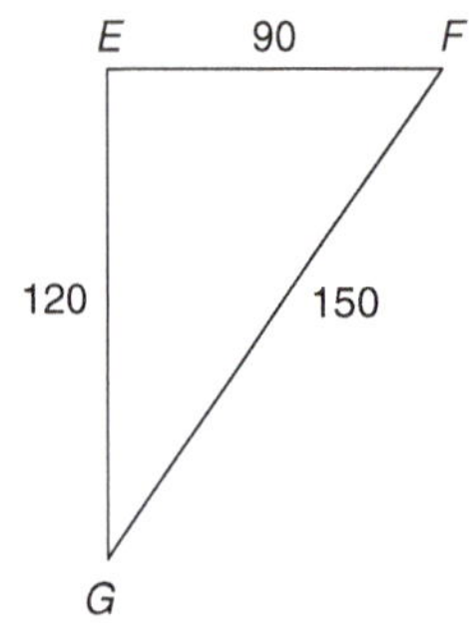

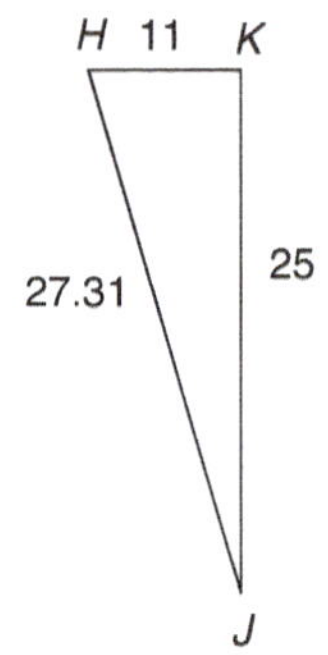

4. a) sin A = ____
   b) sin C = ____
   c) cos A = ____
   d) cos C = ____
   e) tan A = ____
   f) tan C = ____

5. a) sin G = ____
   b) sin F = ____
   c) cos G = ____
   d) cos F = ____
   e) tan G = ____
   f) tan F = ____

6. a) sin H = ____
   b) sin J = ____
   c) cos H = ____
   d) cos J = ____
   e) tan H = ____
   f) tan J = ____

Using your calculator, find the trigonometric value for these angle degree, to the nearest hundredth of a value

7. sin 72° ____

8. tan 20° ____

9. cos 45° ____

10. sin 5° _____

11. tan 45° ____

12. sin 30° ____

13. cos 88° _____

14. tan 68° ____

Using your calculator, find degree measure, to the nearest hundredth of a degree, for these trigonometric ratios

15. arcsin 3/4 _____

16. arccos 2/11 ____

17. arctan 8/9 _____

18. arccos 1/3 _____

19. arctan .111 _____

20. arcsin .456 _____

21. arccos .877 _____

22. arcsin .555 _____

23. arctan 2.678 ____

for 24. and 25.

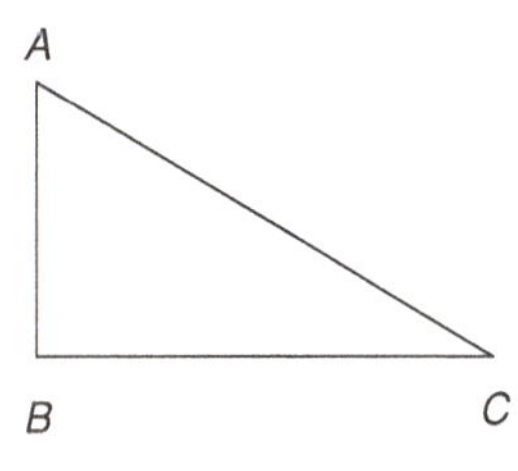

24. Explain: Why, in a right triangle with vertices A, B, C, and angle B is the right angle, is $\sin A = \cos C$ and $\sin C = \cos A$.

25. Explain why cos and sin ratio values can never be greater than 1.

Answers

1. a) 3
   b) 4
   c) 5

2. a) 5
   b) 12
   c) 13

3. a) 7
   b) 24
   c) 25

4. a) 9/12.04
   b) 8/12.04
   c) 8/12.04
   d) 9/12.04
   e) 9/8
   f) 8/9

5. a) 90/150
   b) 120/150
   c) 120/150
   d) 90/150
   e) 90/120
   f) 120/90

6. a) 25/27.31
   b) 11/27.31
   c) 11/27.31
   d) 25/27.31
   e) 25/11
   f) 11/25

7. .95

8. .36

9. .71

10. .09

11. 1

12. .5

13. .03

14. 2.48

15. 48.59°

16. 79.52°

17. 41.63°

18. 70.53°

19. 6.33°

20. 27.13°

21. 28.72°

22. 33.71°

23. 69.52°

6-2 TRIGONOMETRIC RATIOS TO FIND SIDE LENGTHS OF RIGHT TRIANGLES

We can find a side in a right-angled triangle using trigonometric ratios if we know

- One side length, and
- One angle (i.e., apart from the right angle).

Example: Find the other two sides for this right triangle.

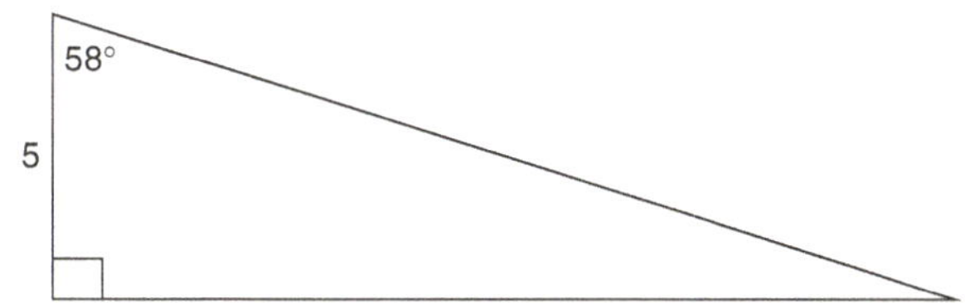

We have the adjacent length, 5, for the angle 58°. If we want to find the leg length opposite the angle, we know that $\tan 58° = \dfrac{\text{opposite leg}}{\text{adjacent leg}} = \dfrac{\text{opposite leg}}{5}$.

So for $\tan 58° = \dfrac{\text{opposite leg}}{5}$ we can multiply both sides by 5 to get

$$5 \cdot \tan 58° = 5 \cdot \frac{\text{opposite leg}}{5}$$

$$5 \cdot \tan 58° = \text{opposite leg}$$

$$8 = \text{opposite leg length}$$

To find the hypotenuse length, we know that $\cos 58° = \dfrac{\text{adjacent leg}}{\text{hypotenuse}} = \dfrac{5}{\text{hypotenuse}}$

So for $\cos 58^{c} = \dfrac{5}{\text{hypotenuse}}$, we multiply both sides by hypotenuse

$$\text{Hypotenuse} \cdot \cos 58° = \text{hypotenuse} \cdot \frac{5}{\text{hypotenuse}}$$

$$\text{Hypotenuse} \cdot \cos 58° = 5 \qquad \text{now we divide both sides by } \cos 58°$$

$$\frac{\text{Hypotenuse} \cdot \cos 58°}{\cos 58°} = \frac{5}{\cos 58°}$$

$$\text{Hypotenuse} = \frac{5}{\cos 58°}$$

$$\text{Hypotenuse length} \approx 9.43$$

You can also find the other non-right angle measure as the two non-right angles degrees add up to 90°. So the other angle measure is $90° - 58° = 32°$.

So the other two side lengths and the other angle measure for this right triangle are

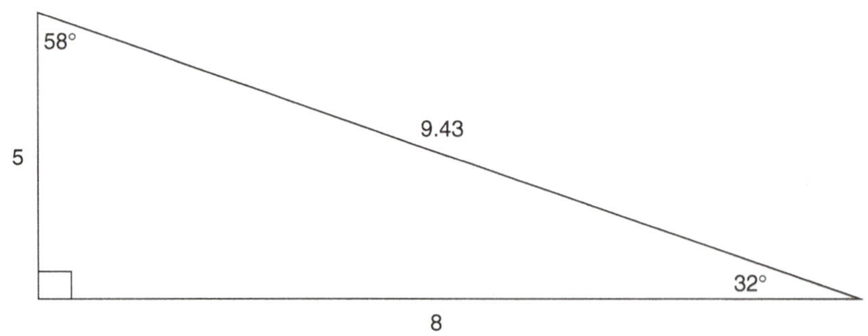

Example: Find the missing side lengths and angle measure for

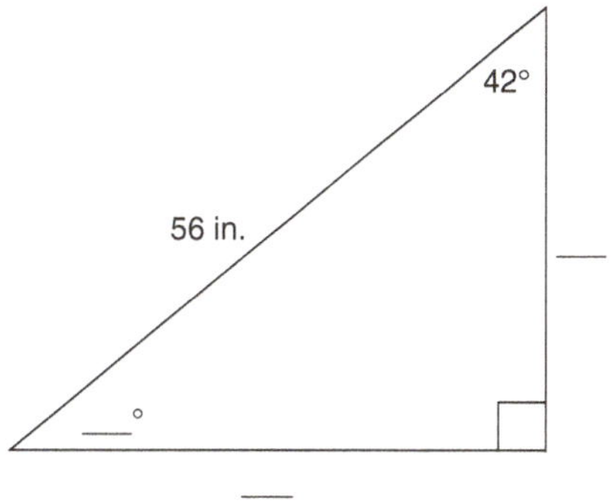

Given is the hypotenuse length of 56 in. The angle measure given is 42°. If we want to find the leg length that is adjacent to the 42° angle, we need $\cos 42° = \frac{\text{adjacent leg}}{\text{hypotenuse}} = \frac{\text{adjacent leg}}{56}$

$$\cos 42° = \frac{\text{adjacent leg}}{56} \qquad \text{multiply both sides by 56}$$

$$56 \cdot \cos 42° = 56 \cdot \frac{\text{adjacent leg}}{56}$$

$$56 \cdot \cos 42° = \text{adjacent leg}$$

$$41.62 \text{ in.} \approx \text{adjacent leg length}$$

To find the opposite leg length to the 42° angle, we can use $\sin 42° = \frac{\text{opposite leg}}{\text{hypotenuse}} = \frac{\text{opposite leg}}{56}$

$$\sin 42° = \frac{\text{opposite leg}}{56} \qquad \text{multiply both sides by 56}$$

$$56 \cdot \sin 42° = 56 \cdot \frac{\text{opposite leg}}{56}$$

$$56 \cdot \sin 42° = \text{opposite leg}$$

$$37.\,47 \text{ in.} \approx \text{opposite leg length}$$

Other angle is $90° - 42° = 48°$.

So missing side lengths and angle measure for

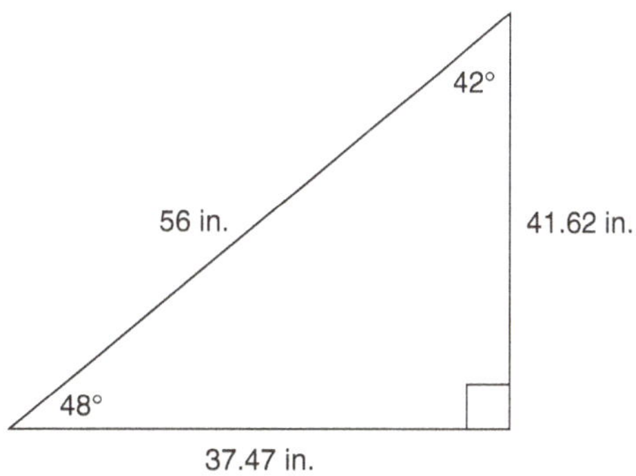

Practice

Find the missing side lengths and angle measures for

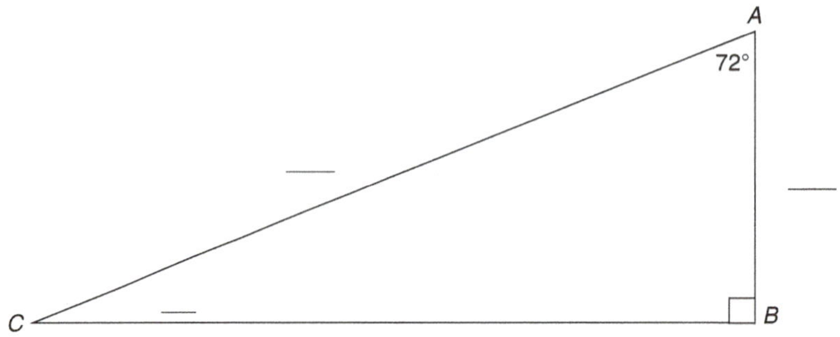

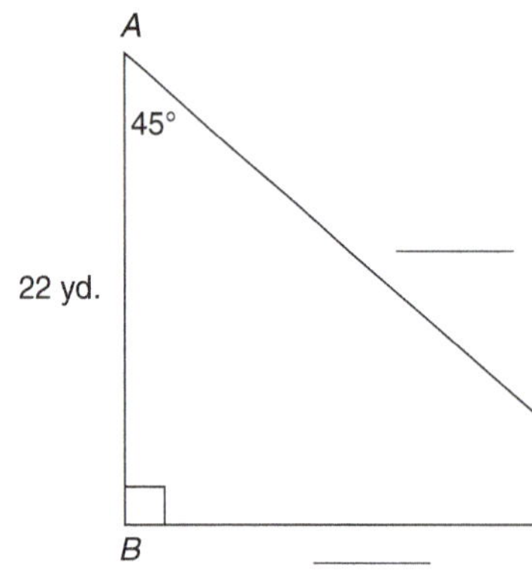

6-2 Assignments

Find the missing sides and angle for the following right triangles using trigonometry ratios

1.

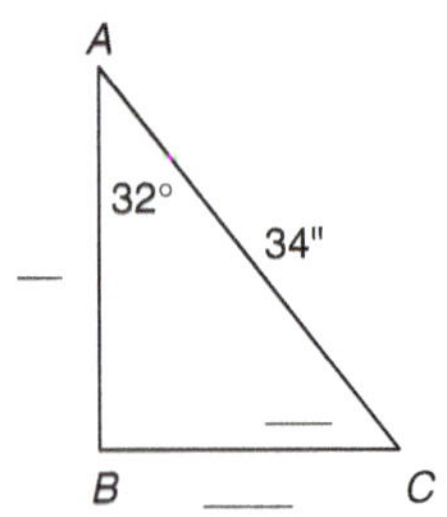

2.

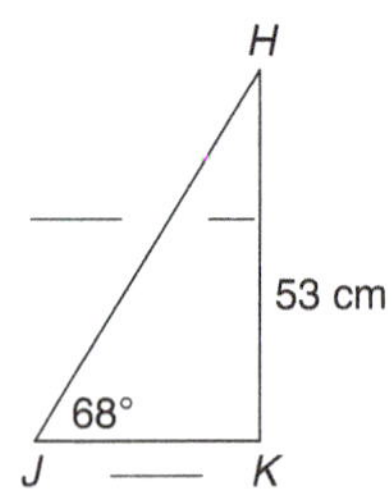

3.

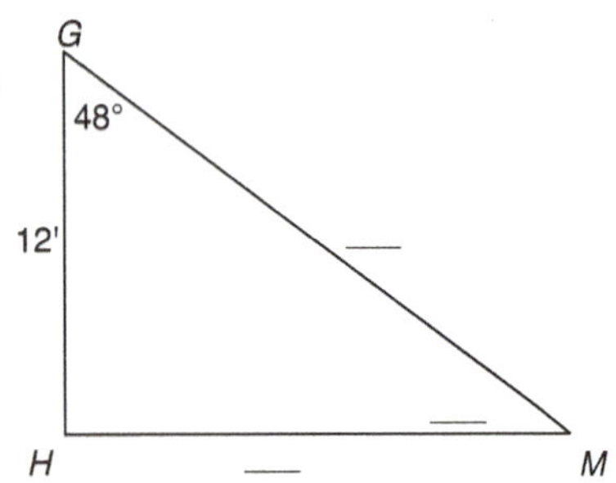

4.

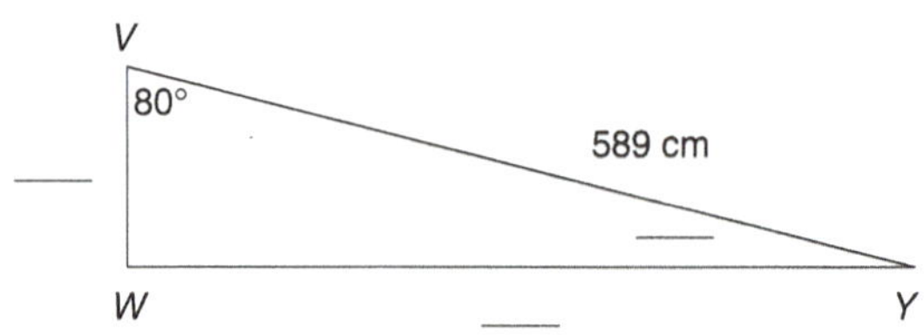

5.

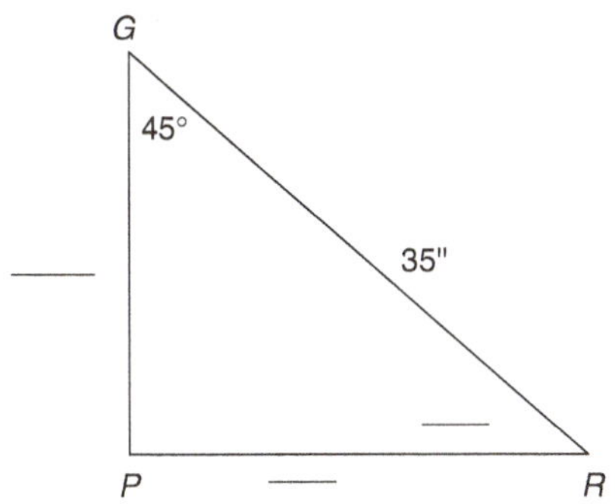

6.

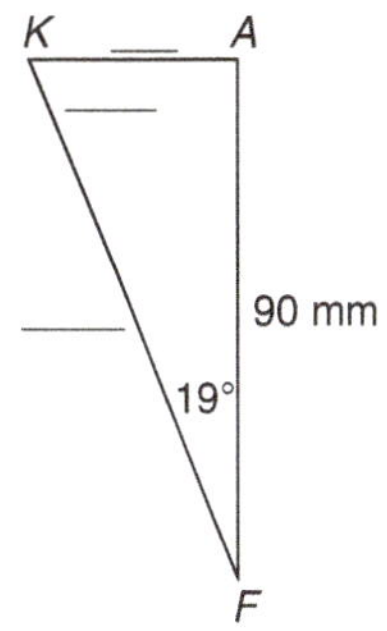

Answers

1.

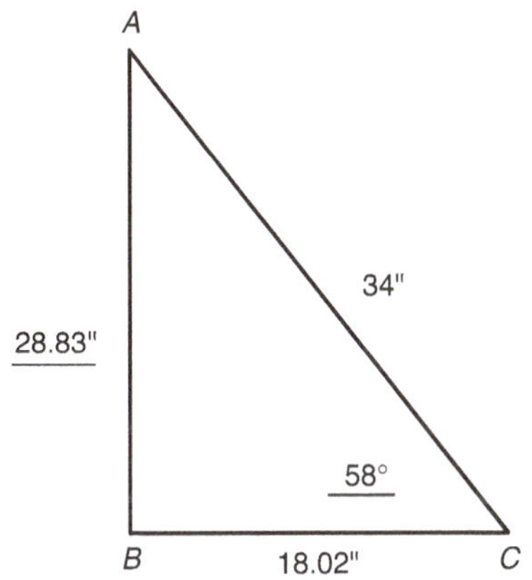

2.

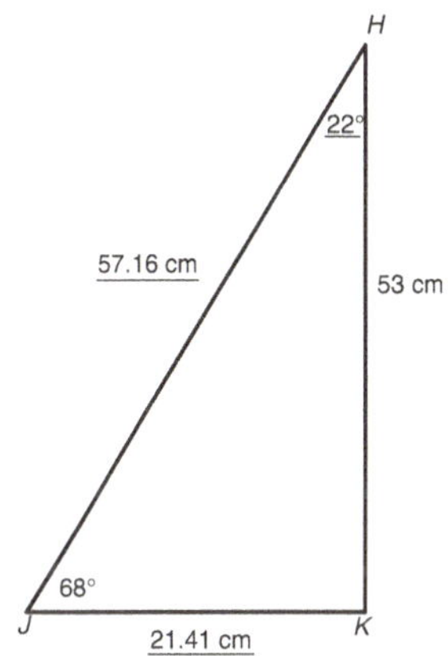

3.

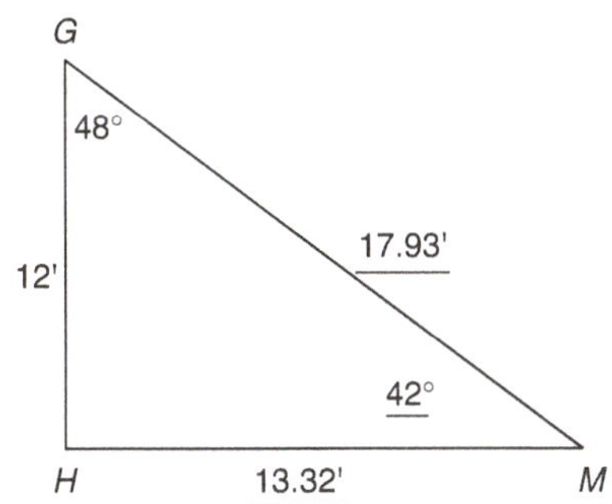

4.

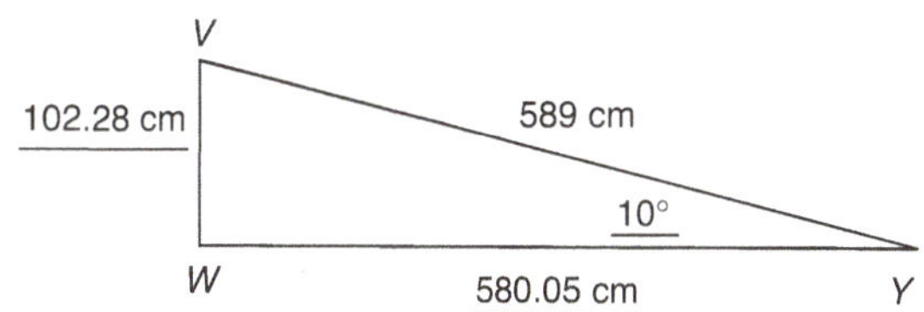

5.

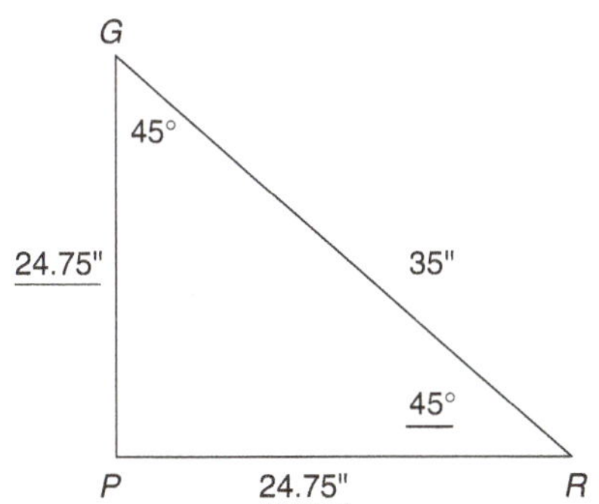

6.

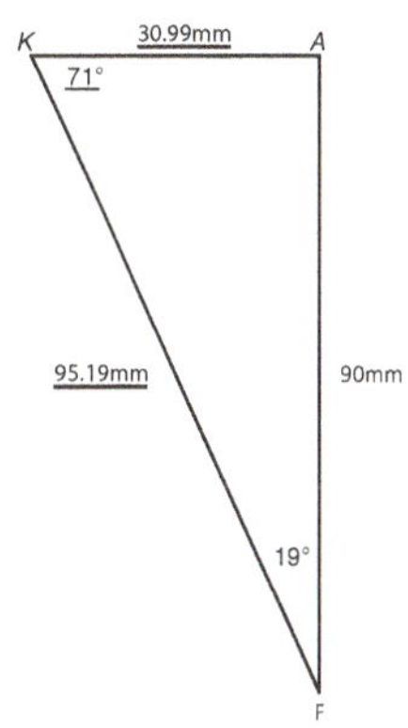

6-3 ARCSIN, ARCCOS, AND ARCTAN TO FIND ANGLES OF RIGHT TRIANGLES

In section 4-3, we found a third side of a right triangle if given two sides using Pythagorean theorem. We can also find the third side in this situation if give two sides using trigonometry. Plus, we can figure the nonright angle measures as well!

Find the missing side and missing angles.

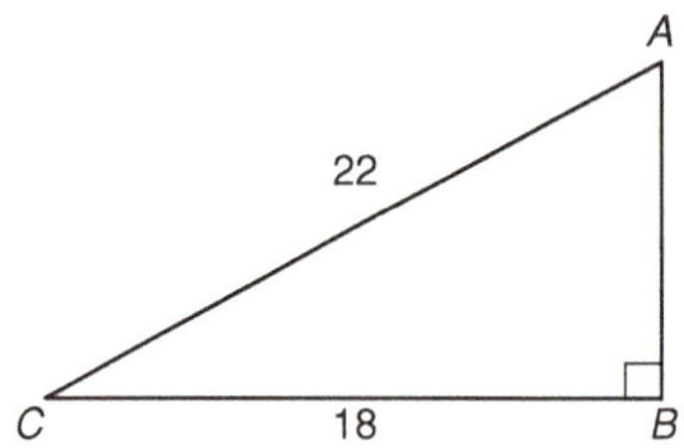

Angle B is the 90° angle. Let first find measure of angle A.

We know the hypotenuse is 22 and the side opposite angle A is 18.

As $\frac{\text{opposite leg}}{\text{hypotenuse}}$ is the sine ratio, then $\sin A = \frac{\text{opposite leg}}{\text{hypotenuse}} = \frac{18}{22}$. Now we can use arcsin to find the angle measure of angle A.

$$\arcsin \frac{18}{22} \approx 54.9 \text{ so angle } A \text{ measure is } 54.9°$$

As the missing side is the adjacent leg to angle A, we can use cosine ratio $\frac{\text{adjacent leg}}{\text{hypotenuse}}$ to find the missing side

$$\cos 54.9° = \frac{\text{adjacent leg}}{\text{hypotenuse}} = \frac{\text{adjacent leg}}{22} \quad \text{multiply both sides by 22}$$

$$22 \cdot \cos 54.9° = 22 \cdot \frac{\text{adjacent leg}}{22}$$

$$22 \cdot \cos 54.9° = \text{adjacent side}$$

$$12.65 \approx \text{adjacent side length}$$

And the third angle is $90° - 54.9° = 35.1°$

so our triangle angles and sides are

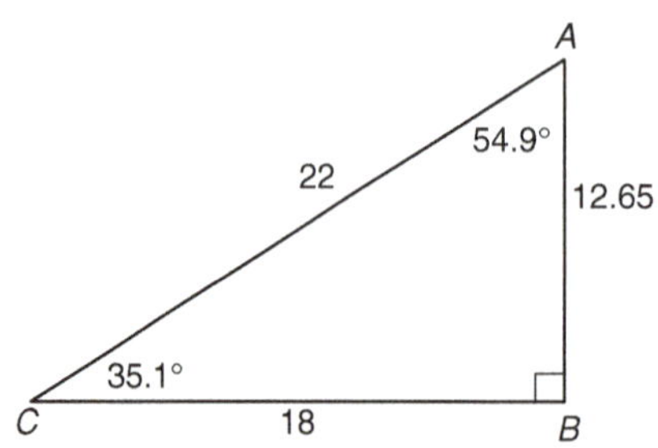

Example: Find the missing side and angles for

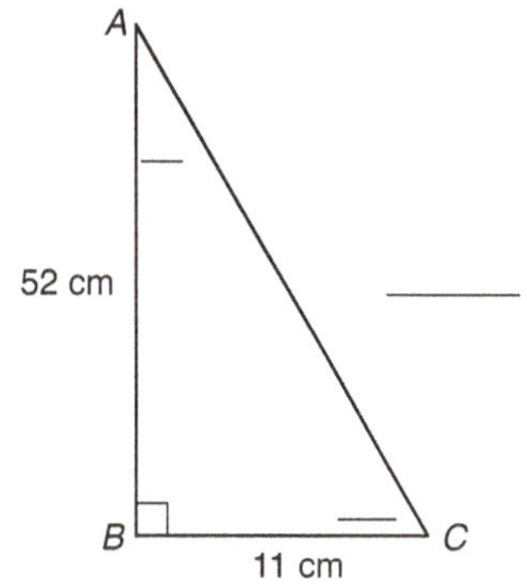

To find angle A measure, use $\tan A = \dfrac{\text{opposite leg}}{\text{hypotenuse}} = \dfrac{11}{52}$

Then take the arctan $\dfrac{11}{52} \approx 11.9$ so angle A measure $= 11.9°$

To find the hypotenuse, use $\sin 11.9° = \dfrac{\text{opposite leg}}{\text{hypotenuse}} = \dfrac{11}{\text{hypotenuse}}$

$$\sin 11.9° = \frac{11}{\text{hypotenuse}} \qquad \text{multiply both sides by } \frac{\text{hypotenuse}}{\sin 11.9°}$$

$$\frac{\text{hypotenuse}}{\sin 11.9°} \cdot \sin 11.9° = \frac{\text{hypotenuse}}{\sin 11.9°} \cdot \frac{11}{\text{hypotenuse}}$$

$$\text{Hypotenuse} = \frac{11}{\sin 11.9°}$$

$$\text{Hypotenuse length} \approx 53.35 \text{ cm}$$

The other angle $= 90° - 11.9° = 78.1°$

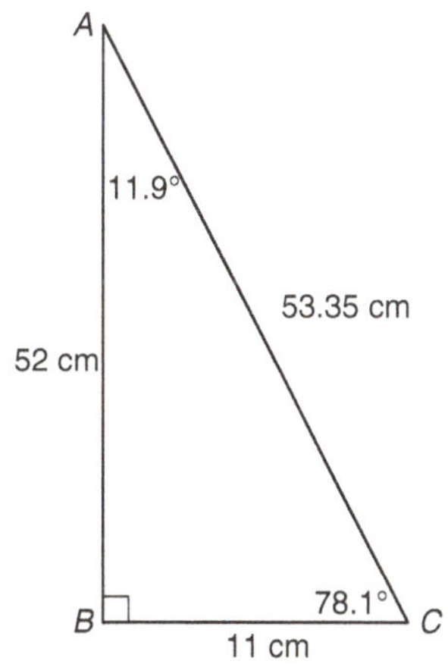

Practice

Using trigonometric functions, find the missing angles and side lengths

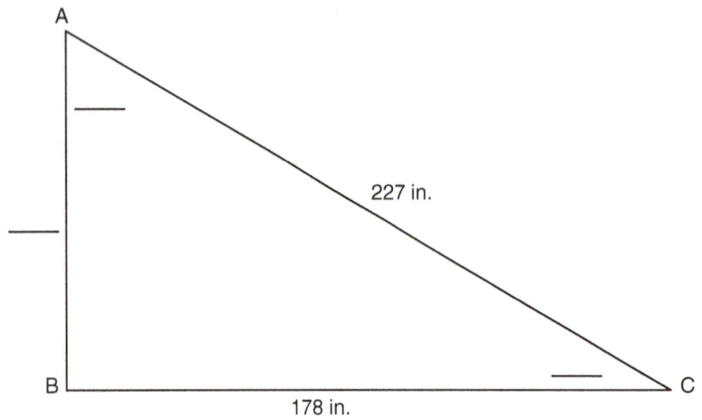

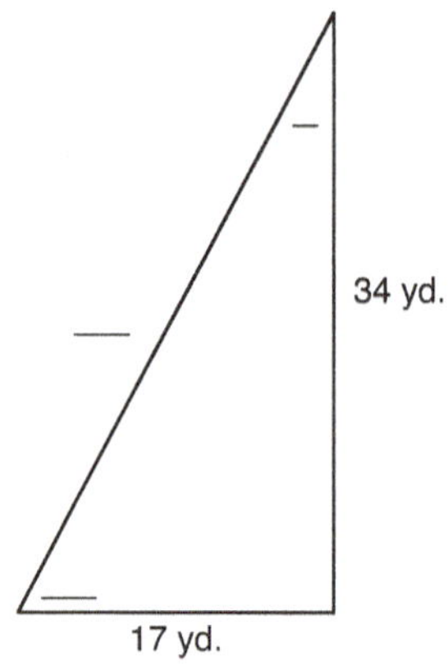

6-3 Assignments

Find the missing angles and side for these triangles

1.

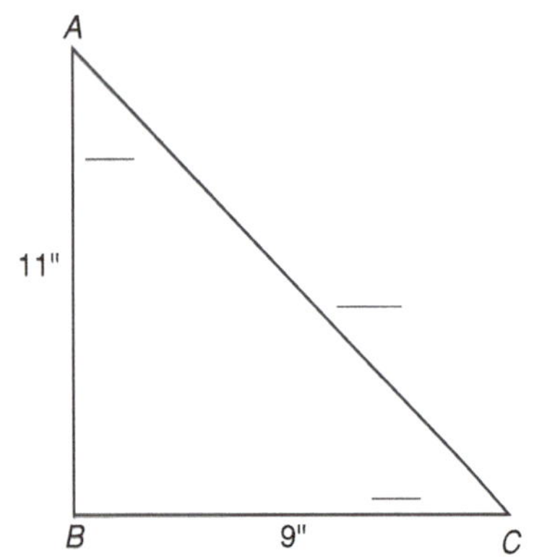

2.

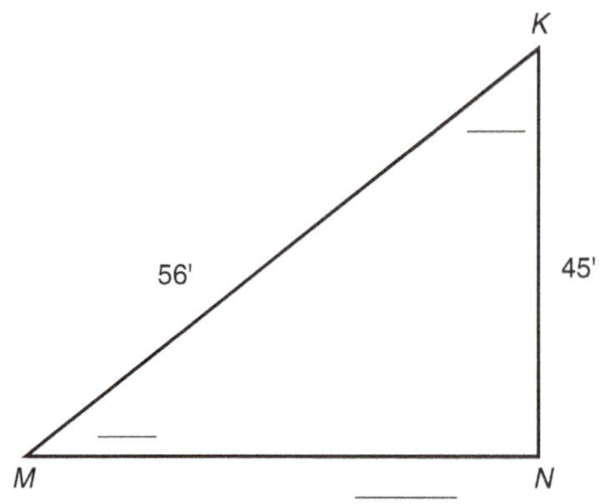

3.

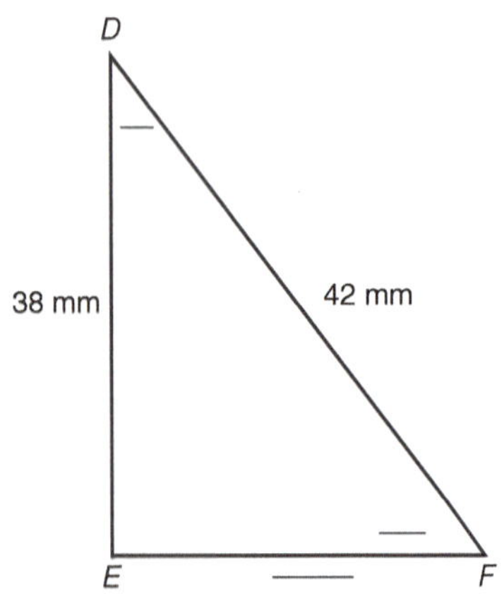

4.

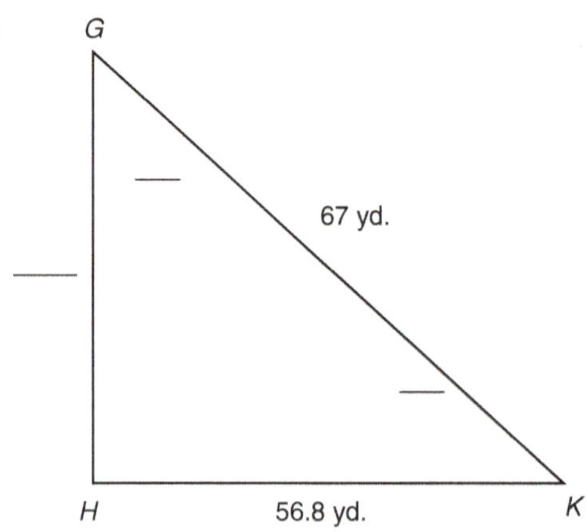

5.

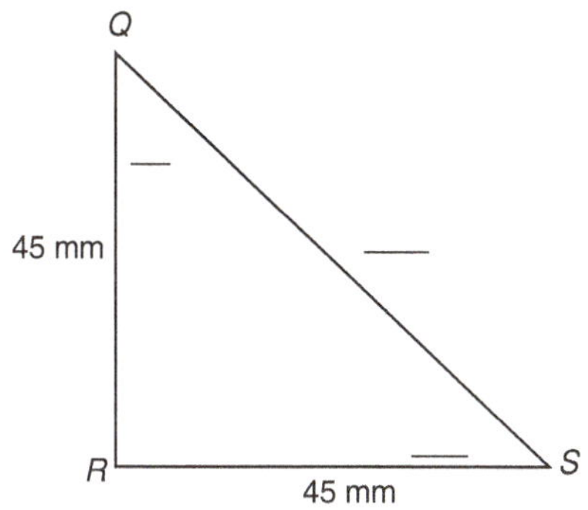

6.

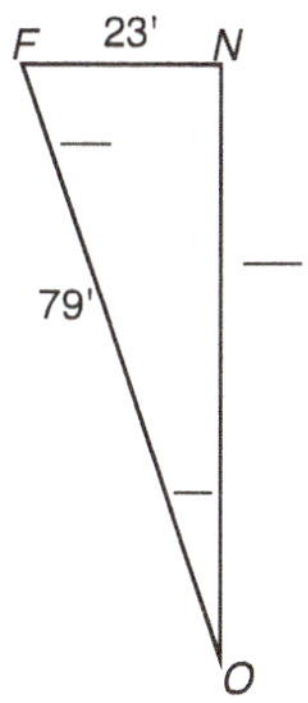

Answers

1.

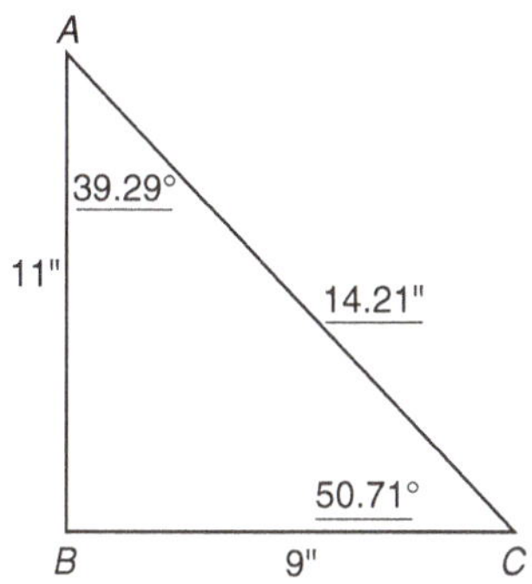

2.

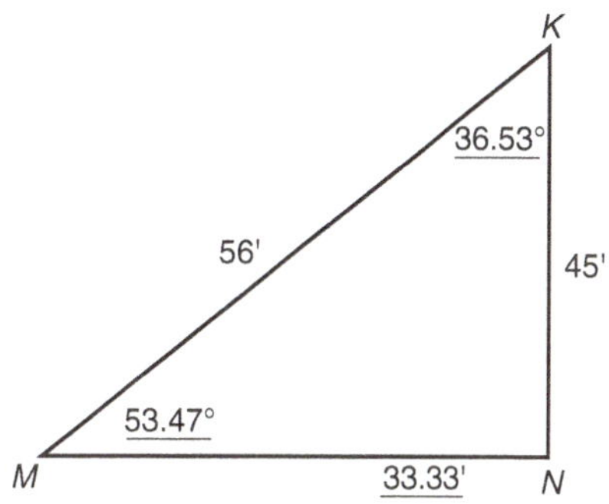

3.

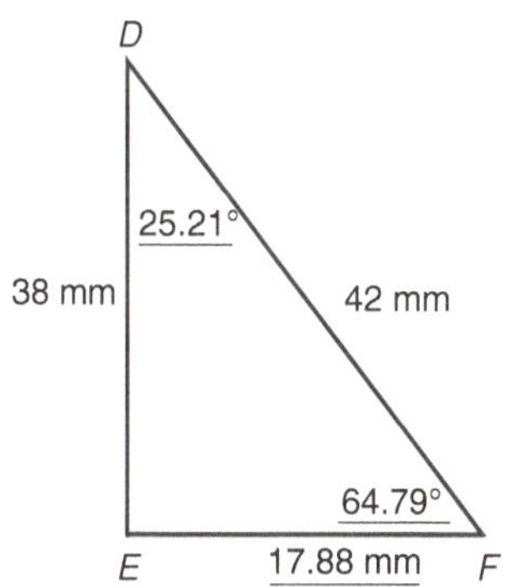

4.

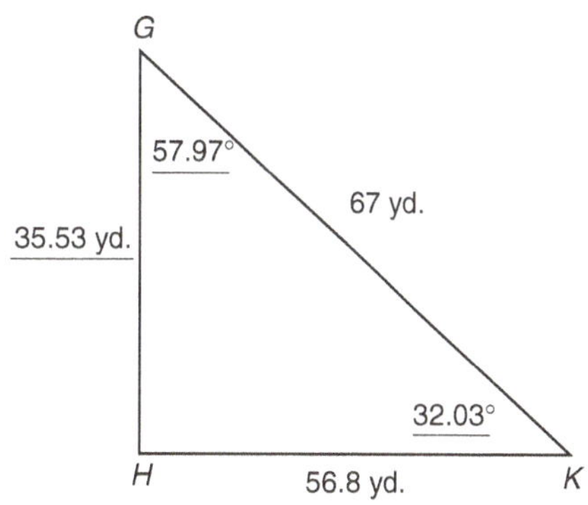

5.

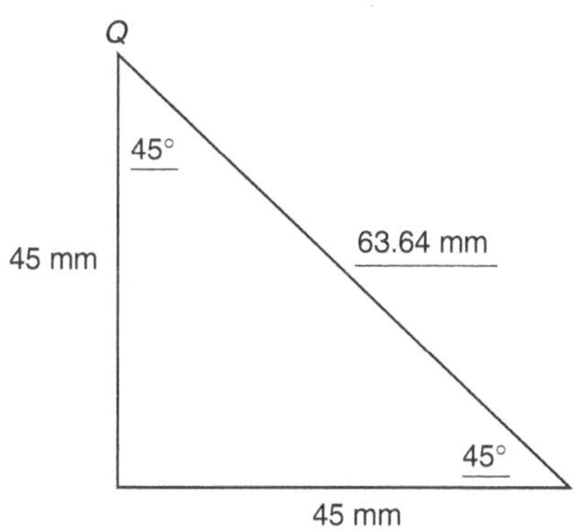

6.

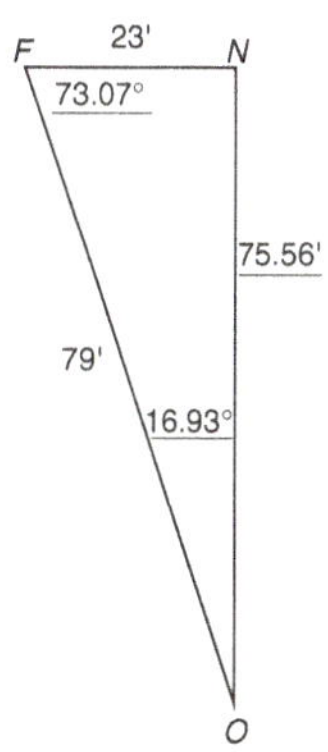

6-4 ANGLE OF ELEVATION AND ANGLE OF DEPRESSION

ANGLES OF ELEVATION go in an upward direction from some horizontal line position.

For example, if looking upward, the angle of elevation is

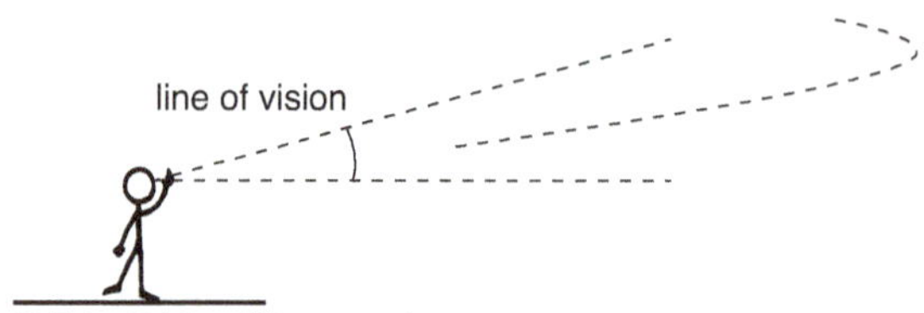

ANGLES OF DEPRESSION go in a downward direction

If looking downward, the angle of depression is

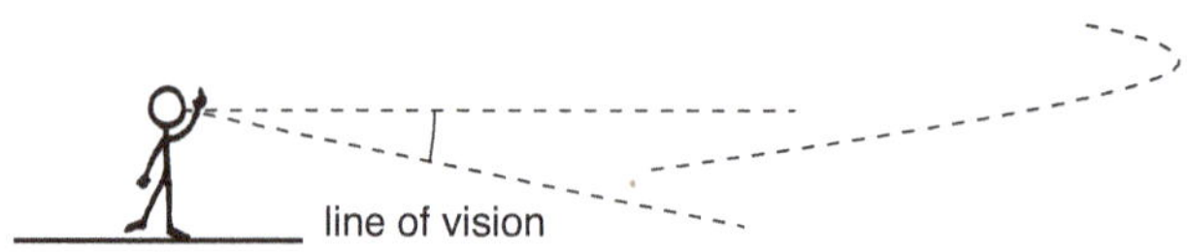

Given angles of elevation and or depression, we can use trigonometry to solve real live problems.

EXAMPLE: A person 6 ft. tall looks up at a bird at the very top of a 30 ft. tree. If the angle of elevation of the persons sight is 39°, the tree is perpendicular to the ground, and the ground is level. How far from the base of the tree is man, and how far from the bird is the man?

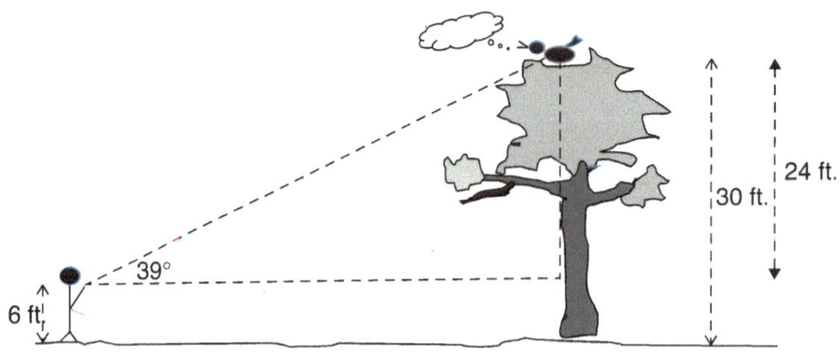

With legs formed by the line from the man to the bird and the line from the man perpendicular to the tree, we can create a right triangle with one angle 39°. As the height of the man's sight is about 6 ft. off the ground, the leg opposite the 39° angle will have a length of 30 ft. – 6 ft. = 24 ft.

Now we can use trigonometry to find the other leg, the distance to the base of the tree, and the hypotenuse, the distance to the bird, to answer our question!

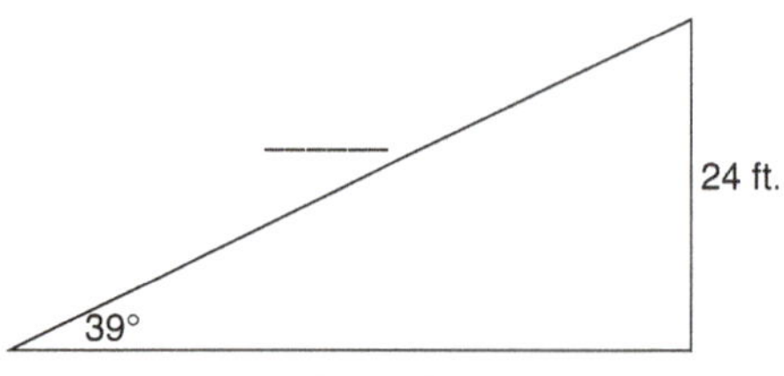

We have the opposite leg for the 39° angle, so to find the hypotenuse, we can use $\sin 39° = \dfrac{\text{opposite leg}}{\text{hypotenuse}}$

$$\sin 39° = \frac{24}{\text{hypotenuse}}$$

$$\frac{\text{hypotenuse}}{\sin 39°} \cdot \sin 39° = \frac{\text{hypotenuse}}{\sin 39°} \cdot \frac{24}{\text{hypotenuse}}$$

$$\text{hypotenuse} = \frac{24}{\sin 39°}$$

$$\text{hypotenuse length} \approx 38.16 \text{ ft.}$$

For the adjacent leg to the 39° angle, we can use tangent, as $\tan 39° = \dfrac{\text{opposite leg}}{\text{adjacent leg}}$ or

$$\tan 39° = \frac{24}{\text{adjacent}}$$

$$\frac{\text{adjacent leg}}{\tan 39°} \cdot \tan 39° = \frac{\text{adjacent leg}}{\tan 39°} \cdot \frac{24}{\text{adjacent leg}}$$

$$\text{adjacent leg} = \frac{24}{\tan 39°}$$

$$29.64 \text{ ft.} = \text{adjacent leg length} \approx 29.64 \text{ ft.}$$

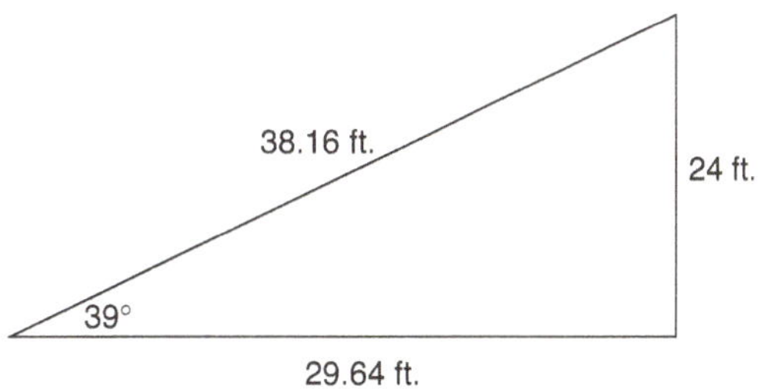

Therefore, the man is 29.64 ft. from the base of the tree and 38.16 ft. from the bird.

Example: An airplane is flying at a <u>height</u> of 2 mi. above the ground. The distance along the ground from the airplane to the airport is 5 mi. What is the <u>angle of depression</u> from the airplane to the airport?

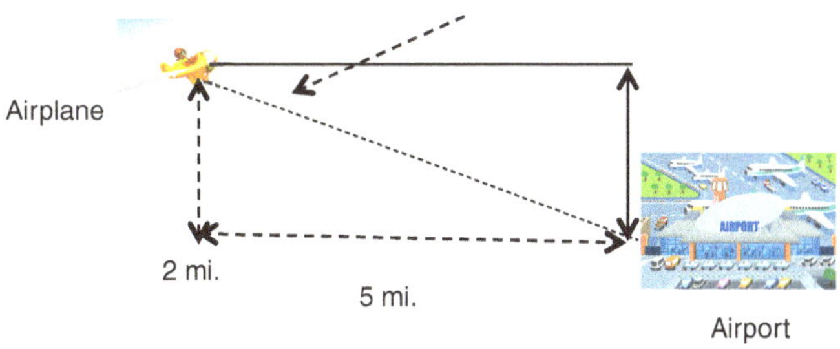

A right triangle is formed with leg lengths of 2 mi and 5 mi.

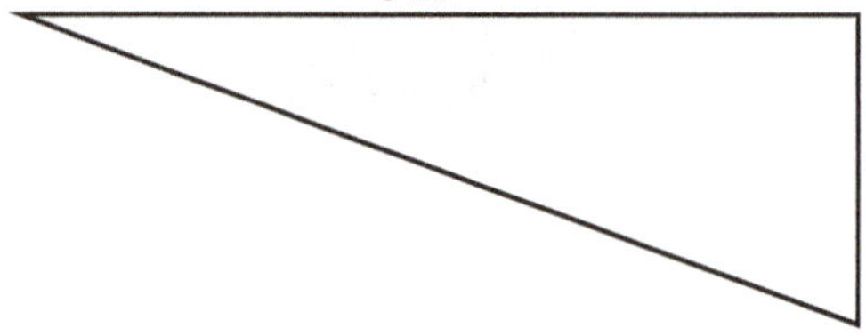

To find the angle of depression, we use the opposite leg length, 2 mi., and the adjacent leg length, 5 mi. and arctan with the ratio $\frac{\text{opposite leg}}{\text{adjacent leg}} = \frac{2}{5}$

So the angle of depression = arctan $\frac{2}{5} \approx 21.80°$

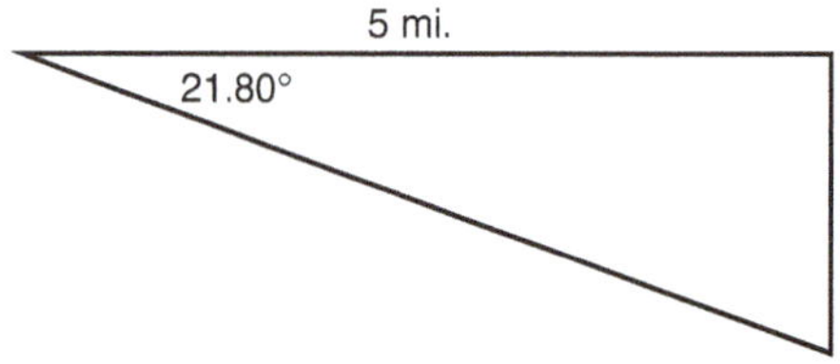

Practice

Curious Correy wants to measure the <u>height</u> of a building. He walks exactly 200 ft. from the <u>base</u> of the building and looks up. The angle from the ground to the top of the building is 33°. How tall is the building?

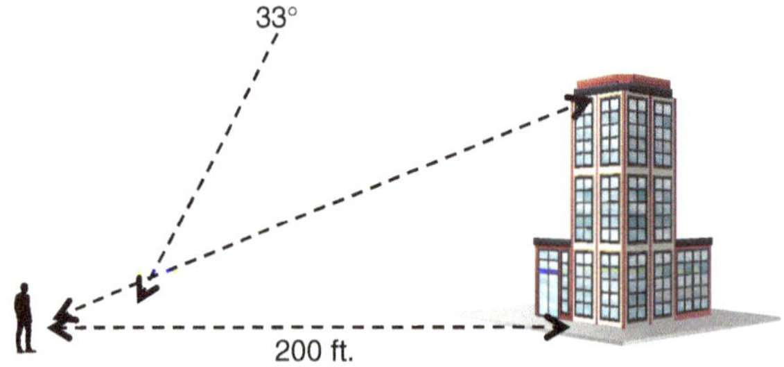

Practice

A Coyote at the top of a cliff 30 m from the ground below, and he spots a roadrunner at an angle of depression of 24^c. How far away is the roadrunner from the coyote, and how far is the roadrunner from the cliff (assuming a shear straight cliff)?

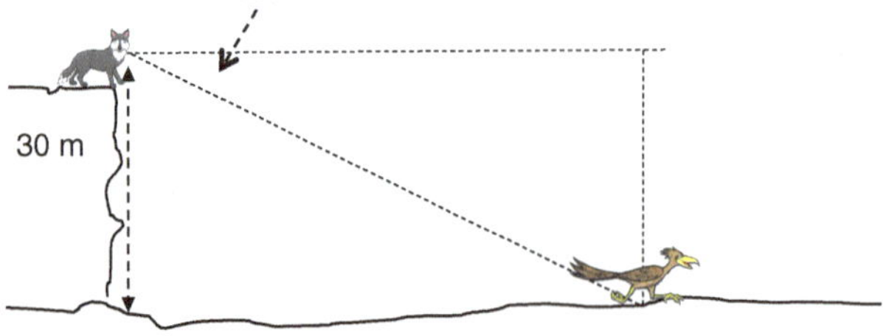

Practice

At a certain time of the day a light post that is 18 ft. tall has a shadow of 5.8 ft. Given the sun's ray at this time, what is the angle of elevation of the sun's ray?

And, given the same time of the day, use this angle to find the shadow's length of a building 300 ft. tall?

6-4 Assignments

1. Matt is at the top of a 100 foot sheer cliff overlooking the ocean and sees a sailboat in the distance. If Matt's angle of depression is 38°,
   a) how far, to the nearest hundredth of a foot, is Matt from the sailboat?
   b) how far, to the nearest hundredth of a foot, is the sailboat form the cliff?

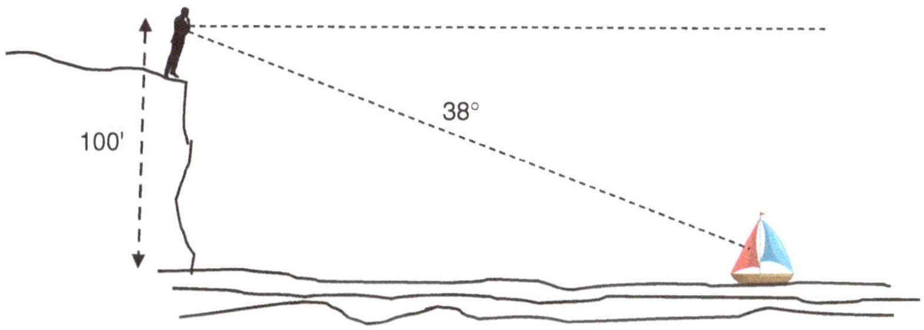

© Michael Sanca/Shutterstock.com (Top Left);
© Studio_G/Shutterstock.com (Bottom Right)

2. Adam looks at the top of a lamppost. He is 300 feet away from the base of the lamppost. If the angle made from the top of the lamppost to Adam with the ground is 25°,
   a) What is the height of the lamppost to the nearest hundredth of a foot?
   b) How far is Adam from the top of the lamppost, to the nearest hundredth of a foot?

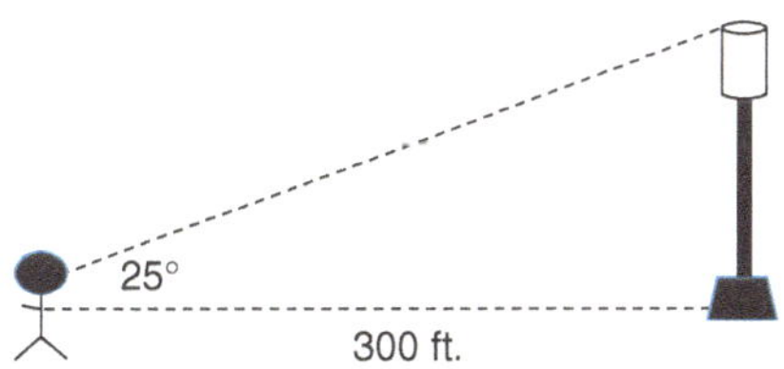

3. A hot air balloon is floating 650 meters above level land. Passengers spots a farm in the distance at an angle of depression of 22°. How far away is the farm from the passengers, to the nearest hundredth of a meter?

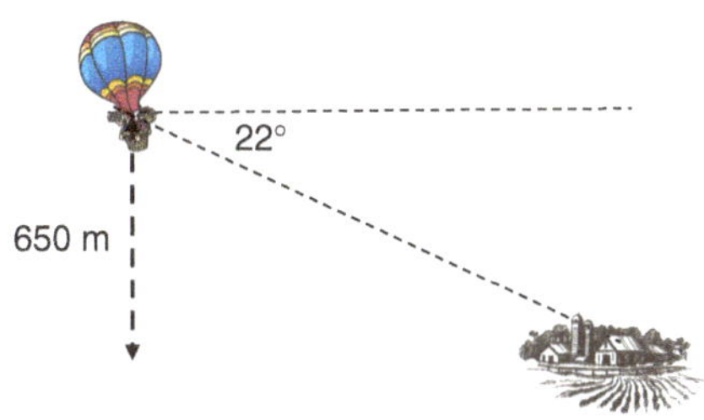

© Memo Angeles/Shutterstock.com (Top Left);
© MorePics/Shutterstock.com (Bottom Right)

4. A flag pole 18 feet tall casts a shadow 12 feet long at a specific time of day. a) Find, to the *nearest hundredth of a degree*, the angle of depression of the sun at this time of day. b) How far is it from the top of the flag pole to the end of its shadow, to the nearest hundredth of a foot?

5. Fabio bought a new beauty shop and wants to order a new sign for the roof of the building. From point P, he finds the angle of elevation of the roof, from ground level, to be 31° and the angle of elevation of the top of the sign to be 38°. If point P is 50 feet from the building, how tall is the sign to the *nearest hundredth of a foot*?

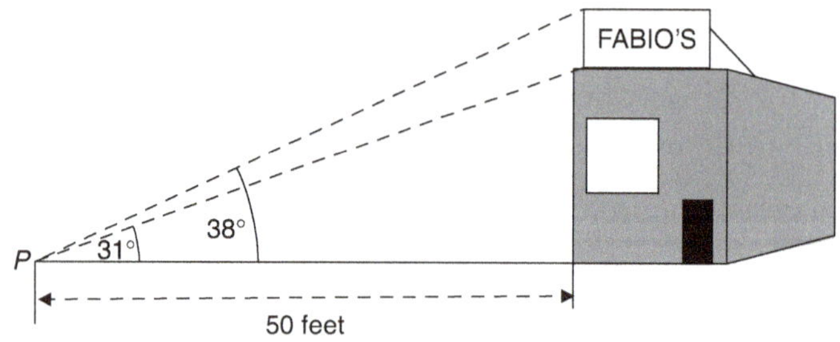

6. Two poles on horizontal ground are 40 meters apart. The shorter pole is 3 meters high. The angle of depression of the top of the shorter pole from the top of the longer pole is 20°.

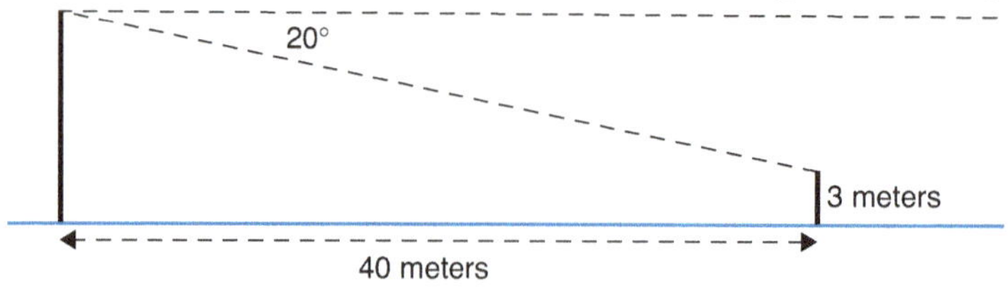

a) Find the height of the other pole, to the nearest hundredth of a meter. b) Find the distance from the top of the two poles, to the nearest hundredth of a meter.

7. A skier is just about to ski down a steep mountain. She is at an elevation of 600 feet from the base of the course. She is headed for a flag at the bottom of the course. If the ski slope is straight and she is 930 feet from the flag, what is her angle of depression with the flag?

8. Buildings A and B are across the street from each other, 100 feet apart. Someone standing on the roof of Building A has an angle of elevation to the top of Building B of 24°, and the angle of depression to the bottom of Building B of 34°. How tall is each building, to the nearest hundredth of a foot?

Answers

1. **a)** 162.45′ **b)** 127.99′

2. **a)** 139.89′ **b)** 331.01′

3. 1735.15 m

4. **a)** 56.31° **b)** 21.63′

5. 9.02′ (height to the top of the building is 30.04′; height from the ground to the top of the sign is 39.06′; 39.06′ – 30.04′ = 9.02′)

6. a) 17.56 m b) 42.57 m

7. 40.18°

8. Building A is 67.45′ tall and Building B is 111.97′ tall.

6-5 LAW OF SINES

We can use the trigonometry functions to find side lengths and angle measures of ANY triangle. One

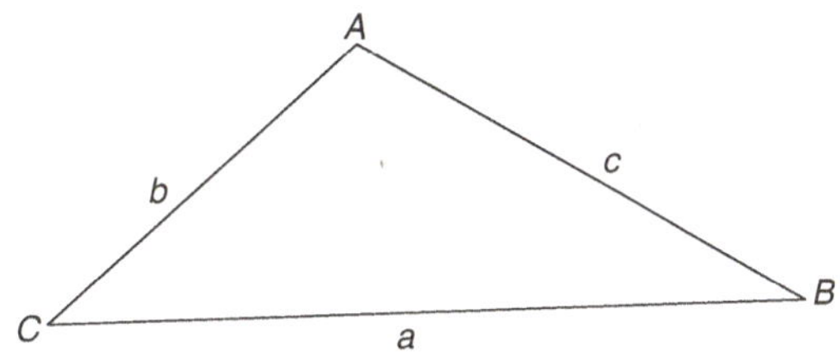

way is using the Law of Sines.

Given

With **ANY** triangle given angles A, B, and C, and with the side opposite to A having length a, the side opposite of B having length b, and the side opposite of C having length c, the following Law of Sines holds true

$$\frac{\sin A}{a} = \frac{\sin B}{b} = \frac{\sin C}{c} \text{ or}$$

$$\frac{a}{\sin A} = \frac{b}{\sin B} = \frac{c}{\sin C}$$

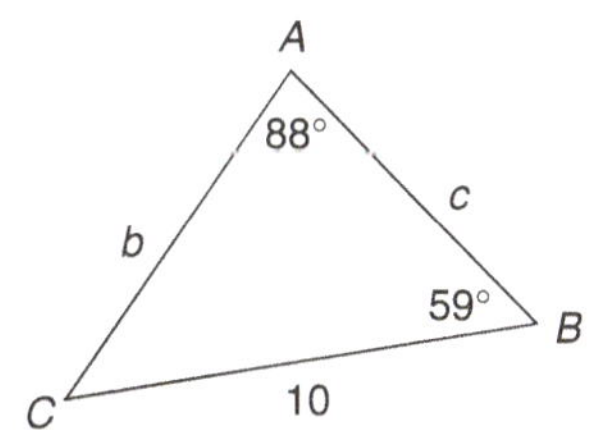

There are three cases where the Law of Sines can be used.

- Case (1) When we know an angle of the triangle, the side opposite the angle, and an additional angle. This situation is known as Angle-Angle-Side or AAS based on the position in the triangle.

  For example:

  Angle A is 88° and the side opposite is $a = 10$. Angle B is 59°. We can find side b using the Law of Sines.

$$\frac{a}{\sin A} = \frac{b}{\sin B} \text{ so } \frac{10}{\sin 88} = \frac{b}{\sin 59} \text{ or } \frac{10(\sin 59)}{\sin 88} \approx 8.58 = b$$

As the sum of interior angles in any triangle is 180°, we can find angle C by Angle $C = 180 - (88 + 59) = 33°$ Now we can use LAW of Sins to find c

$$\frac{a}{\sin A} = \frac{c}{\sin C} \text{ so } \frac{10}{\sin 88} = \frac{c}{\sin 33} \text{ or } \frac{10(\sin 33)}{\sin 88} = 5.45 = c$$

So our triangle is

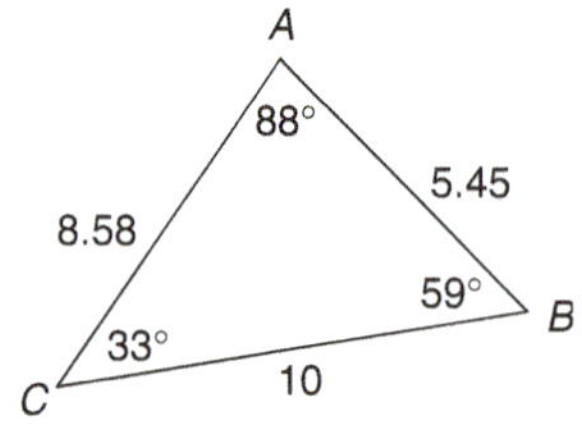

- Case (2) When we know two angles and the side between them. This situation is known as angle-side-angle, or ASA based on the position in the triangle.

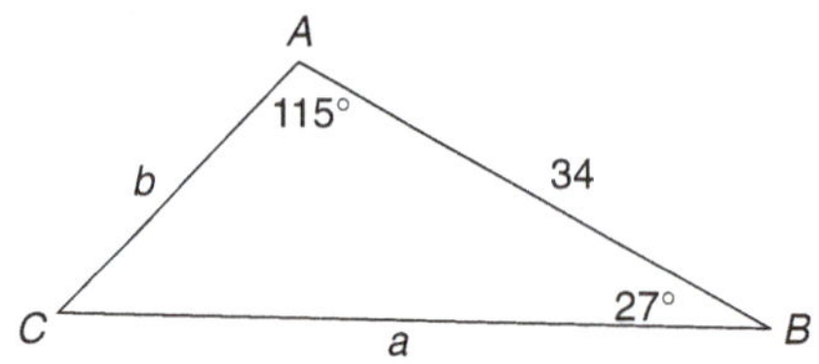

Angle $A = 115°$, side $c = 34$, and Angle $B = 27°$. We can also find angle C, by calculating $180° - (115 + 27) = 38°$. So now we can find side length by Law of Sines

$$\frac{a}{\sin A} = \frac{c}{\sin C} \text{ so } \frac{a}{\sin 115} = \frac{34}{\sin 38} \text{ or } a = \frac{34 \sin 115}{\sin 38} \approx 50.05$$

And to find b

$$\frac{b}{\sin B} = \frac{c}{\sin C} \text{ so } \frac{b}{\sin 27} = \frac{34}{\sin 38} \text{ or } b = \frac{34(\sin 27)}{\sin 38} \approx 25.07$$

our triangle is

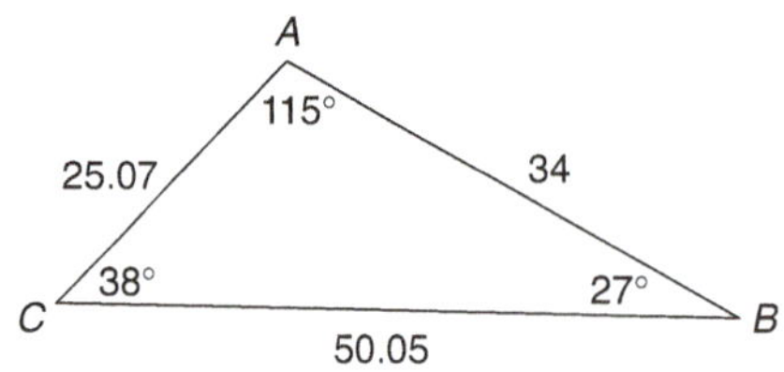

- Case (3) When we know two sides and an angle NOT between them. This situation is known as Side-Side-Angle or SSA.

Example

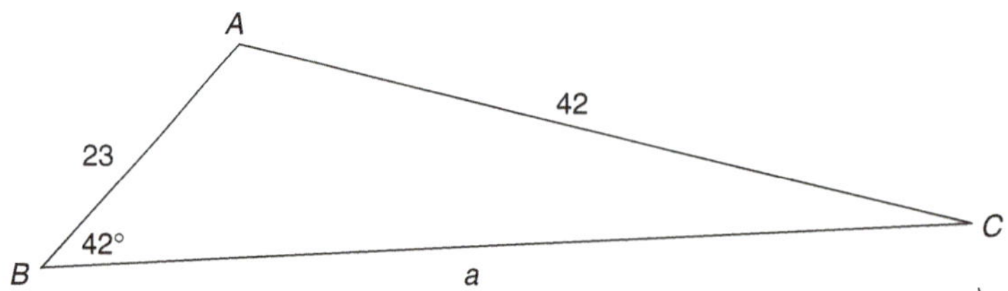

We know angle $B = 42°$, $b = 42$, and $c = 23$. To find angle C, law of sines says

$$\frac{\sin C}{c} = \frac{\sin B}{b} \text{ or } \frac{\sin C}{23} = \frac{\sin 42}{42} \text{ so } \sin C = (23)(\frac{\sin 42}{42}) \approx .366$$

To get the angle C, use the arcsin of the ratio for angle C. arcsin $.366 \approx 21.47°$.

To find angle A, $A = 180° - (42° + 21.47°) = 116.53°$

And to find a, $\frac{a}{\sin A} = \frac{b}{\sin B}$ or $\frac{a}{\sin 116.35} = \frac{42}{\sin 42}$ or $a = \frac{42 \sin 116.35}{\sin 42} \approx 56.25$

So our triangle is

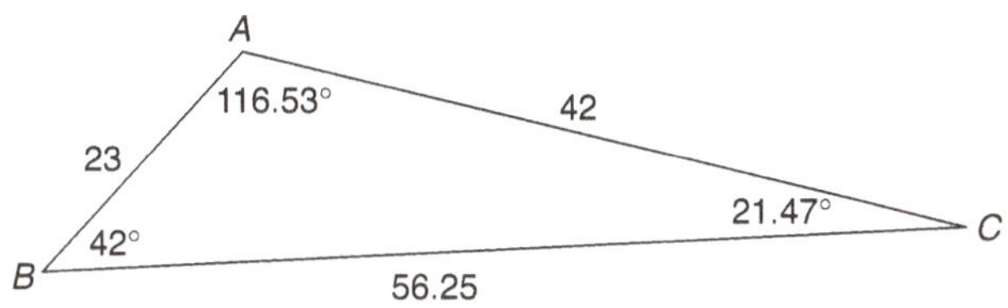

➠ For the remaining CASE for triangles where we are given two SIDES and the ANGLE between them, known as SAS, we CANNOT use the Law of Sines. In Section 6-6, you will learn a technique called the Law of Cosines that will help you find missing side and angle measures for this SAS situation.

Practice

Find the missing sides/angle for

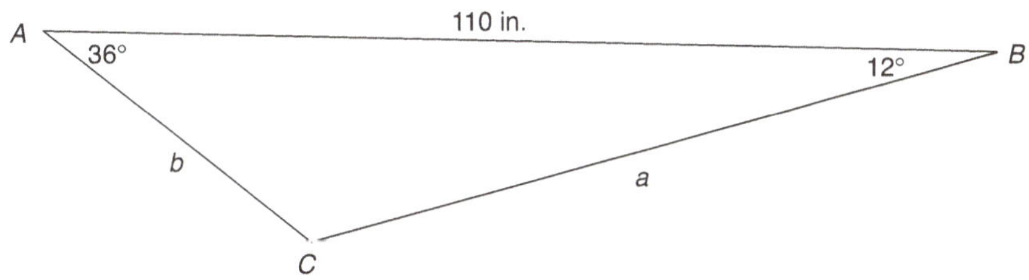

Find the missing side/angles

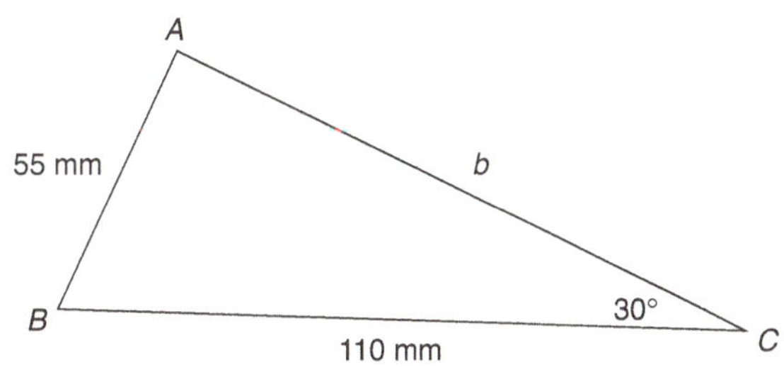

Find the missing side and angles

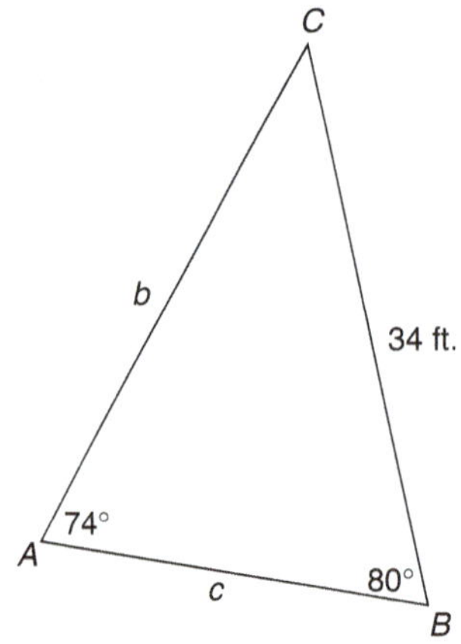

6-5 Assignments

1. Find the missing sides/angle for

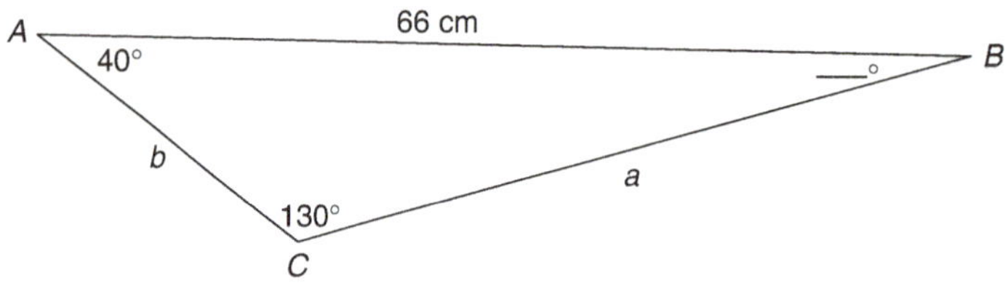

2. Find the missing side/ angles

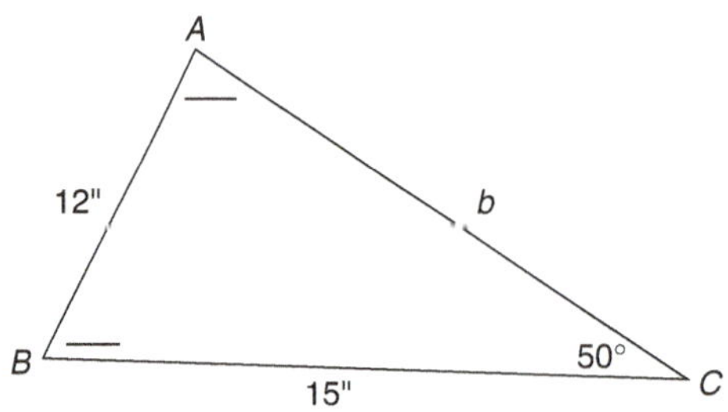

3. Find the missing side and angles

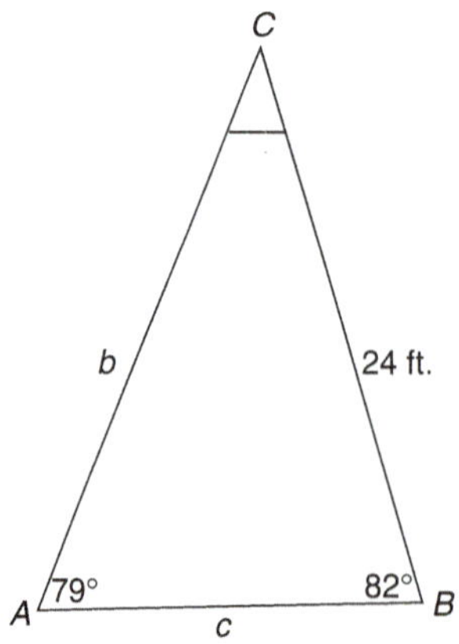

4.

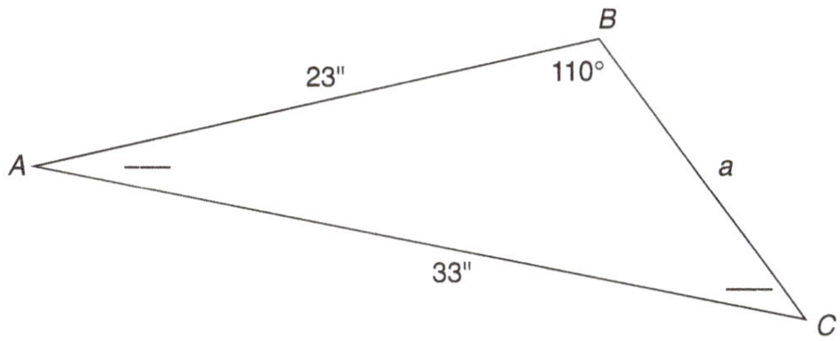

5. In Wild Tacoma Park, a Fire is spotted by park rangers stationed in towers that are 5 miles apart. Using the line between the tower as a baseline, Tower A reports an angle of 58° with the fire and Tower reports an angle of 29° with the fire.

Figure to the nearest hundredth of a mile how far the fire is from a) Tower A and b) Tower B

6. Find the length of a stiff cable from the ground to the top of a culvert according to these measurements.

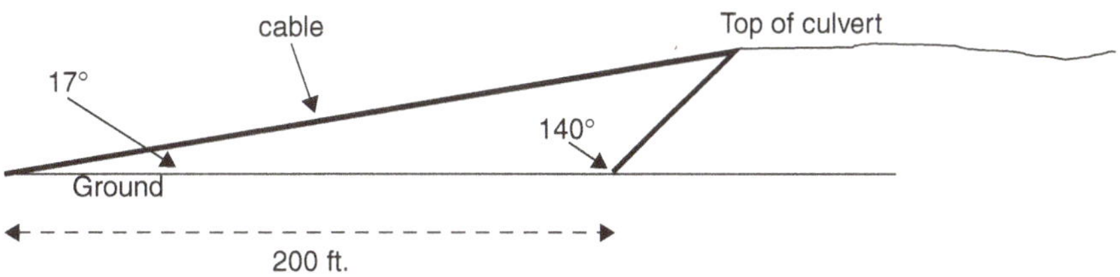

7. A blimp viewed from the south end zone of a football field has an angle of elevation of 62° and at the same time is viewed in the north end zone with an angle of elevation of 70°. If the distance between the two viewpoints is 135 yards, find how far the blimp is from a) south end zone viewer and b) north end zone viewer.

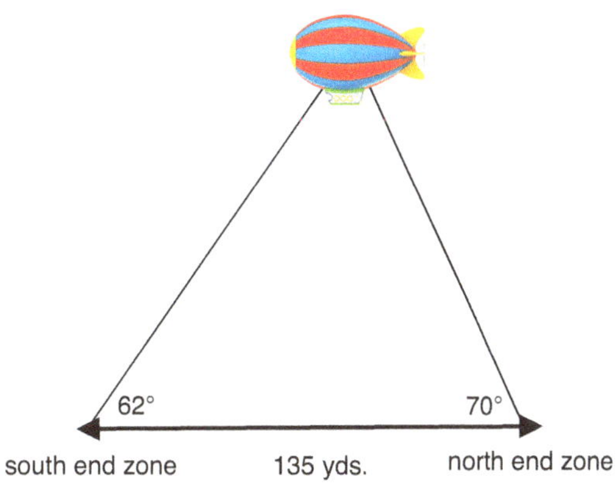

Answers

1. a = 55.38 cm
   b = 14.96 cm
   angle B = 10°

2. b = 13.10″
   angle A = 73.25°
   angle B = 56.75°

3. b = 24.21′
   c = 7.96′
   angle C = 19°

4. a = 17.07″
   angle A = 29.09°
   angle C = 40.91°

5. a) 2.43 miles

   b) 4.25 miles

6. 329.02 feet

7. a) 170.70 feet

   b) 160.40 feet

6-6 The Law of Cosines

In Section 6-5, we learned the Law of Sines can be used to find side lengths and/or angle measures of any triangle given certain sides and angles.

It says given

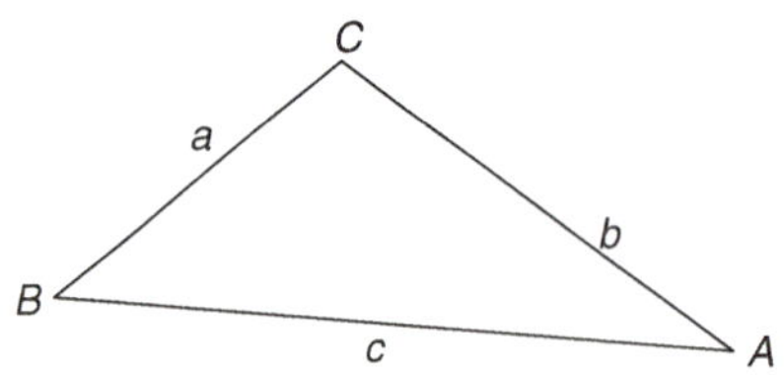

$$\frac{a}{\sin A} = \frac{b}{\sin B} = \frac{c}{\sin C}$$

This law can be used to find the other measure when one has two side lengths and one angle measure or when one has two angle measures and one side length with one EXCEPTION. When you have two adjacent sides and the angle between them, you cannot use the Law of Sines.

However, there is a law that uses cosine that can be used in this case.

THE LAW OF COSINES

Given

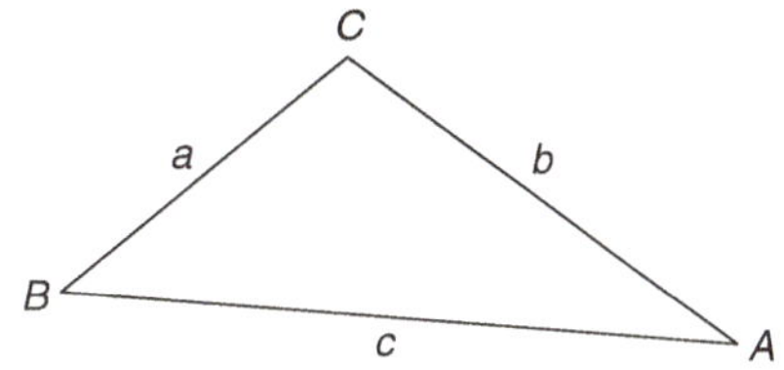

$$c^2 = a^2 + b^2 - 2\ ab\cos C$$

$$a^2 = b^2 + c^2 - 2\ bc\cos A$$

And $$b^2 = a^2 + c^2 - 2\ ac\cos B$$

EXAMPLE: Find c length.

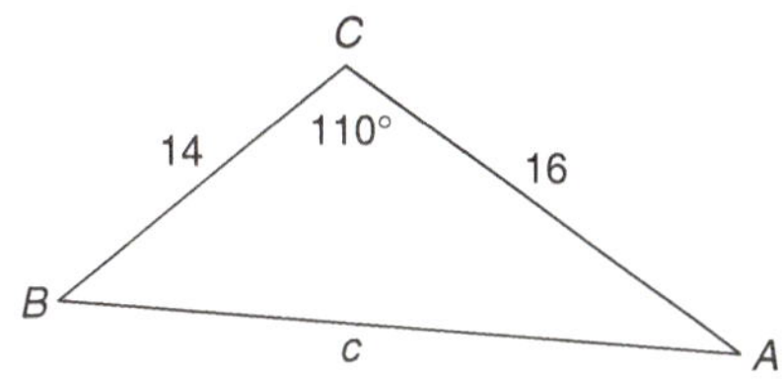

$a = 14$, $b = 16$, angle $C = 110°$

$$c^2 = a^2 + b^2 - 2\ ab\cos C$$

$$c^2 = 14^2 + 16^2 - 2(14)(16)\cos 110°$$

$$c^2 = 14^2 + 16^2 - 2(14)(16)\cos 110°$$

$$c^2 \approx 605.23$$

$$\sqrt{c^2} \approx \sqrt{605.23}$$

$$\mathrm{c} \approx 24.60$$

Now, we can use the Law of Sines to find other angle measures.

$$\frac{a}{\sin A} = \frac{c}{\sin C}$$

$$\frac{14}{\sin A} = \frac{24.60}{\sin 110°}$$

$$\frac{\sin A \cdot \sin 110°}{24.60} \cdot \frac{14}{\sin A} = \frac{\sin A \cdot \sin 110°}{24.60} \cdot \frac{24.60}{\sin 110°}$$

$$\frac{14 \cdot \sin 110°}{24.60} = \sin A$$

$$.53 \approx \sin A$$

Take the arcsin of sin A, arcsin $.53 \approx 32.00° =$ angle A measure

And $180° - (110° + 32°) = 38° =$ angle B measure

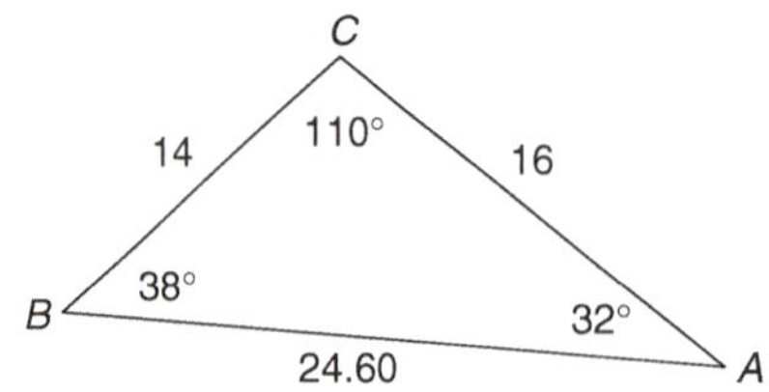

Practice

Find the missing side length and angles measures for

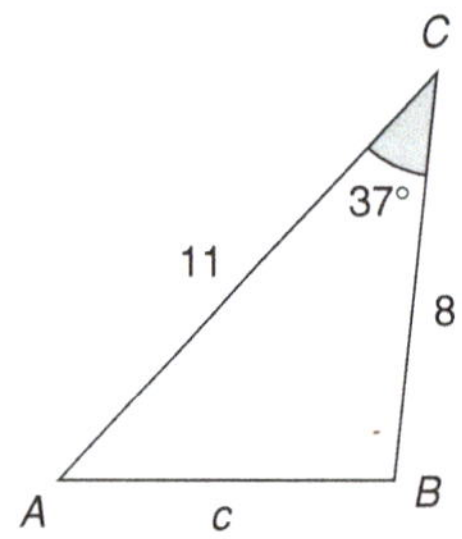

6-6 Assignments

1. Find the missing side and angles to the nearest hundredth.

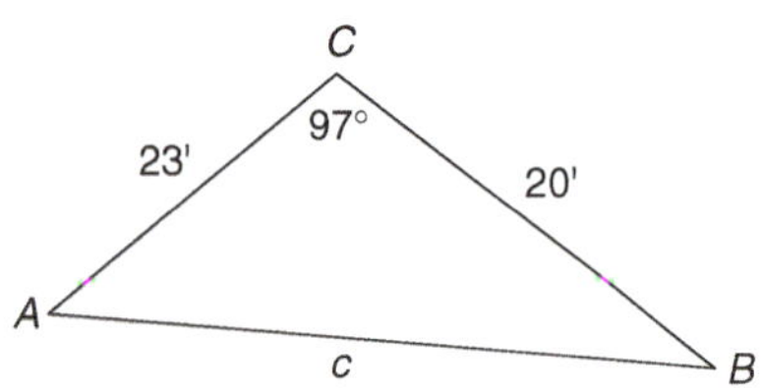

2.

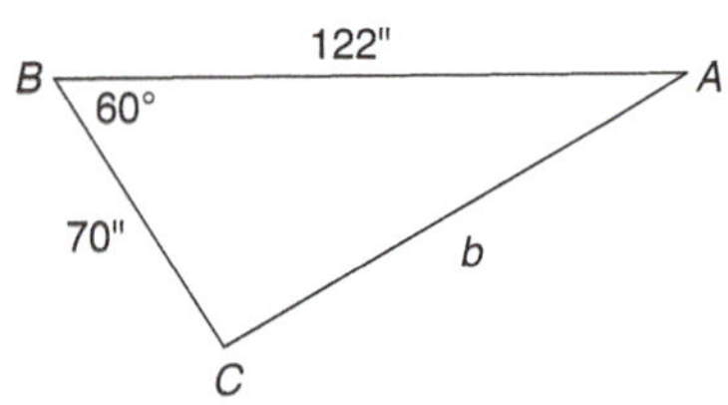

3.

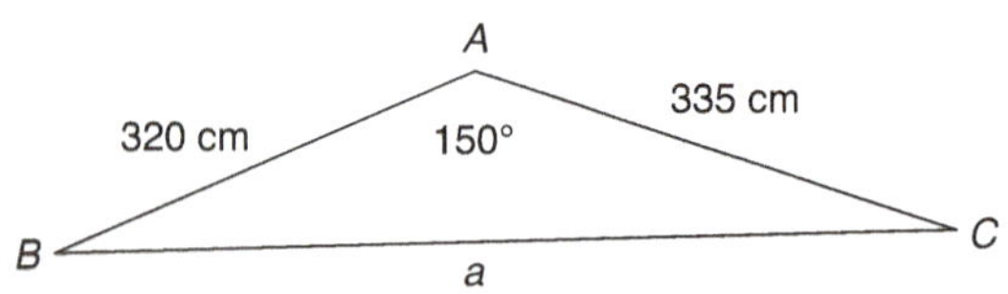

4. Two airplanes leave an airport, and the angle between their flight paths is 40°. An hour later, one plane has traveled 300 miles while the other has traveled 200 miles. How far apart are the planes at this time?

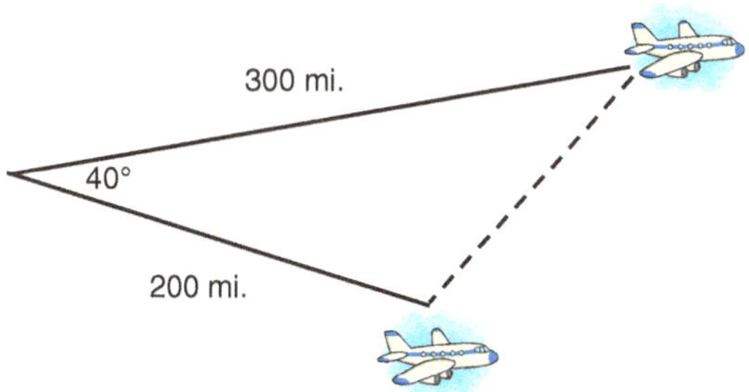

Answers

1. c = 32.27′
   Angle $A = 45.02°$
   Angle $B = 37.90°$

2. b = 106.04″
   Angle $A = 34.88°$
   Angle $C = 85.12°$

3. a = 632.69 cm
   Angle $B = 15.35°$
   Angle $C = 14.65°$

4. 195.13 miles

Practice Test

Practice Test with Answer Key, Test Number 1 Chapters 1 and 2

Use the operations indicated and simplify the following

1. $\frac{4}{7} \div \frac{5}{8}$ ____________

2. $\frac{11}{42} \cdot \frac{6}{55}$ ____________

3. $5\frac{2}{3} + 11\frac{3}{7}$ ____________

4. $\frac{6}{19} + \frac{3}{38}$ ____________

5. $6^2 \div 4 + 5 - 3 \cdot 2 + \frac{8+4 \cdot 2}{8}$ ____________

6. $15 - 16 \div 2 \cdot (1 + 6^2 \div 4 \cdot 2 - 17) + 4$ ____________

7. $2\frac{2}{7} \cdot 3\frac{1}{9}$ ____________

8. $\frac{8}{9} - \frac{3}{5}$ ____________

9. $25\frac{4}{5} - 8\frac{1}{5}$ ____________

10. $\left(7\frac{2}{3}\right)\left(2\frac{1}{11}\right)$ ____________

11. $6\frac{2}{3} \div 1\frac{1}{4}$ ____________

12. $\frac{5}{7}$ of 560 ____________

13. $\sqrt{676}$ ____________

14. $\sqrt{2459}$ to the nearest hundredth place ____________

15. $\sqrt[3]{343}$ ____________

16. $\sqrt[3]{23.15}$ (to nearest hundredth place) ____________

17. $\sqrt[11]{815}$ (to the nearest hundredth place) ____________

18. $[(1 + .08)^4 - 1]$ (to nearest hundredth place) ______________

19. $\dfrac{(5)(9) - 5(6-4)^3}{(5-2)^2 - 2^2}$ ________________

20. Wendy hiked an average of 14 miles per day for five days. Peter hiked an average of $5\frac{2}{5}$ miles per day for the five Days. Pan hiked 3/5th of the total miles Wendy and Peter hiked combined. What were the totals of each for the five days? Wendy ____________

Peter ____________

Pan ____________

21. Solve for y $\dfrac{y}{8} = \dfrac{4}{13}$ ____________

22. Solve for x $\dfrac{4}{11} = \dfrac{20}{3x}$ ________________

23. Solve for x given $3 + \dfrac{x}{6} = 5$ ______________

24. Solve for x $\dfrac{5}{2x} + \dfrac{7}{7x} = 12$ ____________

25. Solve for y $\dfrac{3y}{8} - \dfrac{3}{13} = \dfrac{11y}{13} + \dfrac{1}{8}$ ____________

26. Solve for x, $\dfrac{x+4}{3} = \dfrac{2x-5}{2}$ ________

27. For the formula $y = 3x + 7$ Solve for x, given $y = 2$ ____________

28. Solve for P in $I = Prt$ given $I = 350$ $r = .08$ $t = 4$ ____________

29. For the formula $\dfrac{r+3}{2p-4} = \dfrac{r-5}{p}$ solve for r given $p = 10$ ____________

30. Solve for Y given $610 + .2Y = Y$ ______________

31. Simplify $2x^2 + 11x + 8 - 2x^2 + 12 - 2x$ ______________

32. Solve for w given $8(20 - w) + 15w = 20$ ______________

33. Calculate for A in $A = m[(1+rt)^n - 1]$ given $m = 100$, $r = .006$, $t = \dfrac{1}{12}$ and $n = 180$ (to the nearest hundredth place) ____________

34. Chef Boyardee had 9 pizzas to every 4 calzones in his restaurant. If he had 108 pizzas, how many calzones did Chef have? _________

35. Fab Ulousweathy had 25000 Canadian dollars. He crosses the border for for a whirlwind USA vacation. He exchanges his Canadian dollars to US dollars. The exchange rate is 1.23 Canadian dollars for $1 dollar (US). How much in US Dollars can Fab receive? ____________

36. Speedy Joe drives his motorcycle at an average of 72 miles per hour. How many Kilometers does he travel in 180 minutes, given that there are 1.61 Kilometers to 1 mile? ____________

37. For the formula $\frac{8\text{ wt}}{5\text{ m}} = 32\text{ mx}$ solve the formula for w. ____________

38. For the formula $3y = \frac{3p+t}{5yr^2}$ solve the formula for t ____________

39. Solve the quadratic equation for x $x^2 + 11x - 26 = 0$

40. Solve the quadratic equation for y $2y^2 + 5y = 10$

Answer key

1. $\frac{4}{7} \cdot \frac{8}{5} = \frac{32}{35}$

2. $\frac{11 \cdot 6}{42 \cdot 55} = \frac{1 \cdot 1}{7 \cdot 5} = \frac{1}{35}$

3. $5 + 11 + \frac{2}{3} + \frac{3}{7} = 16 + \frac{14}{21} + \frac{9}{21} = 16 + \frac{23}{21} = 17\frac{2}{21}$

4. $\frac{12}{38} + \frac{3}{38} = \frac{15}{38}$

5. $6^2 \div 4 + 5 - 3 \cdot 2 + \frac{8 + 4 \cdot 2}{8}$

$6^2 \div 4 + 5 - 3 \cdot 2 + \frac{8+8}{8}$

$6^2 \div 4 + 5 - 3 \cdot 2 + \frac{16}{8}$

$36 \div 4 + 5 - 3 \cdot 2 + \frac{16}{8}$

$9 + 5 - 3 \cdot 2 + \frac{16}{8}$

$9 + 5 - 6 + \frac{16}{8}$

$9 + 5 - 6 + 2$

$14 - 6 + 2$

$8 + 2 = 10$

6. $15 - 16 \div 2 \cdot (1 + 36 \div 4 \cdot 2 - 17) + 4$

$15 - 16 \div 2 \cdot (1 + 9 \cdot 2 - 17) + 4$

$15 - 16 \div 2 \cdot (1 + 18 - 17) + 4$

$15 - 16 \div 2 \cdot (19 - 17) + 4$

$15 - 16 \div 2 \cdot 2 + 4$

$15 - 8 \cdot 2 + 4$

$15 - 16 + 4 = -1 + 4 = 3$

7. $\frac{16}{7} \cdot \frac{28}{9} = \frac{16 \cdot 4}{1 \cdot 9} = \frac{64}{9} = 7\frac{1}{9}$

8. $\frac{40}{45} - \frac{27}{45} = \frac{13}{45}$

9. $\frac{129}{5} - \frac{41}{5} = \frac{88}{5} = 17\frac{3}{5}$

10. $\frac{23}{3} \cdot \frac{23}{11} = \frac{529}{33} = 16\frac{1}{33}$

11. $\frac{20}{3} \div \frac{5}{4} = \frac{20}{3} \cdot \frac{4}{5} = \frac{80}{15} = 5\frac{1}{3}$

12. $\frac{5}{7} \cdot 560 = 400$

13. 26

14. 49.59

15. 7

16. 2.85

17. 1.84

18. $[(1 + .08)^4 - 1] = [1.08^4 - 1] = .36$

19. $\frac{(5)(9) - 5(6-4)^3}{(5-2)^2 - 2^2} = \frac{5}{5} = 1$

20. Wendy $= 14 \cdot 5 = 70$ miles, Peter $= 5\frac{2}{5} \cdot 5 = \frac{27}{5} \cdot 5 = 27$ miles, Pan $= \frac{3}{5}(70+27) = \frac{3}{5}(97)$ $= 58.2$ miles

21. $13y = 8(4)$ or $13y = 32$ so $y = \frac{32}{13}$ or $2\frac{6}{13}$

22. $4(3x) = 11(20)$ or $12x = 220$ so $x = \frac{220}{12} = \frac{55}{3}$ or $18\frac{1}{3}$

23. $\frac{x}{6} = 2$ so $x = 6(2) = 12$

24. $14x\left(\frac{5}{2x} + \frac{7}{7x}\right) = 12(14x)$ becomes $35 + 14 = 168x$ or $49 = 168x$, so $x = \frac{49}{168} = \frac{7}{24}$

25. $104\left(\frac{3y}{8}-\frac{3}{13}\right)=\left(\frac{11y}{13}+\frac{1}{8}\right)104$ becomes $39y-24=88y+13$ or $39y=88y+37$ or $-49y=37$ or $y=-\frac{37}{49}$

26. $3(2x-5)=2(x+4)$ or $6x-15=2x+8$ or $6x=2x+23$ or $4x=23$ so $x=\frac{23}{4}$

27. $2=3x+7$ or $-5=3x$ so $x=-\frac{5}{3}$

28. $I=Prt$ so $350=P\,(.08)\,(4)$ or $350=.32P$ so $P=1093.75$

29. $\frac{r+3}{2(10)-4}=\frac{r-5}{10}$ or $\frac{r+3}{16}=\frac{r-5}{10}$ or $10(r+3)=16(r-5)$ or $10r+3=16r-80$ or $10r+110=16r$ or $110=6r$ or $r=\frac{110}{6}=\frac{55}{3}=18\frac{1}{3}$

30. $610=1Y-.2Y$ so $610=.8Y$ and $Y=610\div.8=762.5$

31. $9x+20$

32. $160-8w+15w=20$ or $160+7w=20$ or $+7w=-140$ or $w=\frac{-140}{7}=-20$

33. $A=100\left[\left(1+.006\cdot\frac{1}{12}\right)^{180}-1\right]=9.41$

34. $\frac{9\text{ pizzas}}{4\text{ calzones}}=\frac{108\text{ pizzas}}{x\text{ calzones}}$ so $\frac{9}{4}=\frac{108}{x}$ or

$9x=4(108)$ or $9x=432$ so $x=\frac{432}{9}=48$ so 48 calzones

35. \$20325.20

36. 347.76 Kilometers

37. $w=\frac{20\,m^2x}{t}$

38. $t=15\,y^2w^2-3p$

39. $x=\frac{-11\pm\sqrt{11^2-4(1)(-26)}}{2(1)}=\frac{-11\pm\sqrt{225}}{2}=\frac{-11\pm15}{2}$ so $x=\frac{-11+15}{2}=\frac{4}{2}=2$

or $x=\frac{-11-15}{2}=\frac{-26}{2}=-13$

40. Change to $2y^2+5y-10=0$

$y=\frac{-5\pm\sqrt{5^2-4(2)(-10)}}{2(2)}=\frac{-5\pm\sqrt{105}}{4}$ so $y=\frac{-5+\sqrt{105}}{4}=1.31$

or $y=\frac{-5-\sqrt{105}}{4}=-3.81$

Practice Test for Chapter 3 and 4

1. Convert 4 cups = ____ Quarts
2. Convert 23 Kilogram = ____ grams
3. Convert 1400 milliliters = ______ Liters
4. Convert 34 fluid ounces = _____ milliliters
5. Convert 500 cubic inches = ______ gallons
6. Convert 5 kilometers = _______ centimeters
7. Convert 56 meters/1 hour = _______ feet/1 minute
8. Convert 40 quarts/1 second = _______ cubic feet/1 minute
9. Which is the bigger volume: 32 gallons or 107 Liters ________________
10. What is the longer distance: 170 Kilometers or 100 miles ____________________
11. What is the greater weight: 500 pounds or 230 Kilograms ____________________
12. What is the greater area: 7 square inches or 44 square centimeters ____________
13. Convert 89 Newtons to pounds of force. ____________
14. Convert 30 horsepower to Watts of power ____________
15. Convert 78 BTU to Joules of energy _____________
16. What is more powerful?, 22 horsepower or 17000 Watts ____________
17. What has more energy? 80 Calories or 330 Joules ________________
18. What is a faster acceleration? 56 km per hour^2 or .01 miles per minute^2? _____________
19. It is 50 °F in Toronto Canada. What is the temperature in degrees Celsius _____________
20. 200°C is ________ ° Kelvin

For 21. – 30. Unless otherwise started, find answers to the **nearest hundredth of a unit**.

21. Find the perimeter and area of this rectangle

13 cm

21 cm

22. Find the diagonal length for a square with side length 5.6 inches.
23. An equilateral triangle has a side length of 8.2 feet. What is the triangle's height and area?
24. A regular octagon has a perimeter of 120 meters. Find its side length and its area.
25. For this regular hexagon to the right, find its perimeter, diagonal, and area.

5 mm

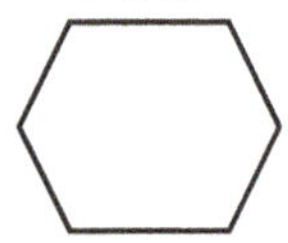

26. A circle has a diameter of 34 inches. Find its circumference and its area.

27. This figure made up of two concentric circles, one with radius 15 inches and the other with radius of 28 inches. Find the area shaded.

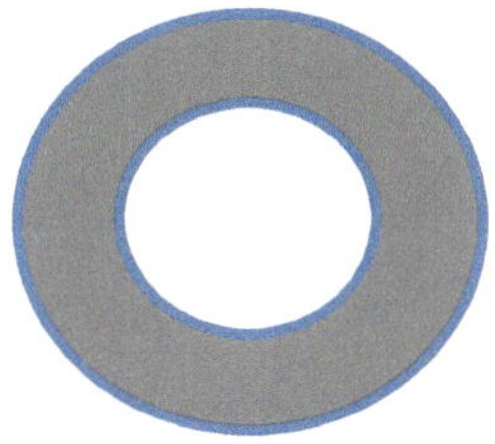

28. This ellipse has short diameter of 24 inches and longer diameter of 52 inches. Find its area and perimeter

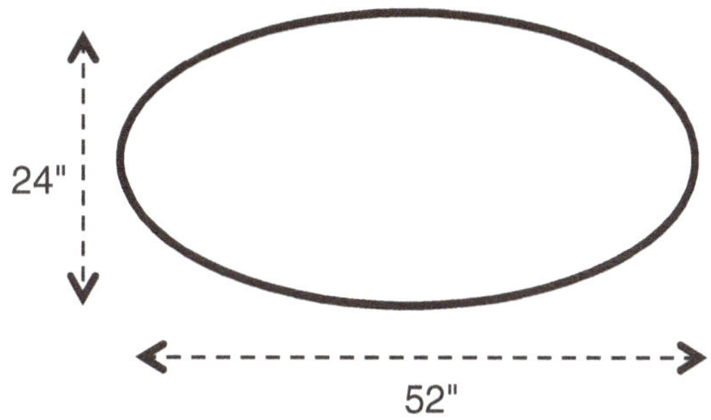

29. Find the perimeter and area for this shape made up of a half circle, rectangle, and right triangle.

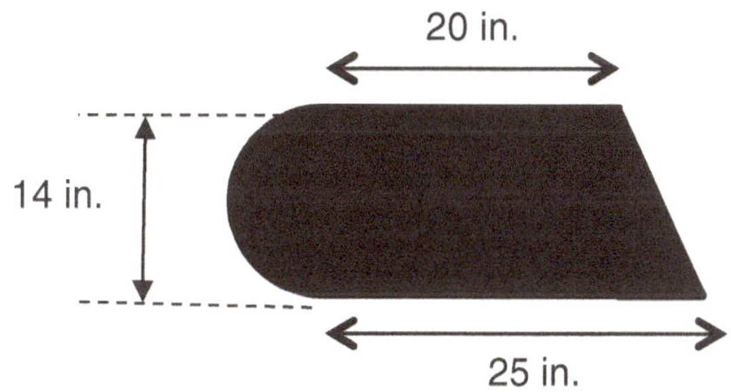

30. For the minor arc from A to B that is intercepted by the 78° central angle circle, find the arc length and find the sector area

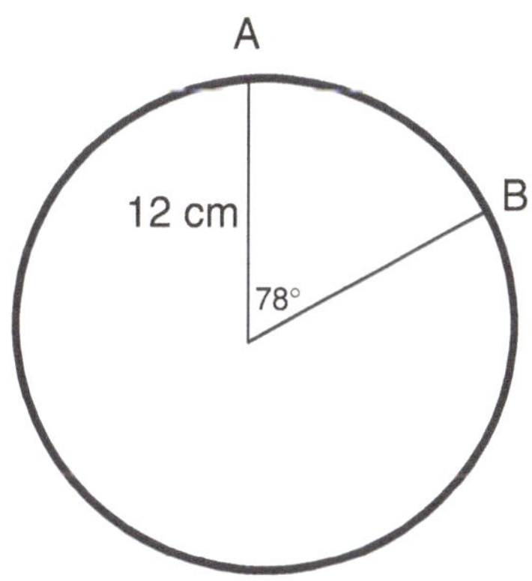

Answer key

1. 1 quart
2. 23000 grams
3. 1.4 Liters
4. 1005.38 milliiters
5. 2.16 gallons
6. 500000 centimeters
7. 3.06 feet/1 minute
8. 80.21 cubic feet/1 minute
9. 32 gallons
10. 170 kilimeters
11. 230 kilograms
12. 7 square inches
13. 20 pounds
14. 22371 watts
15. 82290 Joules
16. 17000 Watts
17. 80 Calories
18. $\frac{.01 \text{ miles}}{\text{minute}^2}$
19. 10°C
20. 473 °K
21. $P = 2(13 \text{ cm} + 21 \text{ cm}) = 68 \text{ cm}$ $A = (13 \text{ cm})(21 \text{ cm}) = 273 \text{ cm}^2$
22. Diagonal $= 1.414\, s = 1.414\, (5.6'') = 7.92''$
23. $h = .866\, s = .866\, (8.2 \text{ ft}) = 7.10$ feet $A = .433\, s^2 = .433\, (8.2\text{ft})^2 = 29.11\text{ft}^2$
24. $S = \frac{p}{8} = \frac{120 \text{ m}}{8} = 15 \text{ m}.$ $A = 4.828\, s^2 = 4.828\,(15 \text{ m})^2 = 1086.30 \text{ m}^2$
25. $P = 6s = 6(5 \text{ mm}) = 30 \text{ mm}.$ $d = 2(s) = 2(5 \text{ mm}) = 10 \text{ mm}$ $A = 2.598\, s^2 = 2.598(5 \text{ mm})^2$ $= 64.95 \text{ mm}^2$
26. $r = 34/2 = 17''$. Circumference $= d\pi = 34''(\pi) = 106.81''$ Area $= \pi r^2 = \pi(17 \text{ in})^2 = 907.92 \text{ in}^2$
27. $\text{area}_{\text{annulus}} = \pi\left(r_o{}^2 - r_i{}^2\right) = \pi\left((28 \text{ in})^2 - (15 \text{ in})^2\right) = 1765.15 \text{ in}^2$
28. $a = 52''/2 = 26''$ $b = 24''/2 = 12''$ Area $= \pi ab = \pi(26'')(12'') = 980.18 \text{ in}^2$

    Perimeter $= \pi\sqrt{2\left(a^2 + b^2\right)} = \pi\sqrt{2\left((26'')^2 + (12'')^2\right)} = 127.22$ in

29. Half circle has diameter of 14" so radius of 7". Rectangle is 20" by 14", right triangle part has legs of (25-20) = 5" and 14".

Perimeter: half circle is $.5(d\pi) = .5(14")(\pi) = 21.99"$; two horizontal lengths are 20" and 25"; the remaining side is hypotenuse of the right triangle, so length is $\sqrt{5^2+14^2}$ = 14.87" so perimeter = 21.99" + 20" + 25" + 14.87" = 81.86"

Area: half circle = $.5(\pi r^2) = .5(\cdot)(7\text{ in})^2 = 76.97\text{ in}^2$; the rest is a trapezoid with base lengths 20" and 25" and height of 14", so Area = $.5(b_1+b_2)$ (h) =.5(20"+ 25") (14") = 315 in^2 so total area = 76.97 + 315 = 391.97 in^2

30. $x = 78°$, $r = 12$ cm so Arc Length $L = \dfrac{\pi rx}{180} = \dfrac{\pi(12\text{ cm})(78)}{180} = 16.33\text{ cm}$

$$\text{Sector area} = \frac{\pi r^2 x}{360} = \frac{\pi 12\text{ cm}^2(78)}{360} = 98.02\text{ cm}^2$$

Practice Test for Chapter 5 and 6

1. For the rectangular prism to the right

Find the surface area and the volume

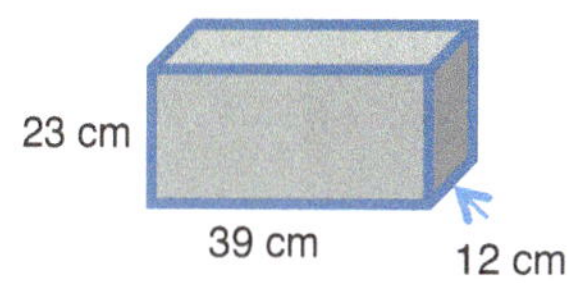

2. For a cylinder with height 6 yards and base radius of 5 yards, find the surface area and the volume.

3. For this sphere, find the surface area and the volume

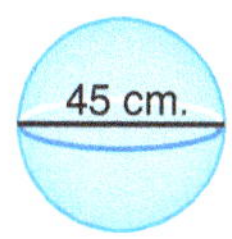

4. Find the volume and surface area of this regular octagonal pyramid with octagonal base side length of 4" and

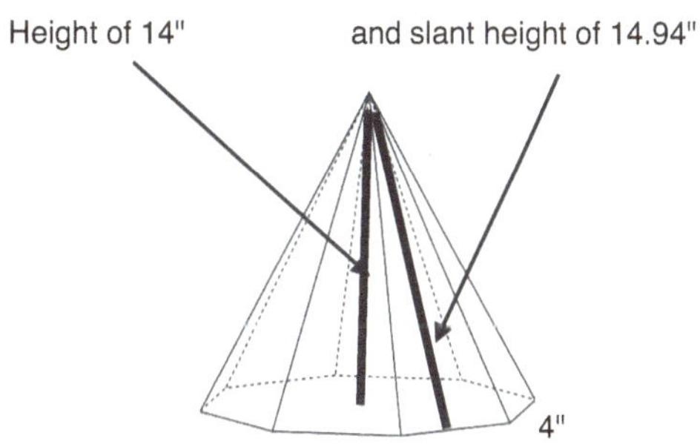

5. A swimming pool was shaped like a right trapezoidal prism, that was 15 feet at the deep end, 4 feet at the shallow end, and the rectangular top was 50 feet long by 25 feet wide. If filled entirely with water, what is the number of gallons of water in the pool's volume? (Remember, there are 231 cubic in to 1 gallon, and 1728 cubic inches to 1 cubic foot.)

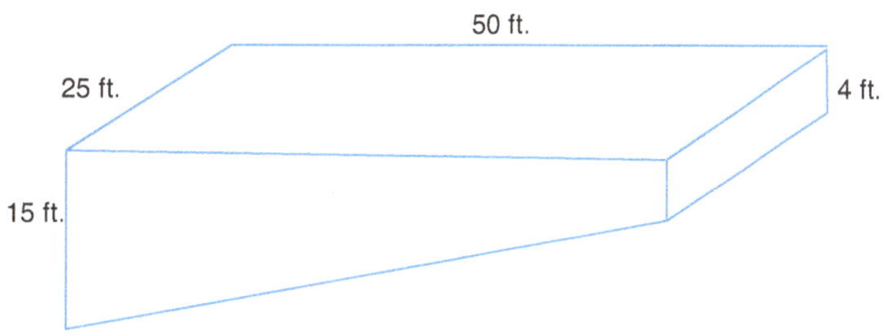

6. A canister holding propane is made up of Hemispherical ends and cylindrical middle. Find its volume

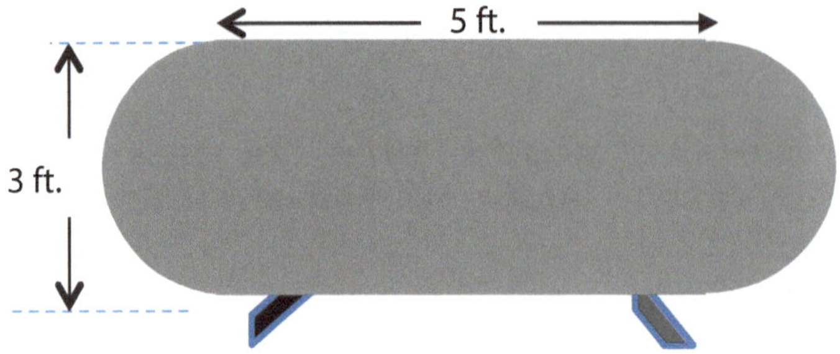

7. Find the volume of a right cone with a 20 foot radius of base and a height of 37 feet.

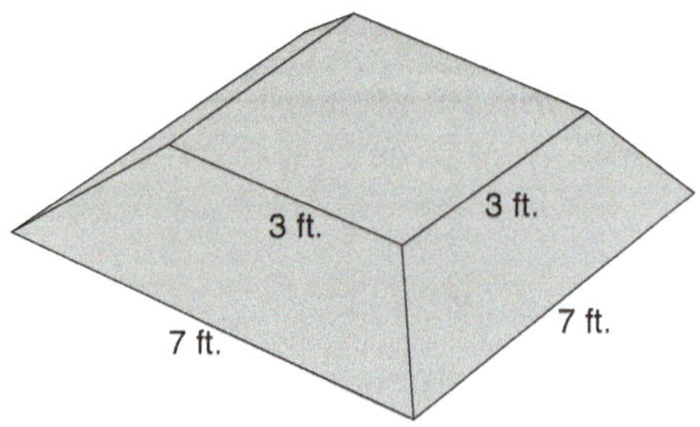

8. For the square pyramidal frustum, find its volume if it's height is 1.5 feet, and its larger base side is 7ft and the smaller base side is 3 ft.

9. How much does 420 cubic centimeters of zinc weigh in kilograms?

10. A silver bar has dimensions of 6 cm by 10 cm by 2 cm. If silver is worth $.51 a gram, what is the total worth of the bar?

11. A lead spherical ball bearing is has a diameter of 1.4 inches. How much does it weigh in grams?

12. A large cherry wood bowl is a hemispherical, with an inner diameter of 50 cm and an outer diameter of 70 cm. What is the bowl's weight in pounds?

13. A sculpture has a right cylinder with a base radius of 4 meters and a height of 9.5 meters sitting on top of a cube that has edge length of 10 meters.

   a) Find its Volume.

   b) If the sculpture is made out of solid granite, what is the sculpture's weight in kilograms?

14. Given a right triangle to the right

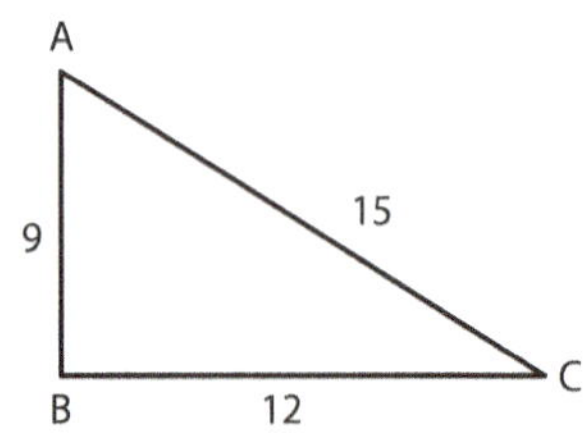

   a) What is the sine ratio for angle A? ____________

   b) What is the cosine ratio for Angle A? _________

   c) What is the tangent ratio for Angle A? ____________

15. For this right triangle find

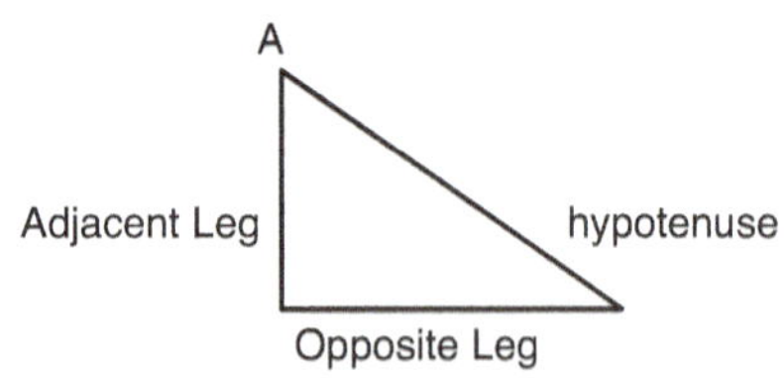

   a) sin A ________

   b) cos C ________

   c) tan A ____________

   d) cos A ____________

   e) tan C ____________

16. Use your calculator to find

a) sin 20° ______

b) cos 69.5° ________

c) tan 45 ° _____

17. For this right triangle, using Trigonometry and your calculator find angle measure for angle C ______

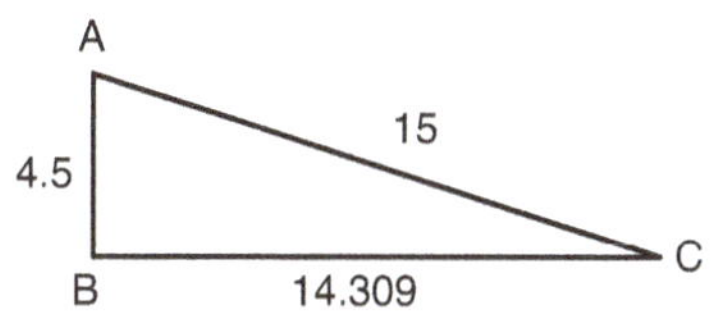

18. Find the other two sides and other angle for this right triangle.

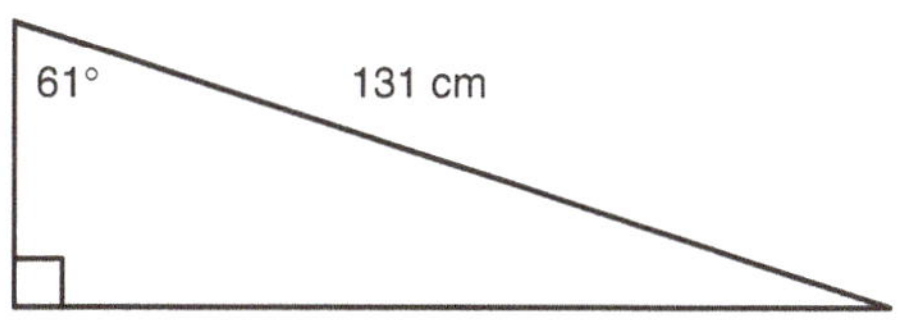

19. Find the missing side and missing angles.

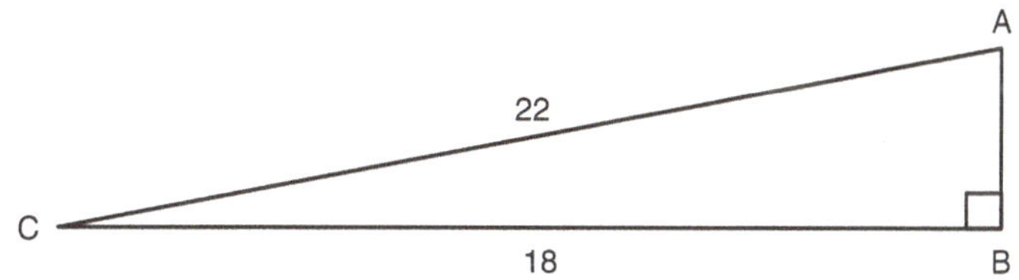

20. A person laying on the ground looks up at a flag at the top of a FLAG POLE. If the person is 25 feet from the flag pole and the angle of elevation for his eyes to the flag is 22.3°, how far is he from the flag, and how tall is the flag pole?

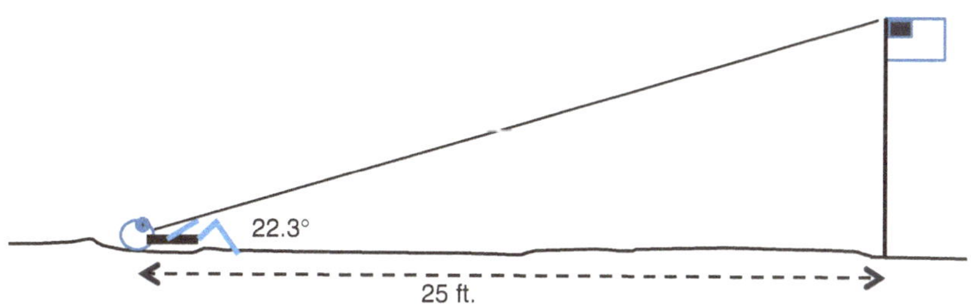

21. A lighthouse sits on a high hill 130 feet above the Bay. How far is the lighthouse from a ship in the bay is the angle of depression is 35°?

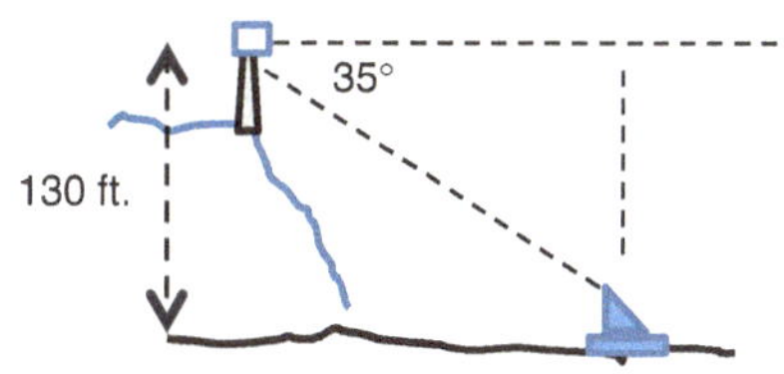

22. A Bird watcher sees a rare purple oriole on a telephone pole. If the bird watcher's eye is 6 feet above the ground, if the pole is 30 feet high, and the bird watcher is 35 feet from the pole, what is the angle of elevation for seeing the bird?

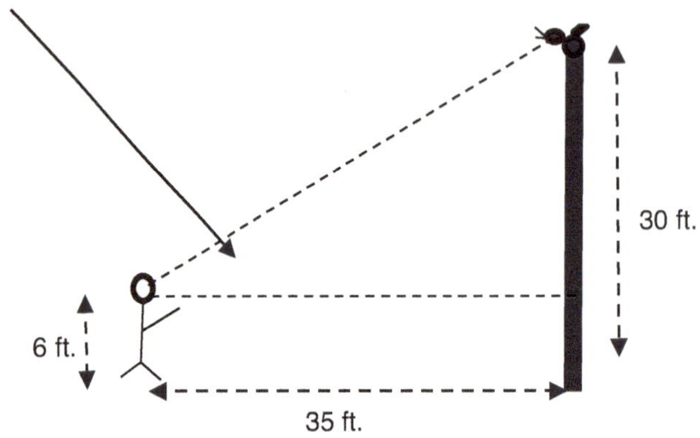

23. Find side b, c and angle C

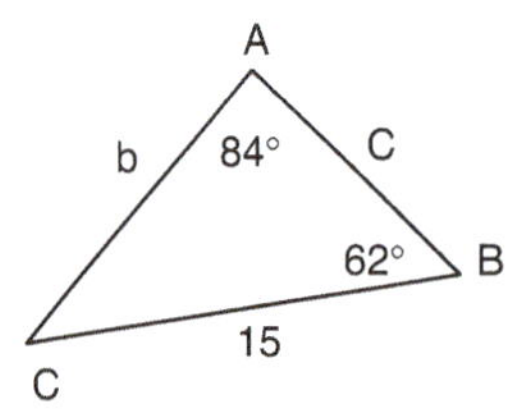

24. Find side a, b, and angle C

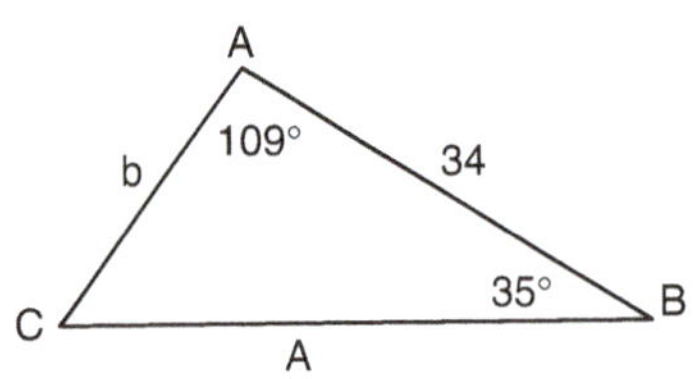

25. Find side side c, angle A and angle B

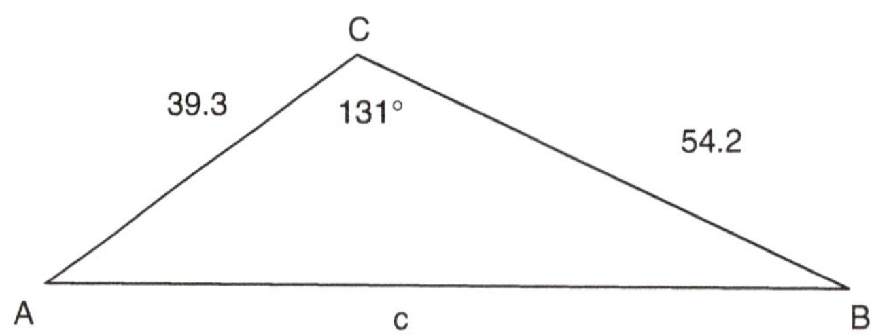

26. Three light towers were arranged in a triangular pattern in a National Park. If straight path lines were drawn from Tower A to Tower B to Tower C then to Tower A, The distance from Tower A to Tower B was 2.6 miles and the angles made by path lines were as below, what is the distance between

Tower A to Tower C ___________ Tower B to Tower C ________

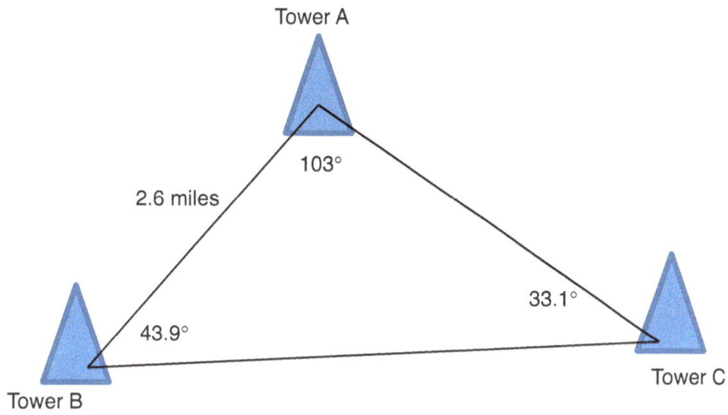

Answer Key

1. Surface area = $2B + (\text{Perimeter}_{base})h = 2(39\text{ cm}\cdot 12\text{ cm}) + (2(39\text{ cm} + 12\text{ cm}))(23\text{ cm})$ $= 3282\text{ cm}^2$

Area = $B\,h = (39\text{ cm}\cdot 12\text{ cm})\,(23\text{ cm}) = 10764\text{ cm}^3$

2. Surface area cylinder = $2\pi r^2 + 2\pi rh = 2\pi(5\text{ yd})^2 + 2\pi(5\text{ yd})(6\text{ yd}) = 345.58\text{ yd}^2$

Volume cylinder = $\pi r^2 h = \pi(5\text{ yd})^2(6\text{ yd}) = 471.24\text{ yd}^3$

3. Radius of sphere = 45 cm/2 = 22.5 cm

Surface Area = $4\pi r^3 = 4\pi(22.5\text{ cm})^2 = 6361.73\text{ in}^2$

Volume = $\frac{4}{3}\pi r^3 = \frac{4}{3}\pi\,(22.5\text{ cm})^3 = 47712.94\text{ cm}^3$

4. Base area of octagon = $4.828\ s^2 = 4.828\ (4\text{ in})^2 = 77.25\text{ in}^2$

Volume = $\frac{1}{3}B\ h = \frac{1}{3}(77.25\text{ in}^2)(14\text{ in}) = 360.49\text{ in}^3$

Surface area = base area + Lateral area = $77.25\text{ in}^2 + 8(.5\cdot l\cdot s) = 77.25\text{ in}^2 + 8(.5\cdot 14.94''\cdot 4'')$

$= 316.26\text{ in}^2$

5. Base area of trapezoid = $.5(15' + 4')(50') = 475\text{ ft}^2$,

Volume = $475\text{ ft}^2\ (25\text{ ft}) = 11875\text{ ft}^3$. $11885\text{ ft}^3\left(\frac{1728\text{ in}^3}{1\text{ ft}^3}\right)\left(\frac{1\text{ gal}}{231\text{ in}^3}\right) = 88831.17$ gallons of water.

6. Hemisphere and cylinder radius is 3 ft/2 = 1.5 ft.

   Hemisphere volume = $.5\left(\frac{4}{3}\pi r^3\right) = \frac{4}{3}\pi(1.5\text{ ft})^3 = 7.07\text{ ft}^3$ Volume of cylinder part = $\pi r^2 h$

   $= \pi(1.5\text{ ft})^2(5\text{ ft}) = 35.34\text{ ft}^3$

   Volume of canister = 2 (hemisphere volume) + cylinder volume) = $2(7.07\text{ ft}^3) + 35.34\text{ ft}^3 = 49.48\text{ ft}^3$

7. Volume cone = $\frac{1}{3}\pi r^2 h = \frac{1}{3}\pi(20\text{ ft})^2\ (37\text{ ft}) = 15498.52\text{ ft}^3$

8. Height = 1.5 ft, $B = (7\text{ ft})^2 = 49\text{ ft}^2$, $b = (3\text{ ft})^2 = 9\text{ ft}^2$

   Volume frustum = $\frac{1}{3}h\left(B + b + \sqrt{B \cdot b}\right) = \frac{1}{3}(1.5\text{ ft})\left(49\text{ ft}^2 + 9\text{ ft}^2 + \sqrt{49\text{ ft}^2 \cdot 9\text{ ft}^2}\right) = 39.5\text{ ft}^3$

9. Mass = Volume · density, density of zinc = $\frac{4.05\text{ g}}{\text{cm}^3}$

   mass = $420\text{ cm}^3\left(\frac{4.05\text{ g}}{\text{cm}^3}\right) = 1701\text{ g}, (1701\text{g})\left(\frac{1\text{ kg}}{1000\text{ g}}\right) = 1.70\text{ kg}$

10. Volume of bar = (6 cm)(10 cm)(2 cm) = 120 cm^3 $\text{Density}_{\text{silver}} = \frac{10.5\text{g}}{\text{cm}^3}$

    Mass = $(120\text{ cm}^3)\left(\frac{10.5\text{ g}}{\text{cm}^3}\right) = 1260$ grams. 1260 g (\$.51/g) = \$642.60

11. The radius of the sphere is 1.4 in/2 = .7 in.

    Volume = $\frac{4}{3}\pi r^3 = \frac{4}{3}\pi(.7\text{ in})^3 = 1.44\text{ in}^3$, $16.39\text{ cm}^3 = 1\text{ in}^3$, convert to cm^3,

    $1.44\text{ in}^3\left(\frac{16.387\text{ cm}^3}{1\text{ in}^3}\right) = 23.60\text{ cm}^3$; $\text{density}_{\text{lead}} = \frac{13.5\text{ g}}{\text{cm}^3}$ so

    Mass = $23.60\text{ cm}^3\left(\frac{13.5\text{ g}}{\text{cm}^3}\right) = 318.60$ grams

12. Outer hemisphere, radius is 70 cm/2 = 35 cm. Inner hemisphere radius is 50 cm/2 = 25 cm.

    Volume of the bowl is volume of outer hemisphere – Volume of inner hemisphere = .5 $\left(\frac{4}{3}\right)\pi(35\text{ cm})^3 - .5\left(\frac{4}{3}\right)\pi(25\text{ cm})^3 = 57072.26\text{ cm}^3$

    $\text{Density}_{\text{cherry wood}} = \frac{.43\text{ g}}{\text{cm}^3}$ so mass = $57072.26\text{ cm}^3\left(\frac{.43\text{ g}}{\text{cm}^3}\right) = 24541.07$ grams

    1000 g = 1 kg, and 2.205 lb = 1 kg.

    $24541.07\text{ g}\left(\frac{1\text{ kg}}{1000\text{ g}}\right)\left(\frac{2.205\text{ lb}}{1\text{ kg}}\right) = 54.11$ pounds

13. Volume cylinder $= \pi r^2 h = \pi(4\text{ m})^2(9.5\text{ m}) = 477.52\ \text{m}^3$

Volume of cube is $(10\text{ m})^3 = 1000\ \text{m}^3$

Total Volume $= 477.52\ \text{m}^3 + 1000\ \text{m}^3 = 1477.52\ \text{m}^3$

As $1000000\ \text{cm}^3 = 1\ \text{m}^3$ Convert to cubic cm, $1477.52\ \text{m}^3 \cdot \dfrac{1000000\ \text{cm}^3}{1\ \text{m}^3} = 1477520000\ \text{cm}^3$

$\text{density}_{\text{granite}} = \dfrac{2.75\text{ g}}{\text{cm}^3}$ so mass $= 1477520000\ \text{cm}^3\left(\dfrac{2.75\text{ g}}{\text{cm}^3}\right) = 4063180000$ grams

$4063180000\text{ g}\left(\dfrac{1\text{ kg}}{1000\text{ g}}\right) = 4063180$ kilograms

14. **a)** $\dfrac{\text{Opposite leg}}{\text{Hypotenuse}}$

b) $\dfrac{\text{Adjacent leg}}{\text{Hypotenuse}}$

c) $\dfrac{\text{Opposite leg}}{\text{Adjacent leg}}$

15. **a)** $\dfrac{12}{15}$ or .8

b) $\dfrac{12}{15}$ or .8

c) $\dfrac{12}{9}$ or 1.33…

d) $\dfrac{9}{15}$ or .6

e) $\dfrac{9}{12} = .75$

16. **a)** .342

b) .350

c) 1

17. $\arcsin\dfrac{4.5}{15} = 17.46°$ $\left(\text{you could also use } \arccos\dfrac{14.309}{15} \text{ or } \arctan\dfrac{4.5}{14.309}\right)$

18. $\cos 61° = \frac{\text{adjacent}}{131}$ so $131(\cos 61) = 63.510$; $\sin 61° = \frac{\text{opposite}}{131}$ so $131 (\sin 61) = 114.575$; $90° - 61° = 29°$

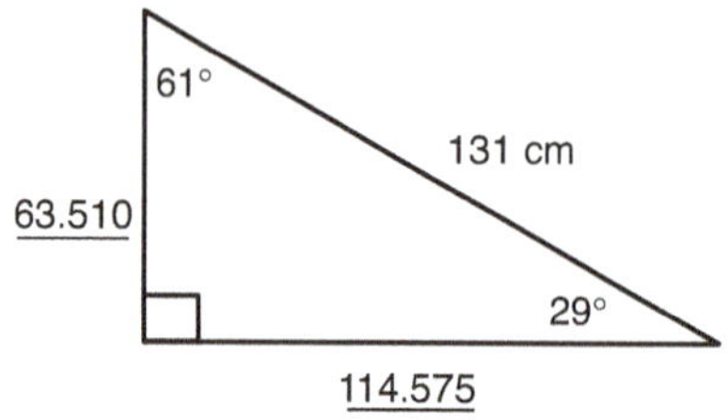

19. $\cos C = \frac{18}{22}$ so $\arccos \frac{18}{22} = 35.10°$; $90 - 35.10 = 54.90°$; $\sin 35.10° = \frac{\text{opposite}}{22}$ so $22(\sin 35.10°) = 12.65$

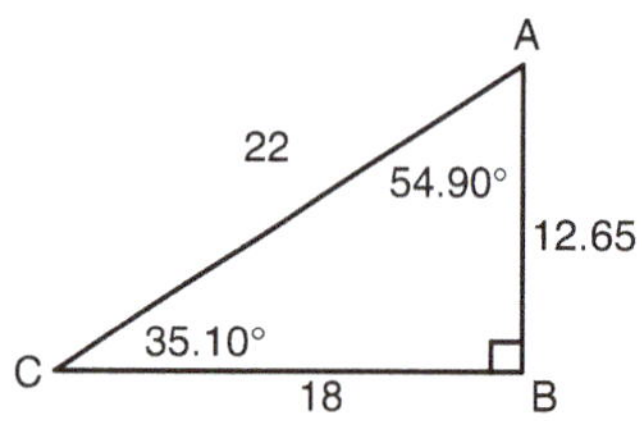

20.

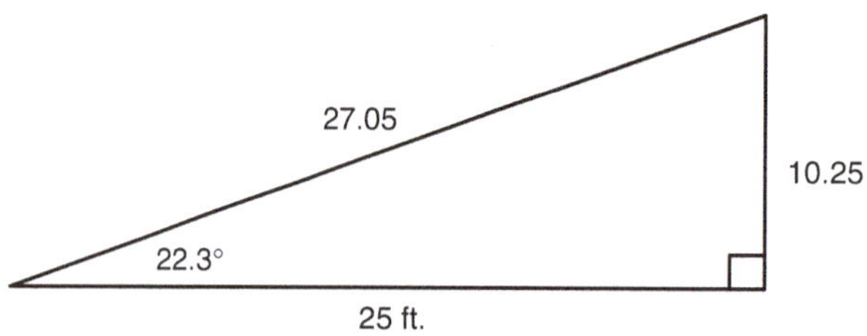

$\cos 22.3° = \frac{25}{\text{hypotenuse}}$ so hypotenuse $= \frac{25}{\cos 22.3} = 27.02$ and

$\tan 22.3° = \frac{\text{opposite}}{25}$ so opposite $= 25(\tan 22.3) = 10.25$

27.02 feet from the flag pole top, the pole is 10.25 ft tall

21.

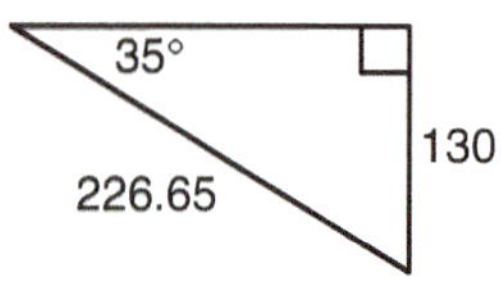

$\sin 35° = \frac{130}{\text{hypotenuse}}$ so hypotenuse $= \frac{130}{\sin 35} = 226.65$

226.65 feet from the lighthouse to the boat.

22.

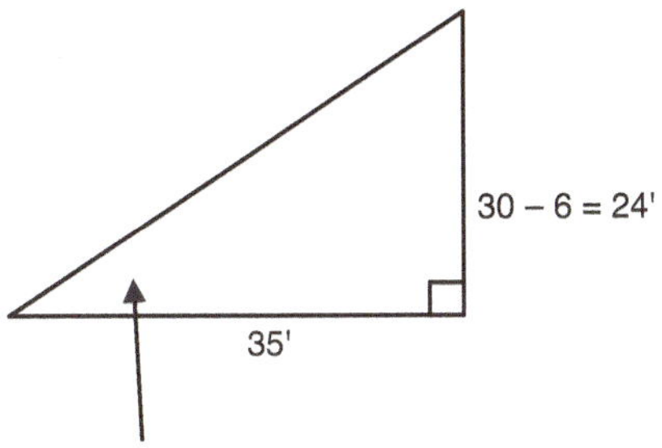

Angle of elevation = Arcsin $\frac{24}{35}$ = 43.29°

23. use law of sines, $\frac{15}{\sin 84} = \frac{b}{\sin 62}$ so $b = \frac{(\sin 62)(15)}{\sin 84} = 13.32$ Angle

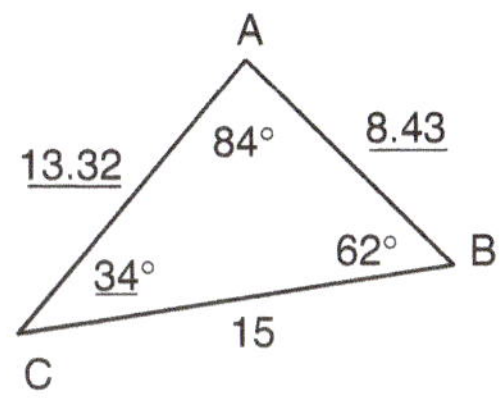

$C = 180 - 84 - 62 = 34°$ $\frac{15}{\sin 84} = \frac{C}{\sin 34}$, so $C = \frac{15(\sin 34)}{\sin 84} = 8.43$

24. angle C = 180 – (109 + 35) = 36°; use law of sines $\frac{34}{\sin 36} = \frac{b}{\sin 35}$, so

$b = \frac{34(\sin 35)}{\sin 36} = 33.18$; $\frac{34}{\sin 36} = \frac{a}{\sin 109}$, so $a = \frac{34(\sin 109)}{\sin 36} = 54.69$

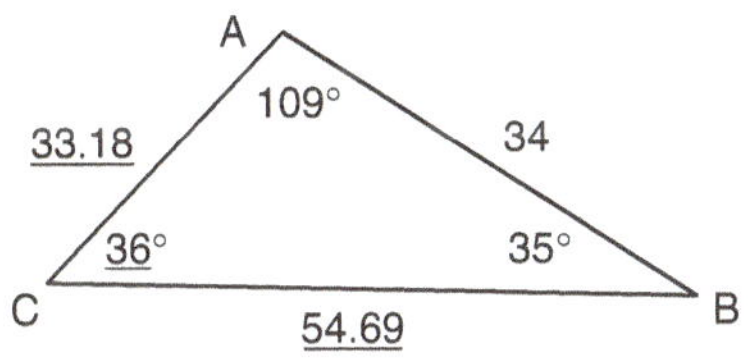

25. use law of cosine $c^2 = a^2 + b^2 - 2\,ab\cos C$

$$c^2 = 39.3^2 + 54.2^2 - 2\,(39.3)(54.2)\cos 131$$

$$c^2 = 7277.020191$$

$$c = \sqrt{7277.020191} = 85.31$$

use law of sines $\frac{54.2}{\sin A} = \frac{85.31}{\sin 131}$ or $\sin A = \frac{54.2 \sin 131}{85.31} = .4794896173$

Angle A = arcsin .4794896173 = 28.65°

Angle B = 180 – (131 + 28.65) = 20.35°

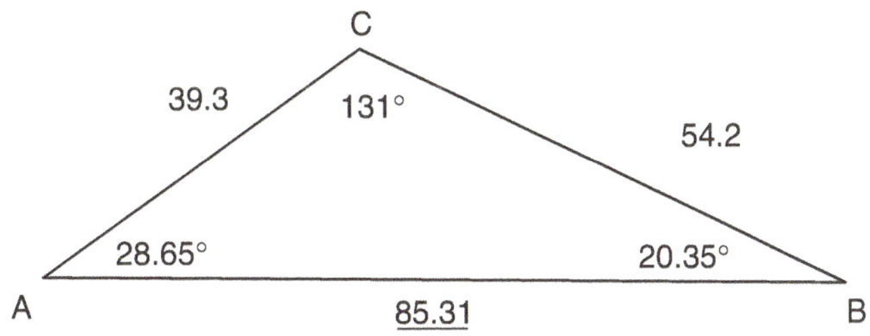

26. law of sines $\dfrac{2.6}{\sin 33.1} = \dfrac{\text{side from tower A to tower C}}{\sin 43.9}$ or

$$\text{side from tower A to tower C} = \frac{2.6\sin 43.9}{\sin 33.1} = 3.30 \text{ miles}$$

$$\frac{2.6}{\sin 33.1} = \frac{\text{side from tower B to tower C}}{\sin 103} \text{ or}$$

$$\text{side from tower B to tower C} = \frac{2.6\sin 103}{\sin 33.1} = 4.64 \text{ miles}$$

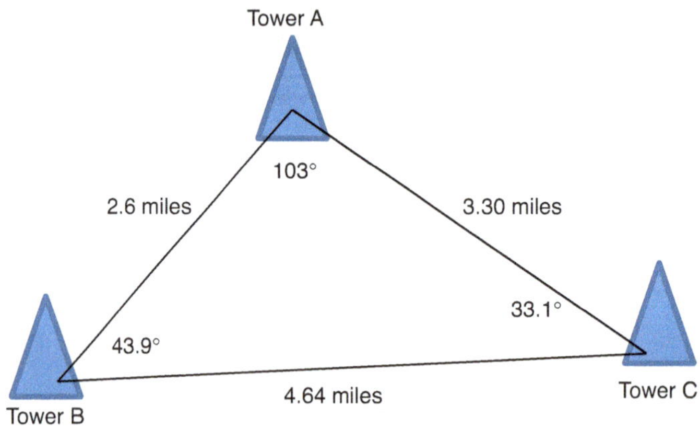